More praise for *The Demon-Haunted World*

"As I close this eloquent and fascinating book, I recall the final chapter title from one of Carl Sagan's earlier works, *Cosmos*. 'Who Speaks for Earth?' is a rhetorical question, but I presume to answer it. My candidate for planetary ambassador can be none other than Carl Sagan himself. He is wise, humane, witty, well read, and incapable of composing a dull sentence. . . . I wish I had written *The Demon-Haunted World*. Having failed to do so the least I can do is press it upon my friends. Please read this book."
— Richard Dawkins
The Times (London)

"Sagan takes no prisoners. . . . Unfailingly respectful of religion in general, he decries the love of ignorance at the heart of fundamentalism. . . . Closely argued . . . Entertaining . . . A major salvo in the batttle against irrationality and superstition."
— *Albany Times Union*

"In Sagan's characteristically elegant prose style, it offers a heavy dose of common sense. . . . Every parent, teacher, clergyman, politician, and high school student should read it immediately."
— *Fort Lauderdale Sun-Sentinel*

"Skeptical books are . . . comparatively rare. Rarer still are those of the caliber of Carl Sagan's new work. . . . Impressive . . . Persuasive . . . Brave."
— *Scientific American*

"Lithe, well-supported, sometimes quite wry, and altogether refreshing."
— *Booklist*

"One man is not content to go gently into the dark night of irrationality, and his lucid and lyrical voice fills the pages of an important new book, *The Demon-Haunted World*. . . . A personal statement of one man's moving belief, wonderfully written."
— *Hackensack Sunday Record*

THE
DEMON-
HAUNTED
WORLD

THE DEMON-HAUNTED WORLD

Science as a Candle in the Dark

CARL SAGAN

BALLANTINE BOOKS • NEW YORK

TO TONIO,
MY GRANDSON.

I WISH YOU A WORLD
FREE OF DEMONS
AND FULL OF LIGHT.

We wait for light, but behold darkness.

ISAIAH 59:9

It is better to light one candle than to curse the darkness.

ADAGE

Preface

MY TEACHERS

It was a blustery fall day in 1939. In the streets outside the apartment building, fallen leaves were swirling in little whirlwinds, each with a life of its own. It was good to be inside and warm and safe, with my mother preparing dinner in the next room. In our apartment there were no older kids who picked on you for no reason. Just the week before, I had been in a fight—I can't remember, after all these years, who it was with; maybe it was Snoony Agata from the third floor— and, after a wild swing, I found I had put my fist through the plate glass window of Schechter's drug store.

Mr. Schechter was solicitous: "It's all right, I'm insured," he said as he put some unbelievably painful antiseptic on my wrist. My mother took me to the doctor whose office was on the ground floor of our building. With a pair of tweezers, he pulled out a fragment of glass. Using needle and thread, he sewed two stitches.

"Two stitches!" my father had repeated later that night. He knew about stitches, because he was a cutter in the garment industry; his job was to use a very scary power saw to cut out patterns—backs, say, or sleeves for ladies' coats and suits—from an enormous stack of cloth. Then the patterns were conveyed to endless rows of women sitting at sewing machines. He was pleased I had gotten angry enough to over- come a natural timidity.

Sometimes it was good to fight back. I hadn't planned to do any- thing violent. It just happened. One moment Snoony was pushing me and the next moment my fist was through Mr. Schechter's window. I had injured my wrist, generated an unexpected medical expense, bro- ken a plate glass window, and no one was mad at me. As for Snoony, he was more friendly than ever.

I puzzled over what the lesson was. But it was much more pleasant to work it out up here in the warmth of the apartment, gazing out through the living room window into Lower New York Bay, than to risk some new misadventure on the streets below.

As she often did, my mother had changed her clothes and made up her face in anticipation of my father's arrival. The Sun was almost setting and together we looked out across the choppy waters.

"There are people fighting out there, killing each other," she said, waving vaguely across the Atlantic. I peered intently.

"I know," I replied. "I can see them."

"No, you can't," she replied, almost severely, before returning to the kitchen. "They're too far away."

How could she know whether I could see them or not? I wondered. Squinting, I had thought I'd made out a thin strip of land at the horizon on which tiny figures were pushing and shoving and dueling with swords as they did in my comic books. But maybe she was right. Maybe it had just been my imagination, a little like the midnight monsters that still, on occasion, awakened me from a deep sleep, my pajamas drenched in sweat, my heart pounding.

How can you tell when someone is only imagining? I gazed out across the gray waters until night fell and I was called to wash my hands for dinner. To my delight, my father swooped me up in his arms. I could feel the cold of the outside world against his one-day growth of beard.

—

On a Sunday in that same year, my father had patiently explained to me about zero as a placeholder in arithmetic, about the wicked-sounding names of big numbers, and about how there's no biggest number. ("You can always add one," he pointed out). Suddenly, I was seized by a childish compulsion to write in sequence all the integers from 1 to 1,000. We had no pads of paper, but my father offered up the stack of gray cardboards he had been saving from when his shirts were sent to the laundry. I started the project eagerly, but was surprised at how slowly it went. When I had gotten no farther than the low hundreds, my mother announced that it was time for me to take my bath. I was disconsolate. I had to get to a thousand. A mediator his whole life, my father intervened: If I would cheerfully submit to the bath, he would continue the sequence. I was overjoyed. By the time I emerged, he was approaching 900, and I was able to reach 1,000 only a little past my ordinary bedtime. The magnitude of large numbers has never ceased to impress me.

Also in 1939 my parents took me to the New York World's Fair. There, I was offered a vision of a perfect future made possible by science and high technology. A time capsule was buried, packed with artifacts of our time for the benefit of those in the far future—who, astonishingly, might not know much about the people of 1939. The "World of Tomorrow" would be sleek, clean, streamlined and, as far as I could tell, without a trace of poor people.

"See sound," one exhibit bewilderingly commanded. And sure enough, when the tuning fork was struck by the little hammer, a beautiful sine wave marched across the oscilloscope screen. "Hear light," another poster exhorted. And sure enough, when the flashlight shone on the photocell, I could hear something like the static on our Motorola radio set when the dial was between stations. Plainly the world held wonders of a kind I had never guessed. How *could* a tone become a picture and light become a noise?

My parents were not scientists. They knew almost nothing about science. But in introducing me simultaneously to skepticism and to wonder, they taught me the two uneasily cohabiting modes of thought that are central to the scientific method. They were only one step out of poverty. But when I announced that I wanted to be an astronomer, I received unqualified support—even if they (as I) had only the most rudimentary idea of what an astronomer does. They never suggested that, all things considered, it might be better to be a doctor or a lawyer.

I wish I could tell you about inspirational teachers in science from my elementary or junior high or high school days. But as I think back on it, there were none. There was rote memorization about the Periodic Table of the Elements, levers and inclined planes, green plant photosynthesis, and the difference between anthracite and bituminous coal. But there was no soaring sense of wonder, no hint of an evolutionary perspective, and nothing about mistaken ideas that everybody had once believed. In high school laboratory courses, there was an answer we were supposed to get. We were marked off if we didn't get it. There was no encouragement to pursue our own interests or hunches or conceptual mistakes. In the backs of textbooks there was material you could tell was interesting. The school year would always end before we got to it. You could find wonderful books on astronomy, say, in the libraries, but not in the classroom. Long division was taught as a set of rules from a cookbook, with no explanation of how this particular sequence of short divisions, multiplications, and subtractions got you the right answer. In high school, extracting square roots was offered reverentially, as if it were a method once handed down from Mt. Sinai. It was our job merely to remember what we had been commanded. Get the right answer, and never mind that you don't understand what you're doing. I had a very capable second-year algebra teacher from whom I learned much mathematics; but he was also a bully who enjoyed reducing young women to tears. My interest in science was maintained through all those school years by reading books and magazines on science fact and fiction.

College was the fulfillment of my dreams: I found teachers who not only understood science, but who were actually able to explain it. I was lucky enough to attend one of the great institutions of learning of the time, the University of Chicago. I was a physics student in a department orbiting around Enrico Fermi; I discovered what true mathematical elegance is from Subrahmanyan Chandrasekhar; I was given the chance to talk chemistry with Harold Urey; over summers I was apprenticed in biology to H. J. Muller at Indiana University; and I learned planetary astronomy from its only full-time practitioner at the time, G. P. Kuiper.

It was from Kuiper that I first got a feeling for what is called a back-of-the-envelope calculation: A possible explanation to a problem occurs to you, you pull out an old envelope, appeal to your knowledge of fundamental physics, scribble a few approximate equations on the envelope, substitute in likely numerical values, and see if your answer comes anywhere near explaining your problem. If not, you look for a different explanation. It cut through nonsense like a knife through butter.

At the University of Chicago I also was lucky enough to go through a general education program devised by Robert M. Hutchins, where science was presented as an integral part of the gorgeous tapestry of human knowledge. It was considered unthinkable for an aspiring physicist not to know Plato, Aristotle, Bach, Shakespeare, Gibbon, Malinowski, and Freud—among many others. In an introductory science class, Ptolemy's view that the Sun revolved around the Earth was presented so compellingly that some students found themselves re-evaluating their commitment to Copernicus. The status of the teachers in the Hutchins curriculum had almost nothing to do with their research; perversely—unlike the American university standard of today—teachers were valued for their teaching, their ability to inform and inspire the next generation.

In this heady atmosphere, I was able to fill in some of the many gaps in my education. Much that had been deeply mysterious, and not just in science, became clearer. I also witnessed at first hand the joy felt by those whose privilege it is to uncover a little about how the Universe works.

I've always been grateful to my mentors of the 1950s, and tried to make sure that each of them knew my appreciation. But as I look back, it seems clear to me that I learned the most essential things not from my school teachers, nor even from my university professors, but from my parents, who knew nothing at all about science, in that single far-off year of 1939.

Contents

* Written with Ann Druyan

Chapter 1

THE
MOST
PRECIOUS
THING

All our science, measured against reality,
is primitive and childlike — and yet it is
the most precious thing we have.

ALBERT EINSTEIN
(1879–1955)

As I got off the plane, he was waiting for me, holding up a scrap of cardboard with my name scribbled on it. I was on my way to a conference of scientists and TV broadcasters devoted to the seemingly hopeless prospect of improving the presentation of science on commercial television. The organizers had kindly sent a driver.

"Do you mind if I ask you a question?" he said as we waited for my bag.

No, I didn't mind.

"Isn't it confusing to have the same name as that scientist guy?"

It took me a moment to understand. Was he pulling my leg? Finally, it dawned on me.

"I *am* that scientist guy," I answered.

He paused and then smiled. "Sorry. That's my problem. I thought it was yours too."

He put out his hand. "My name is William F. Buckley." (Well, he wasn't *exactly* William F. Buckley, but he did bear the name of a contentious and well-known TV interviewer, for which he doubtless took a lot of good-natured ribbing.)

As we settled into the car for the long drive, the windshield wipers rhythmically thwacking, he told me he was glad I was "that scientist guy"—he had so many questions to ask about science. Would I mind?

No, I didn't mind.

And so we got to talking. But not, as it turned out, about science. He wanted to talk about frozen extraterrestrials languishing in an Air Force base near San Antonio, "channeling" (a way to hear what's on the minds of dead people—not much, it turns out), crystals, the prophecies of Nostradamus, astrology, the shroud of Turin . . . He introduced each portentous subject with buoyant enthusiasm. Each time I had to disappoint him:

"The evidence is crummy," I kept saying. "There's a much simpler explanation."

He was, in a way, widely read. He knew the various speculative nuances on, let's say, the "sunken continents" of Atlantis and Lemuria. He had at his fingertips what underwater expeditions were supposedly just setting out to find the tumbled columns and broken minarets of a once-great civilization whose remains were now visited only by deep sea luminescent fish and giant kraken. Except . . . while the ocean keeps many secrets, I knew that there isn't a trace of oceanographic or geophysical support for Atlantis and Lemuria. As far as science can tell, they never existed. By now a little reluctantly, I told him so.

As we drove through the rain, I could see him getting glummer and glummer. I was dismissing not just some errant doctrine, but a precious facet of his inner life.

And yet there's so much in real science that's equally exciting, more mysterious, a greater intellectual challenge—as well as being a lot closer to the truth. Did he know about the molecular building blocks of life sitting out there in the cold, tenuous gas between the stars? Had he heard of the footprints of our ancestors found in 4-million-year-old volcanic ash? What about the raising of the Himalayas when India went crashing into Asia? Or how viruses, built like hypodermic syringes, slip their DNA past the host organism's defenses and subvert the reproductive machinery of cells; or the radio search for extraterrestrial intelligence; or the newly discovered ancient civilization of Ebla that advertised the virtues of Ebla beer? No, he hadn't heard. Nor did he know, even vaguely, about quantum indeterminacy, and he recognized DNA only as three frequently linked capital letters.

Mr. "Buckley"—well-spoken, intelligent, curious—had heard virtually nothing of modern science. He had a natural appetite for the wonders of the Universe. He *wanted* to know about science. It's just that all the science had gotten filtered out before it reached him. Our cultural motifs, our educational system, our communications media had failed this man. What the society permitted to trickle through was mainly pretense and confusion. It had never taught him how to distinguish real science from the cheap imitation. He knew nothing about how science works.

There are hundreds of books about Atlantis—the mythical continent that is said to have existed something like 10,000 years ago in the Atlantic Ocean. (Or somewhere. A recent book locates it in Antarctica.) The story goes back to Plato, who reported it as hearsay coming down to *him* from remote ages. Recent books authoritatively describe the high level of Atlantean technology, morals, and spirituality, and the great tragedy of an entire populated continent sinking beneath the waves. There is a "New Age" Atlantis, "the legendary civilization of advanced sciences," chiefly devoted to the "science" of crystals. In a trilogy called *Crystal Enlightenment*, by Katrina Raphaell—the books mainly responsible for the crystal craze in America—Atlantean crystals read minds, transmit thoughts, are the repositories of ancient history and the model and source of the pyramids of Egypt. Nothing approximating evidence is offered to support these assertions. (A resurgence of crystal mania may follow the recent finding by the real science of seismology that the inner core of the Earth may be composed of a single, huge, nearly perfect crystal—of iron.)

A few books—Dorothy Vitaliano's *Legends of the Earth*, for example—sympathetically interpret the original Atlantis legends in terms of a small island in the Mediterranean that was destroyed by a volcanic eruption, or an ancient city that slid into the Gulf of Corinth after an earthquake. This, for all we know, may be the source of the legend, but it is a far cry from the destruction of a continent on which had sprung forth a preternaturally advanced technical and mystical civilization.

What we almost never find—in public libraries or newsstand magazines or prime time television programs—is the evidence from sea floor spreading and plate tectonics, and from mapping the ocean floor which shows quite unmistakably that there could have been no continent between Europe and the Americas on anything like the timescale proposed.

Spurious accounts that snare the gullible are readily available. Skeptical treatments are much harder to find. Skepticism does not sell well. A bright and curious person who relies entirely on popular culture to be informed about something like Atlantis is hundreds or thousands of times more likely to come upon a fable treated uncritically than a sober and balanced assessment.

Maybe Mr. "Buckley" should know to be more skeptical about what's dished out to him by popular culture. But apart from that, it's hard to see how it's his fault. He simply accepted what the most widely available and accessible sources of information claimed was true. For his naïveté, he was systematically misled and bamboozled.

Science arouses a soaring sense of wonder. But so does pseudoscience. Sparse and poor popularizations of science abandon ecological niches that pseudoscience promptly fills. If it were widely understood that claims to knowledge require adequate evidence before they can be accepted, there would be no room for pseudoscience. But a kind of Gresham's Law prevails in popular culture by which bad science drives out good.

All over the world there are enormous numbers of smart, even gifted, people who harbor a passion for science. But that passion is unrequited. Surveys suggest that some 95 percent of Americans are "scientifically illiterate." That's just the same fraction as those African Americans, almost all of them slaves, who were illiterate just before the Civil War—when severe penalties were in force for anyone who taught a slave to read. Of course there's a degree of arbitrariness about any determination of illiteracy, whether it applies to language or to science. But anything like 95 percent illiteracy is extremely serious.

Every generation worries that educational standards are decaying. One of the oldest short essays in human history, dating from Sumer some 4,000 years ago, laments that the young are disastrously more ignorant than the generation immediately preceding. Twenty-four hundred years ago, the aging and grumpy Plato, in Book VII of the *Laws*, gave his definition of scientific illiteracy:

> Who is unable to count one, two, three, or to distinguish odd from even numbers, or is unable to count at all, or reckon night and day, and who is totally unacquainted with the revolution of the Sun and Moon, and the other stars . . . All freemen, I conceive, should learn as much of these branches of knowledge as every child in Egypt is taught when he learns the alphabet. In that country arithmetical games have been invented for the use of mere children, which they learn as pleasure and amusement . . . I . . . have late in life heard with amazement of our ignorance in these matters; to me we appear to be more like pigs than men, and I am quite ashamed, not only of myself, but of all Greeks.

I don't know to what extent ignorance of science and mathematics contributed to the decline of ancient Athens, but I know that the consequences of scientific illiteracy are far more dangerous in our time than in any that has come before. It's perilous and foolhardy for the average citizen to remain ignorant about global warming, say, or ozone depletion, air pollution, toxic and radioactive wastes, acid rain, topsoil erosion, tropical deforestation, exponential population growth. Jobs and wages depend on science and technology. If our nation can't manufacture, at high quality and low price, products people want to buy, then industries will continue to drift away and transfer a little more prosperity to other parts of the world. Consider the social ramifications of fission and fusion power, supercomputers, data "highways," abortion, radon, massive reductions in strategic weapons, addiction, government eavesdropping on the lives of its citizens, high-resolution TV, airline and airport safety, fetal tissue transplants, health costs, food additives, drugs to ameliorate mania or depression or schizophrenia, animal rights, superconductivity, morning-after pills, alleged hereditary antisocial predispositions, space stations, going to Mars, finding cures for AIDS and cancer.

How can we affect national policy—or even make intelligent decisions in our own lives—if we don't grasp the underlying issues? As I write, Congress is dissolving its own Office of Technology Assessment—the only organization specifically tasked to provide advice to the House and Senate on science and technology. Its competence and integrity over the years have been exemplary. Of the 535 members of the U.S. Congress, rarely in the twentieth century have as many as one percent had any significant background in science. The last scientifically literate President may have been Thomas Jefferson.*

So how do Americans decide these matters? How do they instruct their representatives? Who in fact makes these decisions, and on what basis?

———

Hippocrates of Cos is the father of medicine. He is still remembered 2,500 years later for the Hippocratic Oath (a modified form of which is

---

* Although claims can be made for Theodore Roosevelt, Herbert Hoover and Jimmy Carter. Britain had such a Prime Minister in Margaret Thatcher. Her early studies in chemistry, in part under the tutelage of Nobel Laureate Dorothy Hodgkins, were key to the U.K.'s strong and successful advocacy that ozone-depleting CFCs be banned worldwide.

still here and there taken by medical students upon their graduation). But he is chiefly celebrated because of his efforts to bring medicine out of the pall of superstition and into the light of science. In a typical passage Hippocrates wrote: "Men think epilepsy divine, merely because they do not understand it. But if they called everything divine which they do not understand, why, there would be no end of divine things." Instead of acknowledging that in many areas we are ignorant, we have tended to say things like the Universe is permeated with the ineffable. A God of the Gaps is assigned responsibility for what we do not yet understand. As knowledge of medicine improved since the fourth century B.C., there was more and more that we understood and less and less that had to be attributed to divine intervention—either in the causes or in the treatment of disease. Deaths in childbirth and infant mortality have decreased, lifetimes have lengthened, and medicine has improved the quality of life for billions of us all over the planet.

In the diagnosis of disease, Hippocrates introduced elements of the scientific method. He urged careful and meticulous observation: "Leave nothing to chance. Overlook nothing. Combine contradictory observations. Allow yourself enough time." Before the invention of the thermometer, he charted the temperature curves of many diseases. He recommended that physicians be able to tell, from present symptoms alone, the probable past and future course of each illness. He stressed honesty. He was willing to admit the limitations of the physician's knowledge. He betrayed no embarrassment in confiding to posterity that more than half his patients were killed by the diseases he was treating. His options of course were limited; the drugs available to him were chiefly laxatives, emetics, and narcotics. Surgery was performed, and cauterization. Considerable further advances were made in classical times through the fall of Rome.

While medicine in the Islamic world flourished, what followed in Europe was truly a dark age. Much knowledge of anatomy and surgery was lost. Reliance on prayer and miraculous healing abounded. Secular physicians became extinct. Chants, potions, horoscopes, and amulets were widely used. Dissections of cadavers were restricted or outlawed, so those who practiced medicine were prevented from acquiring firsthand knowledge of the human body. Medical research came to a standstill.

It was very like what the historian Edward Gibbon described for the entire Eastern Empire, whose capital was Constantinople:

> In the revolution of ten centuries, not a single discovery was made to exalt the dignity or promote the happiness of mankind. Not a single idea had been added to the speculative systems of antiquity, and a succession of patient disciples became in their turn the dogmatic teachers of the next servile generation.

Even at its best, pre-modern medical practice did not save many. Queen Anne was the last Stuart monarch of Great Britain. In the last 17 years of the seventeenth century, she was pregnant 18 times. Only five children were born alive. Only one of them survived infancy. He died before reaching adulthood, and before her coronation in 1702. There seems to be no evidence of some genetic disorder. She had the best medical care money could buy.

Diseases that once tragically carried off countless infants and children have been progressively mitigated and cured by science — through the discovery of the microbial world, via the insight that physicians and midwives should wash their hands and sterilize their instruments, through nutrition, public health and sanitation measures, antibiotics, drugs, vaccines, the uncovering of the molecular structure of DNA, molecular biology, and now gene therapy. In the developed world at least, parents today have an enormously better chance of seeing their children live to adulthood than did the heir to the throne of one of the most powerful nations on Earth in the late seventeenth century. Smallpox has been wiped out worldwide. The area of our planet infested with malaria-carrying mosquitoes has dramatically shrunk. The number of years a child diagnosed with leukemia can expect to live has been increasing progressively, year by year. Science permits the Earth to feed about a hundred times more humans, and under conditions much less grim, than it could a few thousand years ago.

We can pray over the cholera victim, or we can give her 500 milligrams of tetracycline every 12 hours. (There is still a religion, Christian Science, that denies the germ theory of disease; if prayer fails, the faithful would rather see their children die than give them antibiotics.) We can try nearly futile psychoanalytic talk therapy on the schizophrenic patient, or we can give him 300 to 500 milligrams a day of clozapine. The scientific treatments are hundreds or thousands of times more effective than the alternatives. (And even when the alternatives seem to work, we don't actually know that they played any role: Spontaneous remissions, even of cholera and schizophrenia, can occur without prayer and without psychoanalysis.) Abandoning science means abandoning much more than air conditioning, CD players, hair dryers, and fast cars.

In hunter-gatherer, pre-agricultural times, the human life expectancy was about 20 to 30 years. That's also what it was in Western Europe in Late Roman and in Medieval times. It didn't rise to 40 years until around the year 1870. It reached 50 in 1915, 60 in 1930, 70 in 1955, and is today approaching 80 (a little more for women, a little less for men). The rest of the world is retracing the European increment in longevity. What is the cause of this stunning, unprecedented, humanitarian transition? The germ theory of disease, public health measures, medicines and medical technology. Longevity is perhaps the best single measure of the physical quality of life. (If you're dead, there's little you can do to be happy.) This is a precious offering from science to humanity—nothing less than the gift of life.

But microorganisms mutate. New diseases spread like wildfire. There is a constant battle between microbial measures and human countermeasures. We keep pace in this competition not just by designing new drugs and treatments, but by penetrating progressively more deeply toward an understanding of the nature of life—basic research.

If the world is to escape the direst consequences of global population growth and 10 or 12 billion people on the planet in the late twenty-first century, we must invent safe but more efficient means of growing food—with accompanying seed stocks, irrigation, fertilizers, pesticides, transportation and refrigeration systems. It will also take widely available and acceptable contraception, significant steps toward political equality of women, and improvements in the standards of living of the poorest people. How can all this be accomplished without science and technology?

I know that science and technology are not just cornucopias pouring gifts out into the world. Scientists not only conceived nuclear weapons; they also took political leaders by the lapels, arguing that *their* nation—whichever it happened to be—had to have one first. Then they manufactured over 60,000 of them. During the Cold War, scientists in the United States, the Soviet Union, China and other nations were willing to expose their own fellow citizens to radiation—in most cases without their knowledge—to prepare for nuclear war. Physicians in Tuskegee, Alabama misled a group of veterans into thinking they were receiving medical treatment for their syphilis, when they were the untreated controls. The atrocious cruelties of Nazi doctors are well-known. Our technology has produced thalidomide, CFCs, Agent Orange, nerve gas, pollution of air and water, species extinctions, and industries so powerful they can ruin the climate of the planet. Roughly half the scientists on Earth work at least part-time for the military. While a few scientists are still perceived as outsiders, courageously criticizing the ills of society and providing early warnings of potential technological catastrophes, many are seen as compliant opportunists, or as the willing source of corporate profits and weapons of mass destruction—never mind the long-term consequences. The technological perils that science serves up, its implicit challenge to received wisdom, and its perceived difficulty, are all reasons for some people to mistrust and avoid it. There's a *reason* people are nervous about science and technology. And so the image of the mad scientist haunts our world—down to the white-coated loonies of Saturday morning children's TV and the plethora of Faustian bargains in popular culture, from the eponymous Dr. Faustus himself to *Dr. Franken-stein, Dr. Strangelove,* and *Jurassic Park.*

But we can't simply conclude that science puts too much power into the hands of morally feeble technologists or corrupt, power-crazed politicians and so decide to get rid of it. Advances in medicine and agriculture have saved vastly more lives than have been lost in all the wars in history.* Advances in transportation, communication, and entertainment have transformed and unified the world. In opinion poll after opinion poll science is rated among the most admired and trusted occupations, despite the misgivings. The sword of science is double-

---

* At a dinner table recently, I asked the assembled guests—ranging in age, I guess, from thirties to sixties—how many of them would be alive today if not for antibiotics, cardiac pacemakers, and the rest of the panoply of modern medicine. Only one hand went up. It was not mine.

edged. Its awesome power forces on all of us, including politicians, but of course especially on scientists, a new responsibility—more attention to the long-term consequences of technology, a global and transgenerational perspective, an incentive to avoid easy appeals to nationalism and chauvinism. Mistakes are becoming too expensive.

—

Do we care what's true? Does it matter?

> *. . . where ignorance is bliss,*
> *'Tis folly to be wise*

wrote the poet Thomas Gray. But is it? Edmund Way Teale in his 1950 book *Circle of the Seasons* understood the dilemma better:

> It is morally as bad not to care whether a thing is true or not, so long as it makes you feel good, as it is not to care how you got your money as long as you have got it.

It's disheartening to discover government corruption and incompetence, for example; but is it better *not* to know about it? Whose interest does ignorance serve? If we humans bear, say, hereditary propensities toward the hatred of strangers, isn't self-knowledge the only antidote? If we long to believe that the stars rise and set for us, that we are the reason there is a Universe, does science do us a disservice in deflating our conceits?

In *The Genealogy of Morals*, Friedrich Nietzsche, as so many before and after, decries the "unbroken progress in the self-belittling of man" brought about by the scientific revolution. Nietzsche mourns the loss of "man's belief in his dignity, his uniqueness, his irreplaceability in the scheme of existence." For me, it is far better to grasp the Universe as it really is than to persist in delusion, however satisfying and reassuring. Which attitude is better geared for our long-term survival? Which gives us more leverage on our future? And if our naïve self-confidence is a little undermined in the process, is that altogether such a loss? Is there not cause to welcome it as a maturing and character-building experience?

To discover that the Universe is some 8 to 15 billion and not 6 to 12 thousand years old* improves our appreciation of its sweep and grandeur; to entertain the notion that we are a particularly complex arrangement of atoms, and not some breath of divinity, at the very least enhances our respect for atoms; to discover, as now seems probable, that our planet is one of billions of other worlds in the Milky Way Galaxy and that our galaxy is one of billions more, majestically expands the arena of what is possible; to find that our ancestors were also the ancestors of apes ties us to the rest of life and makes possible important—if occasionally rueful—reflections on human nature.

Plainly there is no way back. Like it or not, we are stuck with science. We had better make the best of it. When we finally come to terms with it and fully recognize its beauty and its power, we will find, in spiritual as well as in practical matters, that we have made a bargain strongly in our favor.

But superstition and pseudoscience keep getting in the way, distracting all the "Buckleys" among us, providing easy answers, dodging skeptical scrutiny, casually pressing our awe buttons and cheapening the experience, making us routine and comfortable practitioners as well as victims of credulity. Yes, the world *would* be a more interesting place if there were UFOs lurking in the deep waters off Bermuda and eating ships and planes, or if dead people could take control of our hands and write us messages. It would be fascinating if adolescents were able to make telephone handsets rocket off their cradles just by thinking at them, or if our dreams could, more often than can be explained by chance and our knowledge of the world, accurately foretell the future.

These are all instances of pseudoscience. They purport to use the methods and findings of science, while in fact they are faithless to its nature—often because they are based on insufficient evidence or because they ignore clues that point the other way. They ripple with gullibility. With the uninformed cooperation (and often the cynical connivance) of newspapers, magazines, book publishers, radio, television, movie producers, and the like, such ideas are easily and widely

---

* "No thinking religious person believes this. Old hat," writes one of the referees of this book. But many "scientific creationists" not only believe it, but are making increasingly aggressive and successful efforts to have it taught in the schools, museums, zoos, and textbooks. Why? Because adding up the "begats," the ages of patriarchs and others in the Bible, gives such a figure, and the Bible is "inerrant."

available. Far more difficult to come upon, as I was reminded by my encounter with Mr. "Buckley," are the alternative, more challenging and even more dazzling findings of science.

Pseudoscience is easier to contrive than science, because distracting confrontations with reality—where we cannot control the outcome of the comparison—are more readily avoided. The standards of argument, what passes for evidence, are much more relaxed. In part for these same reasons, it is much easier to present pseudoscience to the general public than science. But this isn't enough to explain its popularity.

Naturally people try various belief systems on for size, to see if they help. And if we're desperate enough, we become all too willing to abandon what may be perceived as the heavy burden of skepticism. Pseudoscience speaks to powerful emotional needs that science often leaves unfulfilled. It caters to fantasies about personal powers we lack and long for (like those attributed to comic book superheroes today, and earlier, to the gods). In some of its manifestations, it offers satisfaction of spiritual hungers, cures for disease, promises that death is not the end. It reassures us of our cosmic centrality and importance. It vouchsafes that we are hooked up with, tied to, the Universe.* Sometimes it's a kind of halfway house between old religion and new science, mistrusted by both.

At the heart of some pseudoscience (and some religion also, New Age and Old) is the idea that wishing makes it so. How satisfying it would be, as in folklore and children's stories, to fulfill our heart's desire just by wishing. How seductive this notion is, especially when compared with the hard work and good luck usually required to achieve our hopes. The enchanted fish or the genie from the lamp will grant us three wishes—anything we want except more wishes. Who has not pondered—just to be on the safe side, just in case we ever come upon and accidentally rub an old, squat brass oil lamp— what to ask for?

I remember, from childhood comic strips and books, a top-hatted,

---

* Although it's hard for me to see a more profound cosmic connection than the astonishing findings of modern nuclear astrophysics: Except for hydrogen, all the atoms that make each of us up—the iron in our blood, the calcium in our bones, the carbon in our brains—were manufactured in red giant stars thousands of light-years away in space and billions of years ago in time. We are, as I like to say, starstuff.

mustachioed magician who brandished an ebony walking stick. His name was Zatara. He could make anything happen, anything at all. How did he do it? Easy. He uttered his commands backwards. So if he wanted a million dollars, he would say "srallod noillim a em evig." That's all there was to it. It was something like prayer, but much surer of results.

I spent a lot of time at age eight experimenting in this vein, commanding stones to levitate: "esir, enots." It never worked. I blamed my pronunciation.

—

Pseudoscience is embraced, it might be argued, in exact proportion as real science is misunderstood—except that the language breaks down here. If you've never heard of science (to say nothing of how it works), you can hardly be aware you're embracing pseudoscience. You're simply thinking in one of the ways that humans always have. Religions are often the state-protected nurseries of pseudoscience, although there's no reason why religions have to play that role. In a way, it's an artifact from times long gone. In some countries nearly everyone believes in astrology and precognition, including government leaders. But this is not simply drummed into them by religion; it is drawn out of the enveloping culture in which everyone is comfortable with these practices, and affirming testimonials are everywhere.

Most of the case histories I will relate in this book are American—because these are the cases I know best, not because pseudoscience and mysticism are more prominent in the United States than elsewhere. But the psychic spoon bender and extraterrestrial channeler Uri Geller hails from Israel. As tensions rise between Algerian secularists and Moslem fundamentalists, more and more people are discreetly consulting the country's 10,000 soothsayers and clairvoyants (about half of whom operate with a license from the government). High French officials, including a former President of France, arranged for millions of dollars to be invested in a scam (the Elf-Aquitaine scandal) to find new petroleum reserves from the air. In Germany, there is concern about carcinogenic "Earth rays" undetectable by science; they can be sensed only by experienced dowsers brandishing forked sticks. "Psychic surgery" flourishes in the Philippines. Ghosts are something of a national obsession in Britain. Since

World War II, Japan has spawned enormous numbers of new religions featuring the supernatural. An estimated 100,000 fortune-tellers flourish in Japan; the clientele are mainly young women. Aum Shinrikyo, a sect thought to be involved in the release of the nerve gas sarin in the Tokyo subway system in March 1995, features levitation, faith healing and ESP among its main tenets. Followers, at a high price, drank the "miracle pond" water—from the bath of Asahara, their leader. In Thailand, diseases are treated with pills manufactured from pulverized sacred Scripture. "Witches" are today being burned in South Africa. Australian peace-keeping forces in Haiti rescue a woman tied to a tree; she is accused of flying from rooftop to rooftop, and sucking the blood of children. Astrology is rife in India, geomancy widespread in China.

Perhaps the most successful recent global pseudoscience—by many criteria, already a religion—is the Hindu doctrine of transcendental meditation (TM). The soporific homilies of its founder and spiritual leader, the Maharishi Mahesh Yogi, can be seen on television. Seated in the yogi position, his white hair here and there flecked with black, surrounded by garlands and floral offerings, he has a *look*. One day while channel surfing we came upon this visage. "You know who that is?" asked our four-year-old son. "God." The worldwide TM organization has an estimated valuation of $3 billion. For a fee they promise through meditation to be able to walk you through walls, to make you invisible, to enable you to fly. By thinking in unison they have, they say, diminished the crime rate in Washington, D.C., and caused the collapse of the Soviet Union, among other secular miracles. Not one smattering of real evidence has been offered for any such claims. TM sells folk medicine, runs trading companies, medical clinics and "research" universities, and has unsuccessfully entered politics. In its oddly charismatic leader, its promise of community, and the offer of magical powers in exchange for money and fervent belief, it is typical of many pseudosciences marketed for sacerdotal export.

At each relinquishing of civil controls and scientific education another little spurt in pseudoscience occurs. Leon Trotsky described it for Germany on the eve of the Hitler takeover (but in a description that might equally have applied to the Soviet Union of 1933):

> Not only in peasant homes, but also in city skyscrapers, there lives along side the twentieth century the thirteenth. A hundred mil-

lion people use electricity and still believe in the magic powers of signs and exorcisms. . . Movie stars go to mediums. Aviators who pilot miraculous mechanisms created by man's genius wear amulets on their sweaters. What inexhaustible reserves they possess of darkness, ignorance and savagery!

Russia is an instructive case. Under the Tsars, religious superstition was encouraged, but scientific and skeptical thinking—except by a few tame scientists—was ruthlessly expunged. Under Communism, both religion and pseudoscience were systematically suppressed—except for the superstition of the state ideological religion. It was advertised as scientific, but fell as far short of this ideal as the most unselfcritical mystery cult. Critical thinking—except by scientists in hermetically sealed compartments of knowledge—was recognized as dangerous, was not taught in the schools, and was punished where expressed. As a result, post-Communism, many Russians view science with suspicion. When the lid was lifted, as was also true of virulent ethnic hatreds, what had all along been bubbling subsurface was exposed to view. The region is now awash in UFOs, poltergeists, faith healers, quack medicines, magic waters, and old-time superstition. A stunning decline in life expectancy, increasing infant mortality, rampant epidemic disease, subminimal medical standards, and ignorance of preventative medicine all work to raise the threshold at which skepticism is triggered in an increasingly desperate population. As I write, the electorally most popular member of the Duma, a leading supporter of the ultranationalist Vladimir Zhirinovsky, is one Anatoly Kashpirovsky—a faith healer who remotely cures diseases ranging from hernias to AIDS by glaring at you out of your television set. His face starts stopped clocks.

A somewhat analogous situation exists in China. After the death of Mao Zedong and the gradual emergence of a market economy, UFOs, channeling and other examples of Western pseudoscience emerged, along with such ancient Chinese practices as ancestor worship, astrology and fortune telling—especially that version that involves throwing yarrow sticks and working through the hoary hexagrams of the *I Ching*. The government newspaper lamented that "the superstition of feudal ideology is reviving in our countryside." It was (and remains) a rural, not primarily an urban, affliction.

Individuals with "special powers" gained enormous followings.

They could, they said, project Qi, the "energy field of the Universe," out of their bodies to change the molecular structure of a chemical 2000 kilometers away, to communicate with aliens, to cure diseases. Some patients died under the ministrations of one of these "masters of Qi Gong" who was arrested and convicted in 1993. Wang Hongcheng, an amateur chemist, claimed to have synthesized a liquid, small amounts of which, when added to water, would convert it to gasoline or the equivalent. For a time he was funded by the army and the secret police, but when his invention was found to be a scam he was arrested and imprisoned. Naturally the story spread that his misfortune resulted not from fraud, but from his unwillingness to reveal his "secret formula" to the government. (Similar stories have circulated in America for decades, usually with the government role replaced by a major oil or auto company.) Asian rhinos are being driven to extinction because their horns, when pulverized, are said to prevent impotence; the market encompasses all of East Asia.

The government of China and the Chinese Communist Party were alarmed by certain of these developments. On December 5, 1994, they issued a joint proclamation that read in part:

> [P]ublic education in science has been withering in recent years. At the same time, activities of superstition and ignorance have been growing, and antiscience and pseudoscience cases have become frequent. Therefore, effective measures must be applied as soon as possible to strengthen public education in science. The level of public education in science and technology is an important sign of the national scientific accomplishment. It is a matter of overall importance in economic development, scientific advance, and the progress of society. We must be attentive and implement such public education as part of the strategy to modernize our socialist country and to make our nation powerful and prosperous. Ignorance is never socialist, nor is poverty.

So pseudoscience in America is part of a global trend. Its causes, dangers, diagnosis and treatment are likely to be similar everywhere. Here, psychics ply their wares on extended television commercials, personally endorsed by entertainers. They have their own channel, the "Psychic Friends Network"; a million people a year sign on and use such guidance in their everyday lives. For the CEOs of major corporations, for financial analysts, for lawyers and bankers there is a species of astrologer/soothsayer/psychic ready to advise on any matter. "If people knew how many people, especially the very rich and powerful ones, went to psychics, their jaws would drop through the floor," says a psychic from Cleveland, Ohio. Royalty has traditionally been vulnerable to psychic frauds. In ancient China and Rome astrology was the exclusive property of the emperor; any private use of this potent art was considered a capital offense. Emerging from a particularly credulous Southern California culture, Nancy and Ronald Reagan relied on an astrologer in private and public matters—unknown to the voting public. Some portion of the decision-making that influences the future of our civilization is plainly in the hands of charlatans. If anything, the practice is comparatively muted in America; its venue is worldwide.

—

As amusing as some of pseudoscience may seem, as confident as we may be that we would never be so gullible as to be swept up by such a doctrine, we know it's happening all around us. Transcendental Meditation and Aum Shinrikyo seem to have attracted a large number of accomplished people, some with advanced degrees in physics or engineering. These are not doctrines for nitwits. Something else is going on.

What's more, no one interested in what religions are and how they begin can ignore them. While vast barriers may seem to stretch between a local, single-focus contention of pseudoscience and something like a world religion, the partitions are very thin. The world presents us with nearly insurmountable problems. A wide variety of solutions are offered, some of very limited worldview, some of portentous sweep. In the usual Darwinian natural selection of doctrines, some thrive for a time, while most quickly vanish. But a few—sometimes, as history has shown, the most scruffy and least prepossessing among them—may have the power to profoundly change the history of the world.

The continuum stretching from ill-practiced science, pseudoscience, and superstition (New Age or Old), all the way to respectable mystery religion, based on revelation, is indistinct. I try not to use the word "cult" in this book in its usual meaning of a religion the speaker dislikes, but try to reach for the headstone of knowledge—do they really know what they claim to know? Everyone, it turns out, has relevant expertise.

In certain passages of this book I will be critical of the excesses of theology, because at the extremes it is difficult to distinguish pseudoscience from rigid, doctrinaire religion. Nevertheless, I want to acknowledge at the outset the prodigious diversity and complexity of religious thought and practice over the millennia; the growth of liberal religion and ecumenical fellowship during the last century; and the fact that—as in the Protestant Reformation, the rise of Reform Judaism, Vatican II, and the so-called higher criticism of the Bible—religion has fought (with varying degrees of success) its own excesses. But in parallel to the many scientists who seem reluctant to debate or even publicly discuss pseudoscience, many proponents of mainstream religions are reluctant to take on extreme conservatives and fundamentalists. If the trend continues, eventually the field is theirs; they can win the debate by default.

One religious leader writes to me of his longing for "disciplined integrity" in religion:

> We have grown far too sentimental. . . Devotionalism and cheap psychology on one side, and arrogance and dogmatic intolerance on the other distort authentic religious life almost beyond recognition. Sometimes I come close to despair, but then I live tenaciously and always with hope. . . Honest religion, more familiar than its critics with the distortions and absurdities perpetrated in its name, has an active interest in encouraging a healthy skepticism for its own purposes. . . There is the possibility for religion and science to forge a potent partnership against pseudo-science. Strangely, I think it would soon be engaged also in opposing pseudo-religion.

Pseudoscience differs from erroneous science. Science thrives on errors, cutting them away one by one. False conclusions are drawn all the time, but they are drawn tentatively. Hypotheses are framed so they are capable of being disproved. A succession of alternative hypotheses is confronted by experiment and observation. Science gropes and staggers toward improved understanding. Proprietary feelings are of course offended when a scientific hypothesis is disproved, but such disproofs are recognized as central to the scientific enterprise.

Pseudoscience is just the opposite. Hypotheses are often framed precisely so they are invulnerable to any experiment that offers a prospect of disproof, so even in principle they cannot be invalidated. Practitioners are defensive and wary. Skeptical scrutiny is opposed. When the pseudoscientific hypothesis fails to catch fire with scientists, conspiracies to suppress it are deduced.

Motor ability in healthy people is almost perfect. We rarely stumble and fall, except in young and old age. We can learn tasks such as riding a bicycle or skating or skipping, jumping rope or driving a car, and retain that mastery for the rest of our lives. Even if we've gone a decade without doing it, it comes back to us effortlessly. The precision and retention of our motor skills may, however, give us a false sense of confidence in our other talents. Our perceptions are fallible. We sometimes see what isn't there. We are prey to optical illusions. Occasionally we hallucinate. We are error-prone. A most illuminating book called *How We Know What Isn't So: The Fallibility of Human Reason in Everyday Life*, by Thomas Gilovich, shows how people systematically err in understanding numbers, in rejecting unpleasant evidence, in being influenced by the opinions of others. We're good in some things, but not in everything. Wisdom lies in understanding our limitations. "For Man is a giddy thing," teaches William Shakespeare. That's where the stuffy skeptical rigor of science comes in.

Perhaps the sharpest distinction between science and pseudoscience is that science has a far keener appreciation of human imperfections and fallibility than does pseudoscience (or "inerrant" revelation). If we resolutely refuse to acknowledge where we are liable to fall into error, then we can confidently expect that error—even serious error, profound mistakes—will be our companion forever. But if we are capable of a little courageous self-assessment, whatever rueful reflections they may engender, our chances improve enormously.

If we teach only the findings and products of science—no matter how useful and even inspiring they may be—without communicating its critical method, how can the average person possibly distinguish science from pseudoscience? Both then are presented as unsupported assertion. In Russia and China, it used to be easy. Authoritative science was what the authorities taught. The distinction between science and pseudoscience was made *for* you. No perplexities needed to be muddled through. But when profound political changes occurred and strictures on free thought were loosened, a host of confident or charismatic claims—especially those that told us what we wanted to hear—gained a vast following. Every notion, however improbable, became authoritative.

It is a supreme challenge for the popularizer of science to make clear the actual, tortuous history of its great discoveries and the misapprehensions and occasional stubborn refusal by its practitioners to change course. Many, perhaps most, science textbooks for budding scientists tread lightly here. It is enormously easier to present in an appealing way the wisdom distilled from centuries of patient and collective interrogation of Nature than to detail the messy distillation apparatus. The method of science, as stodgy and grumpy as it may seem, is far more important than the findings of science.

Chapter 2

———

SCIENCE
AND HOPE

Two men came to a hole in the sky.
One asked the other to lift him up . . .
But so beautiful was it in heaven that
the man who looked in over the edge
forgot everything, forgot his companion
whom he had promised to help up
and simply ran off into all the
splendor of heaven.

from an Iglulik Inuit prose poem, early
twentieth century, told by I N U G P A S U G J U K
to K N U D   R A S M U S S E N , the Greenlandic
arctic explorer

I was a child in a time of hope. I wanted to be a scientist from my earliest school days. The crystallizing moment came when I first caught on that the stars are mighty suns, when it first dawned on me how staggeringly far away they must be to appear as mere points of light in the sky. I'm not sure I even knew the meaning of the word "science" then, but I wanted somehow to immerse myself in all that grandeur. I was gripped by the splendor of the Universe, transfixed by the prospect of understanding how things really work, of helping to uncover deep mysteries, of exploring new worlds—maybe even literally. It has been my good fortune to have had that dream in part fulfilled. For me, the romance of science remains as appealing and new as it was on that day, more than half a century ago, when I was shown the wonders of the 1939 World's Fair.

Popularizing science—trying to make its methods and findings accessible to non-scientists—then follows naturally and immediately. *Not* explaining science seems to me perverse. When you're in love, you want to tell the world. This book is a personal statement, reflecting my lifelong love affair with science.

But there's another reason: Science is more than a body of knowledge; it is a way of thinking. I have a foreboding of an America in my children's or grandchildren's time—when the United States is a service and information economy; when nearly all the key manufacturing industries have slipped away to other countries; when awesome technological powers are in the hands of a very few, and no one representing the public interest can even grasp the issues; when the people have lost the ability to set their own agendas or knowledgeably question those in authority; when, clutching our crystals and nervously consulting our horoscopes, our critical faculties in decline, unable to distinguish between what feels good and what's true, we slide, almost without noticing, back into superstition and darkness.

The dumbing down of America is most evident in the slow decay of substantive content in the enormously influential media, the 30-second sound bites (now down to 10 seconds or less), lowest common denominator programming, credulous presentations on pseudoscience and superstition, but especially a kind of celebration of ignorance. As I write, the number-one videocassette rental in America is the movie *Dumb and Dumber*. "Beavis and Butthead" remain popular (and influential) with young TV viewers. The plain lesson is that study and learning—not just of science, but of anything—are avoidable, even undesirable.

We've arranged a global civilization in which most crucial elements—transportation, communications, and all other industries; agriculture, medicine, education, entertainment, protecting the environment; and even the key democratic institution of voting—profoundly depend on science and technology. We have also arranged things so that almost no one understands science and technology. This is a prescription for disaster. We might get away with it for a while, but sooner or later this combustible mixture of ignorance and power is going to blow up in our faces.

A Candle in the Dark is the title of a courageous, largely Biblically based, book by Thomas Ady, published in London in 1656, attacking the witch hunts then in progress as a scam "to delude the people." Any illness or storm, anything out of the ordinary, was popularly attributed to witchcraft. Witches must exist, Ady quoted the "witchmongers" as arguing—"else how should these things be, or come to pass?" For much of our history, we were so fearful of the outside world, with its unpredictable dangers, that we gladly embraced anything that promised to soften or explain away the terror. Science is an attempt, largely successful, to understand the world, to get a grip on things, to get hold of ourselves, to steer a safe course. Microbiology and meteorology now explain what only a few centuries ago was considered sufficient cause to burn women to death.

Ady also warned of the danger that "the Nations [will] perish for lack of knowledge." Avoidable human misery is more often caused not so much by stupidity as by ignorance, particularly our ignorance about ourselves. I worry that, especially as the Millennium edges nearer, pseudoscience and superstition will seem year by year more tempting, the siren song of unreason more sonorous and attractive. Where have we heard it before? Whenever our ethnic or national prejudices are aroused, in times of scarcity, during challenges to national self-esteem or nerve, when we agonize about our diminished cosmic place and purpose, or when fanaticism is bubbling up around us—then, habits of thought familiar from ages past reach for the controls.

The candle flame gutters. Its little pool of light trembles. Darkness gathers. The demons begin to stir.

—

There is much that science doesn't understand, many mysteries still to be resolved. In a Universe tens of billions of light-years across and some ten or fifteen billion years old, this may be the case forever. We are constantly stumbling on surprises. Yet some New Age and religious writers assert that scientists believe that "what they find is all there is." Scientists may reject mystic revelations for which there is no evidence except somebody's say-so, but they hardly believe their knowledge of Nature to be complete.

Science is far from a perfect instrument of knowledge. It's just the best we have. In this respect, as in many others, it's like democracy. Science by itself cannot advocate courses of human action, but it can certainly illuminate the possible consequences of alternative courses of action.

The scientific way of thinking is at once imaginative and disciplined. This is central to its success. Science invites us to let the facts in, even when they don't conform to our preconceptions. It counsels us to carry alternative hypotheses in our heads and see which best fit the facts. It urges on us a delicate balance between no-holds-barred openness to new ideas, however heretical, and the most rigorous skeptical scrutiny of everything—new ideas and established wisdom. This kind of thinking is also an essential tool for a democracy in an age of change.

One of the reasons for its success is that science has built-in, error-correcting machinery at its very heart. Some may consider this an overbroad characterization, but to me every time we exercise self-criticism, every time we test our ideas against the outside world, we are doing science. When we are self-indulgent and uncritical, when we confuse hopes and facts, we slide into pseudoscience and superstition.

Every time a scientific paper presents a bit of data, it's accompanied by an error bar—a quiet but insistent reminder that no knowledge is complete or perfect. It's a calibration of how much we trust what we think we know. If the error bars are small, the accuracy of our empirical knowledge is high; if the error bars are large, then so is the uncertainty in our knowledge. Except in pure mathematics, nothing is known for certain (although much is certainly false).

Moreover, scientists are usually careful to characterize the veridical status of their attempts to understand the world—ranging from conjectures and hypotheses, which are highly tentative, all the way up to laws of Nature which are repeatedly and systematically confirmed through many interrogations of how the world works. But even laws of Nature are not absolutely certain. There may be new circumstances never before examined—inside black holes, say, or within the electron, or close to the speed of light—where even our vaunted laws of Nature break down and, however valid they may be in ordinary circumstances, need correction.

Humans may crave absolute certainty; they may aspire to it; they may pretend, as partisans of certain religions do, to have attained it. But the history of science—by far the most successful claim to knowledge accessible to humans—teaches that the most we can hope for is successive improvement in our understanding, learning from our mistakes, an asymptotic approach to the Universe, but with the proviso that absolute certainty will always elude us.

We will always be mired in error. The most each generation can hope for is to reduce the error bars a little, and to add to the body of data to which error bars apply. The error bar is a pervasive, visible self-assessment of the reliability of our knowledge. You can often see error bars in public opinion polls ("an uncertainty of plus or minus 3 percent," say). Imagine a society in which every speech in the *Congressional Record*, every television commercial, every sermon had an accompanying error bar or its equivalent.

One of the great commandments of science is, "Mistrust arguments from authority." (Scientists, being primates, and thus given to dominance hierarchies, of course do not always follow this commandment.) Too many such arguments have proved too painfully wrong. Authorities must prove their contentions like everybody else. This independence of science, its occasional unwillingness to accept conventional wisdom, makes it dangerous to doctrines less self-critical, or with pretensions to certitude.

Because science carries us toward an understanding of how the world is, rather than how we would wish it to be, its findings may not in all cases be immediately comprehensible or satisfying. It may take a little work to restructure our mindsets. Some of science is very simple. When it gets complicated, that's usually because the world is complicated—or because *we're* complicated. When we shy away from it because it seems too difficult (or because we've been taught so poorly), we surrender the ability to take charge of our future. We are disenfranchised. Our self-confidence erodes.

But when we pass beyond the barrier, when the findings and methods of science get through to us, when we understand and put this knowledge to use, many feel deep satisfaction. This is true for everyone, but especially for children—born with a zest for knowledge, aware that they must live in a future molded by science, but so often convinced in their adolescence that science is not for them. I know personally, both from having science explained to me and from my attempts to explain it to others, how gratifying it is when we get it, when obscure terms suddenly take on meaning, when we grasp what all the fuss is about, when deep wonders are revealed.

In its encounter with Nature, science invariably elicits a sense of reverence and awe. The very act of understanding is a celebration of joining, merging, even if on a very modest scale, with the magnificence of the Cosmos. And the cumulative worldwide buildup of knowledge over time converts science into something only a little short of a transnational, transgenerational meta-mind.

"Spirit" comes from the Latin word "to breathe." What we breathe is air, which is certainly matter, however thin. Despite usage to the contrary, there is no necessary implication in the word "spiritual" that we are talking of anything other than matter (including the matter of which the brain is made), or anything outside the realm of science. On occasion, I will feel free to use the word. Science is not only compatible with spirituality; it is a profound source of spirituality. When we recognize our place in an immensity of light-years and in the passage of ages, when we grasp the intricacy, beauty, and subtlety of life, then that soaring feeling, that sense of elation and humility combined, is surely spiritual. So are our emotions in the presence of great art or music or literature, or of acts of exemplary selfless courage such as those of Mohandas Gandhi or Martin Luther King, Jr. The notion that science and spirituality are somehow mutually exclusive does a disservice to both.

—

Science may be hard to understand. It may challenge cherished beliefs. When its products are placed at the disposal of politicians or industrialists, it may lead to weapons of mass destruction and grave threats to the environment. But one thing you have to say about it: It delivers the goods.

Not every branch of science can foretell the future—paleontology can't—but many can and with stunning accuracy. If you want to know when the next eclipse of the Sun will be, you might try magicians or mystics, but you'll do much better with scientists. They will tell you where on Earth to stand, when you have to be there, and whether it will be a partial eclipse, a total eclipse, or an annular eclipse. They can routinely predict a solar eclipse, to the minute, a millennium in advance. You can go to the witch doctor to lift the spell that causes your pernicious anemia, or you can take vitamin B_{12}. If you want to save your child from polio, you can pray or you can inoculate. If you're interested in the sex of your unborn child, you can consult plumb-bob danglers all you want (left-right, a boy; forward-back, a girl—or maybe it's the other way around), but they'll be right, on average, only one time in two. If you want real accuracy (here, 99 percent accuracy), try amniocentesis and sonograms. Try science.

Think of how many religions attempt to validate themselves with prophecy. Think of how many people rely on these prophecies, however vague, however unfulfilled, to support or prop up their beliefs. Yet has there ever been a religion with the prophetic accuracy and reliability of science? There isn't a religion on the planet that doesn't long for a comparable ability—precise, and repeatedly demonstrated before committed skeptics—to foretell future events. No other human institution comes close.

Is this worshiping at the altar of science? Is this replacing one faith by another, equally arbitrary? In my view, not at all. The directly observed success of science is the reason I advocate its use. If something else worked better, I would advocate the something else. Does science insulate itself from philosophical criticism? Does it define itself as having a monopoly on the "truth"? Think again of that eclipse a thousand years in the future. Compare as many doctrines as you can think of, note what predictions they make of the future, which ones are vague, which ones are precise, and which doctrines—every one of them subject to human fallibility—have error-correcting mechanisms built in. Take account of the fact that not one of them is perfect. Then simply pick the one that in a fair comparison works (as opposed to feels) best. If different doctrines are superior in quite separate and independent fields, we are of course free to choose several—but not if they contradict one another. Far from being idolatry, this is the means by which we can distinguish the false idols from the real thing.

Again, the reason science works so well is partly that built-in error-correcting machinery. There are no forbidden questions in science, no matters too sensitive or delicate to be probed, no sacred truths. That openness to new ideas, combined with the most rigorous, skeptical scrutiny of all ideas, sifts the wheat from the chaff. It makes no difference how smart, august, or beloved you are. You must prove your case in the face of determined, expert criticism. Diversity and debate are valued. Opinions are encouraged to contend—substantively and in depth.

The process of science may sound messy and disorderly. In a way, it is. If you examine science in its everyday aspect, of course you find that scientists run the gamut of human emotion, personality, and character. But there's one facet that is really striking to the outsider, and that is the gauntlet of criticism considered acceptable or even desirable. There is much warm and inspired encouragement of apprentice scientists by their mentors. But the poor graduate student at his or her Ph.D. oral exam is subjected to a withering crossfire of questions from the very professors who have the candidate's future in their grasp. Naturally the students are nervous; who wouldn't be? True, they've prepared for it for years. But they understand that at this critical moment, they have to be able to answer searching questions posed by experts. So in preparing to defend their theses, they must practice a very useful habit of thought: They must anticipate questions; they have to ask: Where in my dissertation is there a weakness that someone else might find? I'd better identify it before they do.

You sit in at contentious scientific meetings. You find university colloquia in which the speaker has hardly gotten 30 seconds into the talk before there are devastating questions and comments from the audience. You examine the conventions in which a written report is submitted to a scientific journal, for possible publication, then is conveyed by the editor to anonymous referees whose job it is to ask: Did the author do anything stupid? Is there anything in here that is sufficiently interesting to be published? What are the deficiencies of this paper? Have the main results been found by anybody else? Is the argument adequate, or should the paper be resubmitted after the author has actually demonstrated what is here only speculated on? And it's anonymous: The author doesn't know who the critics are. This is the everyday expectation in the scientific community.

Why do we put up with it? Do we like to be criticized? No, no scientist enjoys it. Every scientist feels a proprietary affection for his or her ideas and findings. Even so, you don't reply to critics, Wait a minute; this is a really good idea; I'm very fond of it; it's done you no harm; please leave it alone. Instead, the hard but just rule is that if the ideas don't work, you must throw them away. Don't waste neurons on what doesn't work. Devote those neurons to new ideas that better explain the data. The British physicist Michael Faraday warned of the powerful temptation

> to seek for such evidence and appearances as are in the favour of our desires, and to disregard those which oppose them. . . We receive as friendly that which agrees with [us], we resist with dislike that which opposes us; whereas the very reverse is required by every dictate of common sense.

Valid criticism does you a favor.

Some people consider science arrogant—especially when it purports to contradict beliefs of long standing or when it introduces bizarre concepts that seem contradictory to common sense. Like an earthquake that rattles our faith in the very ground we're standing on, challenging our accustomed beliefs, shaking the doctrines we have grown to rely upon can be profoundly disturbing. Nevertheless, I maintain that science is part and parcel humility. Scientists do not seek to impose their needs and wants on Nature, but instead humbly interrogate Nature and take seriously what they find. We are aware that revered scientists have been wrong. We understand human imperfection. We insist on independent and—to the extent possible—quantitative verification of proposed tenets of belief. We are constantly prodding, challenging, seeking contradictions or small, persistent residual errors, proposing alternative explanations, encouraging heresy. We give our highest rewards to those who convincingly disprove established beliefs.

Here's one of many examples: The laws of motion and the inverse square law of gravitation associated with the name of Isaac Newton are properly considered among the crowning achievements of the human species. Three hundred years later we use Newtonian dynamics to predict those eclipses. Years after launch, billions of miles from Earth (with only tiny corrections from Einstein), the spacecraft beautifully arrives at a predetermined point in the orbit of the target world, just as the world comes ambling by. The accuracy is astonishing. Plainly, Newton knew what he was doing.

But scientists have not been content to leave well enough alone. They have persistently sought chinks in the Newtonian armor. At high speeds and strong gravities, Newtonian physics breaks down. This is one of the great findings of Albert Einstein's Special and General Relativity, and is one of the reasons his memory is so greatly honored. Newtonian physics is valid over a wide range of conditions including those of everyday life. But in certain circumstances highly unusual for human beings—we are not, after all, in the habit of traveling near light speed—it simply doesn't give the right answer; it does not conform to observations of Nature. Special and General Relativity are indistinguishable from Newtonian physics in its realm of validity, but make very different predictions—predictions in excellent accord with observation—in those other regimes (high speed, strong gravity). Newtonian physics turns out to be an approximation to the truth, good in circumstances with which we are routinely familiar, bad in others. It is a splendid and justly celebrated accomplishment of the human mind, but it has its limitations.

However, in accord with our understanding of human fallibility, heeding the counsel that we may asymptotically approach the truth but will never fully reach it, scientists are today investigating regimes in which General Relativity may break down. For example, General Relativity predicts a startling phenomenon called gravitational waves. They have never been detected directly. But if they do not exist, there is something fundamentally wrong with General Relativity. Pulsars are rapidly rotating neutron stars whose flicker rates can now be measured to fifteen decimal places. Two very dense pulsars in orbit around each other are predicted to radiate copious quantities of gravitational waves—which will in time slightly alter the orbits and rotation periods of the two stars. Joseph Taylor and Russell Hulse of Princeton University have used this method to test the predictions of General Relativity in a wholly novel way. For all they knew, the results would be inconsistent with General Relativity and they would have overturned one of the chief pillars of modern physics. Not only were they willing to challenge General Relativity, they were widely encouraged to do so. As it turns out, the observations of binary pulsars give a precise verification of the predictions of General Relativity, and for this Taylor and Hulse were co-recipients of the 1993 Nobel Prize in Physics. In diverse ways, many other physicists are testing General Relativity—for example by attempting directly to detect the elusive gravitational waves. They hope to strain the theory to the breaking point and discover whether a regime of Nature exists in which Einstein's great advance in understanding in turn begins to fray.

These efforts will continue as long as there are scientists. General Relativity is certainly an inadequate description of Nature at the quantum level, but even if that were not the case, even if General Relativity were everywhere and forever valid, what better way of convincing ourselves of its validity than a concerted effort to discover its failings and limitations?

This is one of the reasons that the organized religions do not inspire me with confidence. Which leaders of the major faiths acknowledge that their beliefs might be incomplete or erroneous and establish institutes to uncover possible doctrinal deficiencies? Beyond the test of everyday living, who is systematically testing the circumstances in which traditional religious teachings may no longer apply? (It is certainly conceivable that doctrines and ethics that may have worked fairly well in patriarchal or patristic or medieval times might be thoroughly invalid in the very different world we inhabit today.) What sermons even-handedly examine the God hypothesis? What rewards are religious skeptics given by the established religions—or, for that matter, social and economic skeptics by the society in which they swim?

Science, Ann Druyan notes, is forever whispering in our ears, "Remember, you're very new at this. You might be mistaken. You've been wrong before." Despite all the talk of humility, show me something comparable in religion. Scripture is said to be divinely inspired—a phrase with many meanings. But what if it's simply made up by fallible humans? Miracles are attested, but what if they're instead some mix of charlatanry, unfamiliar states of consciousness, misapprehensions of natural phenomena, and mental illness? No contemporary religion and no New Age belief seems to me to take sufficient account of the grandeur, magnificence, subtlety and intricacy of the Universe revealed by science. The fact that so little of the findings of modern science is prefigured in Scripture to my mind casts further doubt on its divine inspiration.

But of course I might be wrong.

—

Read the following two paragraphs—not to understand the science described, but to get a feeling for the author's style of thinking. He is facing anomalies, apparent paradoxes in physics; "asymmetries" he calls them. What can we learn from them?

It is known that Maxwell's electrodynamics—as usually understood at the present time—when applied to moving bodies, leads to asymmetries which do not appear to be inherent in the phenomena. Take, for example, the reciprocal electrodynamic action of a magnet and a conductor. The observable phenomenon here depends only on the relative motion of the conductor and the magnet, whereas the customary view draws a sharp distinction between the two cases in which either the one or the other of these bodies is in motion. For if the magnet is in motion and the conductor at rest, there arises in the neighbourhood of the magnet an electric field with a certain definite energy, producing a current at the places where parts of the conductor are situated. But if the magnet is stationary and the conductor in motion, no electric field arises in the neighbourhood of the magnet. In the conductor, however, we find an electromotive force, to which in itself there is no corresponding energy, but which gives rise—assuming equality of relative motion in the two cases discussed—to electric currents of the same path and intensity as those produced by the electric forces in the former case.

Examples of this sort, together with the unsuccessful attempts to discover any motion of the earth relative to the "ether," suggest that the phenomena of electrodynamics as well as of mechanics possess no properties corresponding to the idea of absolute rest. They suggest rather that, as has already been shown to the first order of small quantities, the same laws of electrodynamics and optics will be valid for all frames of reference for which the equations of mechanics hold good.

What is the author trying to tell us here? I'll try to explain the background later in this book. For now, we can perhaps recognize that the language is spare, technical, cautious, clear, and not a jot more complicated than it need be. You would not offhand guess from how it's phrased (or from its unostentatious title, "On the Electrodynamics of Moving Bodies") that this article represents the crucial arrival of the theory of Special Relativity into the world, the gateway to the triumphant announcement of the equivalence of mass and energy, the deflation of the conceit that our small world occupies some "privileged reference frame" in the Universe, and in several different ways an epochal event in human history. The opening words of Albert Einstein's 1905 paper are characteristic of the scientific report. It is refreshingly unself-serving, circumspect, understated. Contrast its restrained tone with, say, the products of modern advertising, political speeches, authoritative theological pronouncements—or for that matter the blurb on the cover of this book.

Notice how Einstein's paper begins by trying to make sense of experimental results. Wherever possible, scientists experiment. Which experiments suggest themselves often depends on which theories currently prevail. Scientists are intent on testing those theories to the breaking point. They do not trust what is intuitively obvious. That the Earth is flat was once obvious. That heavy bodies fall faster than light ones was once obvious. That bloodsucking leeches cure most diseases was once obvious. That some people are naturally and by divine decree slaves was once obvious. That there is such a place as the center of the Universe, and that the Earth sits in that exalted spot was once obvious. That there is an absolute standard of rest was once obvious. The truth may be puzzling or counterintuitive. It may contradict deeply held beliefs. Experiment is how we get a handle on it.

At a dinner many decades ago, the physicist Robert W. Wood was asked to respond to the toast, "To physics and metaphysics." By "meta-physics," people then meant something like philosophy, or truths you could recognize just by thinking about them. They could also have included pseudoscience. Wood answered along these lines:

The physicist has an idea. The more he thinks it through, the more sense it seems to make. He consults the scientific literature. The more he reads, the more promising the idea becomes. Thus prepared, he goes to the laboratory and devises an experiment to test it. The experiment is painstaking. Many possibilities are checked. The accuracy of measurement is refined, the error bars reduced. He lets the chips fall where they may. He is devoted only to what the experiment teaches. At the end of all this work, through careful experimentation, the idea is found to be worthless. So the physicist discards it, frees his mind from the clutter of error, and moves on to something else.*

The difference between physics and metaphysics, Wood concluded as he raised his glass high, is not that the practitioners of one are smarter than the practitioners of the other. The difference is that the metaphysicist has no laboratory.

—

For me, there are four main reasons for a concerted effort to convey science—in radio, TV, movies, newspapers, books, computer programs, theme parks, and classrooms—to every citizen. In all uses of science, it is insufficient—indeed it is dangerous—to produce only a small, highly competent, well-rewarded priesthood of professionals. Instead, some fundamental understanding of the findings and methods of science must be available on the broadest scale.

• Despite plentiful opportunities for misuse, science can be the golden road out of poverty and backwardness for emerging nations. It makes national economies and the global civilization run. Many nations understand this. It is why so many graduate students in science and engineering at American universities—still the best in the

---

* As the pioneering physicist Benjamin Franklin put it, "In going on with these experiments, how many pretty systems do we build, which we soon find ourselves obliged to destroy?" At the very least, he thought, the experience sufficed to "help to make a vain Man humble."

world—are from other countries. The corollary, one that the United States sometimes fails to grasp, is that abandoning science is the road back into poverty and backwardness.

• Science alerts us to the perils introduced by our world-altering technologies, especially to the global environment on which our lives depend. Science provides an essential early warning system.

• Science teaches us about the deepest issues of origins, natures, and fates—of our species, of life, of our planet, of the Universe. For the first time in human history we are able to secure a real understanding of some of these matters. Every culture on Earth has addressed such issues and valued their importance. All of us feel goosebumps when we approach these grand questions. In the long run, the greatest gift of science may be in teaching us, in ways no other human endeavor has been able, something about our cosmic context, about where, when, and who we are.

• The values of science and the values of democracy are concordant, in many cases indistinguishable. Science and democracy began—in their civilized incarnations—in the same time and place, Greece in the seventh and sixth centuries B.C. Science confers power on anyone who takes the trouble to learn it (although too many have been systematically prevented from doing so). Science thrives on, indeed requires, the free exchange of ideas; its values are antithetical to secrecy. Science holds to no special vantage points or privileged positions. Both science and democracy encourage unconventional opinions and vigorous debate. Both demand adequate reason, coherent argument, rigorous standards of evidence and honesty. Science is a way to call the bluff of those who only pretend to knowledge. It is a bulwark against mysticism, against superstition, against religion misapplied to where it has no business being. If we're true to its values, it can tell us when we're being lied to. It provides a mid-course correction to our mistakes. The more widespread its language, rules, and methods, the better chance we have of preserving what Thomas Jefferson and his colleagues had in mind. But democracy can also be subverted more thoroughly through the products of science than any pre-industrial demagogue ever dreamed.

Finding the occasional straw of truth awash in a great ocean of confusion and bamboozle requires vigilance, dedication, and courage. But if we don't practice these tough habits of thought, we cannot hope to solve the truly serious problems that face us—and we risk becoming a nation of suckers, a world of suckers, up for grabs by the next charlatan who saunters along.

—

An extraterrestrial being, newly arrived on Earth—scrutinizing what we mainly present to our children in television, radio, movies, newspapers, magazines, the comics, and many books—might easily conclude that we are intent on teaching them murder, rape, cruelty, superstition, credulity, and consumerism. We keep at it, and through constant repetition many of them finally get it. What kind of society could we create if, instead, we drummed into them science and a sense of hope?

Chapter 3

THE MAN
IN THE MOON
AND THE FACE
ON MARS

The moon leaps
In the Great River's current . . .
Floating on the wind,
What do I resemble?

DU FU,
"Traveling at Night"
(China, Tang Dynasty, 765)

Each field of science has its own complement of pseudoscience. Geophysicists have flat Earths, hollow Earths, Earths with wildly bobbing axes to contend with, rapidly rising and sinking continents, plus earthquake prophets. Botanists have plants whose passionate emotional lives can be monitored with lie detectors, anthropologists have surviving ape-men, zoologists have extant dinosaurs, and evolutionary biologists have Biblical literalists snapping at their flanks. Archaeologists have ancient astronauts, forged runes, and spurious statuary. Physicists have perpetual motion machines, an army of amateur relativity disprovers, and perhaps cold fusion. Chemists still have alchemy. Psychologists have much of psychoanalysis and almost all of parapsychology. Economists have long-range economic forecasting. Meteorologists, so far, have long-range weather forecasting, as in the sunspot-oriented *Farmer's Almanac* (although long-term climate forecasting is another matter). Astronomy has, as its most prominent pseudoscience, astrology—the discipline out of which it emerged. The pseudosciences sometimes intersect, compounding the confusion—as in telepathic searches for buried treasures from Atlantis, or astrological economic forecasting.

But because I work mainly with planets, and because I've been interested in the possibility of extraterrestrial life, the pseudosciences that most often park themselves on my doorstep involve other worlds and what we have come so easily in our time to call "aliens." In the chapters immediately following, I want to lay out two recent, somewhat related pseudoscientific doctrines. They share the possibility that human perceptual and cognitive imperfections play a role in deceiving us on matters of great import. The first contends that a giant stone face from ages past is staring expressionlessly up at the sky from the sands of Mars. The second maintains that alien beings from distant worlds visit the Earth with casual impunity.

Even when summarized so baldly, isn't there a kind of thrill in contemplating these claims? What if such hoary science fiction ideas—resonant surely with deep human fears and longings—actually were coming to pass? Whose interest can fail to be aroused? Immersed in such material, even the crassest cynic is stirred. Are we absolutely sure, beyond the shadow of a doubt, that we can dismiss these claims? And if hardened debunkers can sense the appeal, what must those untutored in scientific skepticism, like Mr. "Buckley," feel?

—

For most of history—before spacecraft, before telescopes, when we were still largely immersed in magical thinking—the Moon was an enigma. Almost no one thought of it as a world.

What do we actually see when we look up at the Moon with the naked eye? We make out a configuration of irregular bright and dark markings—not a close representation of any familiar object. But, almost irresistibly, our eyes connect the markings, emphasizing some, ignoring others. We seek a pattern, and we find one. In world myth and folklore, many images are seen: a woman weaving, stands of laurel trees, an elephant jumping off a cliff, a girl with a basket on her back, a rabbit, the lunar intestines spilled out on its surface after evisceration by an irritable flightless bird, a woman pounding tapa cloth, a four-eyed jaguar. People of one culture have trouble understanding how such bizarre things could be seen by the people of another.

The most common image is the Man in the Moon. Of course, it doesn't really look like a man. Its features are lopsided, warped, drooping. There's a beefsteak or something over the left eye. And what expression does that mouth convey? An "O" of surprise? A hint of sadness, even lamentation? Doleful recognition of the travails of life on Earth? Certainly the face is too round. The ears are missing. I guess he's bald on top. Nevertheless, every time I look at it, I see a human face.

World folklore depicts the Moon as something prosaic. In the pre-Apollo generation, children were told that the Moon was made of green (that is, smelly) cheese, and for some reason this was thought not marvelous but hilarious. In children's books and editorial cartoons, the Man in the Moon is often drawn simply as a face set in a circle, not too different from the bland "happy face" of a pair of dots and an upturned arc. Benignly, he looks down on the nocturnal frolics of animals and children, of the knife and the spoon.

Consider again the two categories of terrain we recognize when we examine the Moon with the naked eye: the brighter forehead, cheeks, and chin; and the darker eyes and mouth. Through a telescope, the bright features are revealed to be ancient cratered highlands, dating back, we now know (from the radioactive dating of samples returned by the Apollo astronauts), to almost 4.5 billion years ago. The dark features are somewhat younger flows of basaltic lava called maria (singular, mare—both from the Latin word for ocean, although the Moon, we now know, is dry as a bone). The maria welled up in the first few hundred million years of lunar history, partly induced by the high-speed impact of enormous asteroids and comets. The right eye is Mare Imbrium, the beefsteak drooping over the left eye is the combination of Mare Serenitatis and Mare Tranquilitatis (where *Apollo 11* landed), and the off-center open mouth is Mare Humorum. (No craters can be made out by ordinary, unaided human vision.)

The Man in the Moon is in fact a record of ancient catastrophes—most of which took place before humans, before mammals, before vertebrates, before multicelled organisms, and probably even before life arose on Earth. It is a characteristic conceit of our species to put a human face on random cosmic violence.

—

Humans, like other primates, are a gregarious lot. We enjoy one another's company. We're mammals, and parental care of the young is essential for the continuance of the hereditary lines. The parent smiles at the child, the child smiles back, and a bond is forged or strengthened. As soon as the infant can see, it recognizes faces, and we now know that this skill is hardwired in our brains. Those infants who a million years ago were unable to recognize a face smiled back less, were less likely to win the hearts of their parents, and less likely to prosper. These days, nearly every infant is quick to identify a human face, and to respond with a goony grin.

As an inadvertent side effect, the pattern-recognition machinery in our brains is so efficient in extracting a face from a clutter of other detail that we sometimes see faces where there are none. We assemble disconnected patches of light and dark and unconsciously try to see a face. The Man in the Moon is one result. Michelangelo Antonioni's film *Blowup* describes another. There are many other examples.

Sometimes it's a geological formation, such as the Old Man of the Mountains at Franconia Notch, New Hampshire. We recognize that, rather than some supernatural agency or an otherwise undiscovered ancient civilization in New Hampshire, this is the product of erosion and collapse of a rock face. Anyway, it doesn't look much like a face anymore. There's the Devil's Head in North Carolina, the Sphinx Rock in Wastwater, England, the Old Woman in France, the Vartan Rock in Armenia. Sometimes it's a reclining woman, as Mt. Ixtacci-huatl in Mexico. Sometimes it's other body parts, like the Grand Tetons in Wyoming—approached from the West, a pair of mountain peaks named by French explorers. (Actually there are three.) Sometimes it's changing patterns in the clouds. In late medieval and Renaissance Spain, visions of the Virgin Mary were "confirmed" by people seeing saints in cloud forms. (While sailing out of Suva, Fiji, I once saw the head of a truly terrifying monster, jaws agape, set in a brooding storm cloud.)

Occasionally, a vegetable or a pattern of wood grain or the hide of a cow resembles a human face. There was a celebrated eggplant that closely resembled Richard M. Nixon. What shall we deduce from this fact? Divine or extraterrestrial intervention? Republican meddling in eggplant genetics? No. We recognize that there are large numbers of eggplants in the world and that, given enough of them, sooner or later we'll come upon one that looks like a human face, even a very particular human face.

When the face is of a religious personage—as, for example, a tortilla purported to exhibit the face of Jesus—believers tend quickly to deduce the hand of God. In an age more skeptical than most, they crave reassurance. Still, it seems unlikely that a miracle is being worked on so evanescent a medium. Considering how many tortillas have been pounded out since the beginning of the world, it would be surprising if a few didn't have at least vaguely familiar features.*

Magical properties have been ascribed to ginseng and mandrake roots, in part because of vague resemblances to the human form.

---

* These cases are very different from that of the so-called Shroud of Turin, which shows something too close to a human form to be a misapprehended natural pattern and which is now suggested by carbon-14 dating to be not the death shroud of Jesus, but a pious hoax from the fourteenth century—a time when the manufacture of fraud-ulent religious relics was a thriving and profitable home handicraft industry.

Some chestnut shoots show smiling faces. Some corals look like hands. The ear fungus (also unpleasantly called "Jew's ear") indeed looks like an ear, and something rather like enormous eyes can be seen on the wings of certain moths. Some of this may not be mere coincidence; plants and animals that suggest a face may be less likely to be gobbled up by creatures with faces—or creatures who are afraid of predators with faces. A "walking stick" is an insect spectacularly well disguised as a twig. Naturally, it tends to live on and around trees. Its mimicry of the plant world saves it from birds and other predators, and is almost certainly the reason that its extraordinary form was slowly molded by Darwinian natural selection. Such crossings of the boundaries between kingdoms of life are unnerving. A young child viewing a walking stick can easily imagine an army of sticks, branches, and trees marching for some ominous planty purpose.

Many instances of this sort are described and illustrated in a 1979 book called *Natural Likeness*, by John Michell, a British enthusiast of the occult. He takes seriously the claims of Richard Shaver, who—as described below—played a role in the origin of the UFO excitement in America. Shaver cut open rocks on his Wisconsin farm and discovered, written in a pictographic language that only he could see, much less understand, a comprehensive history of the world. Michell also accepts at face value the claims of the dramatist and Surrealist theoretician Antonin Artaud, who, in part under the influence of peyote, saw in the patterns on the outsides of rocks erotic images, a man being tortured, ferocious animals, and the like. "The whole landscape revealed itself," Michell says, "as the creation of a single thought." But a key question: was that thought inside or outside Artaud's head? Artaud concluded, and Michell agrees, that the patterns so apparent in the rocks were manufactured by an ancient civilization, rather than by Artaud's partly hallucinogen-induced altered state of consciousness. When Artaud returned from Mexico to Europe, he was diagnosed as mad. Michell decries the "materialist outlook" that greeted Artaud's patterns skeptically.

Michell shows us a photograph of the Sun taken in X-ray light which looks vaguely like a face and informs us that "followers of Gurdjieff see the face of their Master" in the solar corona. Innumerable faces in trees, mountains, and boulders all over the world are inferred to be the product of ancient wisdom. Perhaps some are: It's a good practical joke, as well as a tempting religious symbol, to pile stones so from afar they look like a giant face.

The view that most of these forms are patterns natural to rock-forming processes and the bilateral symmetry of plants and animals, plus a little natural selection—all processed through the human-biased filter of our perception—Michell describes as "materialism" and a "nineteenth-century delusion." "Conditioned by rationalist beliefs, our view of the world is duller and more confined than nature intended." By what process he has plumbed the intentions of Nature is not revealed.

Of the images he presents, Michell concludes that

their mystery remains essentially untouched, a constant source of wonder, delight and speculation. All we know for sure is that nature created them and at the same time gave us the apparatus to perceive them and minds to appreciate their endless fascination. For the greatest profit and enjoyment they should be viewed as nature intended, with the eye of innocence, unclouded by theories and preconceptions, with the manifold vision, innate in all of us, that enriches and dignifies human life, rather than with the cultivated single vision of the dull and opinionated.

—

Perhaps the most famous spurious claim of a portentous pattern involves the canals of Mars. First observed in 1877, they were seemingly confirmed by a succession of dedicated professional astronomers peering through large telescopes all over the world. A network of single and double straight lines was reported, crisscrossing the Martian surface and with such uncanny geometrical regularity that they could only be of intelligent origin. Evocative conclusions were drawn about a parched and dying planet populated by an older and wiser technical civilization dedicated to conservation of water resources. Hundreds of canals were mapped and named. But, oddly, they avoided showing up on photographs. The human eye, it was suggested, could remember the brief instants of perfect atmospheric transparency, while the undiscriminating photographic plate averaged the few clear with the many blurry moments. Some astronomers saw the canals. Many did not. Perhaps certain observers were more skilled at seeing canals. Or perhaps the whole business was some kind of perceptual delusion.

Much of the idea of Mars as an abode of life, as well as the prevalence of "Martians" in popular fiction, derives from the canals. I myself grew up steeped in this literature, and when I found myself an experimenter on the *Mariner* 9 mission to Mars—the first spacecraft to orbit the red planet—naturally I was interested to see what the real circumstances were. With *Mariner* 9 and with *Viking*, we were able to map the planet pole-to-pole, detecting features hundreds of times smaller than the best that could be seen from Earth. I found, not altogether to my surprise, not a trace of canals. There were a few more or less linear features that had been made out through the telescope—for example, a 5,000-kilometer-long rift valley that would have been hard to miss. But the hundreds of "classical" canals carrying water from the polar caps through the arid deserts to the parched equatorial cities simply did not exist. They were an illusion, some malfunction of the human hand-eye-brain combination at the limit of resolution when we peer through an unsteady and turbulent atmosphere.

Even a succession of professional scientists—including famous astronomers who had made other discoveries that are confirmed and now justly celebrated—can make serious, even profound errors in pattern recognition. Especially where the implications of what we think we are seeing seem to be profound, we may not exercise adequate self-discipline and self-criticism. The Martian canal myth constitutes an important cautionary tale.

For the canals, spacecraft missions provided the means of correcting our misapprehensions. But it is also true that some of the most haunting claims of unexpected patterns emerge from spacecraft exploration. In the early 1960s, I urged that we be attentive to the possibility of finding the artifacts of ancient civilizations—either those indigenous to a given world, or those constructed by visitors from elsewhere. I didn't imagine that this would be easy or probable, and I certainly did not suggest that, on so important a matter, anything short of iron-clad evidence would be worth considering.

Beginning with John Glenn's evocative report of "fireflies" surrounding his space capsule, every time an astronaut reported seeing something not immediately understood, there were those who deduced "aliens." Prosaic explanations—specks of paint flecking off the ship in the space environment, say—were dismissed with contempt. The lure of the marvelous blunts our critical faculties. (As if a man become a moon is not marvel enough.)

Around the time of the Apollo lunar landings, many non-experts—owners of small telescopes, flying saucer zealots, writers for aerospace magazines—pored over the returned photographs seeking anomalies that NASA scientists and astronauts had overlooked. Soon there were reports of gigantic Latin letters and Arabic numerals inscribed on the lunar surface, pyramids, highways, crosses, glowing UFOs. Bridges were reported on the Moon, radio antennas, the tracks of enormous crawling vehicles, and the devastation left by machines able to slice craters in two. Every one of these claims, though, turns out to be a natural lunar geological formation misjudged by amateur analysts, internal reflections in the optics of the astronauts' Hasselblad cameras, and the like. Some enthusiasts discerned the long shadows of ballistic missiles—Soviet missiles it was ominously confided, aimed at America. The rockets, also described as "spires," turn out to be low hills casting long shadows when the Sun is near the lunar horizon. A little trigonometry dispels the mirage.

These experiences also provide fair warning: For a complex terrain sculpted by unfamiliar processes, amateurs (and sometimes even professionals) examining photographs, especially near the limit of resolution, may get into trouble. Their hopes and fears, the excitement of possible discoveries of great import, may overwhelm the usual skeptical and cautious approach of science.

If we examine available surface images of Venus, occasionally a peculiar landform swims into view—as, for example, a rough portrait of Joseph Stalin discovered by American geologists analyzing Soviet orbital radar imagery. No one maintains, I gather, that unreconstructed Stalinists had doctored the magnetic tapes, or that the former Soviets were engaged in engineering activities of unprecedented and hitherto unrevealed scale on the surface of Venus—where every spacecraft to land has been fried in an hour or two. The odds are overwhelming that this feature, whatever it is, is due to geology. The same is true of what seems to be a portrait of the cartoon character Bugs Bunny on the Uranian moon Ariel. A Hubble Space Telescope image of Titan in the near-infrared shows clouds roughly configured to make a world-sized smiling face. Every planetary scientist has a favorite example.

The astronomy of the Milky Way also is replete with imagined likenesses—for example, the Horsehead, Eskimo, Owl, Homunculus, Tarantula, and North American Nebulae, all irregular clouds of gas and dust, illuminated by bright stars and each on a scale that dwarfs our solar system. When astronomers mapped the distribution of galaxies out to a few hundred million light-years, they found themselves outlining a crude human form which has been called "the Stickman." The configuration is understood as something like enormous adjacent soap bubbles, the galaxies formed on the surface of adjacent bubbles and almost no galaxies in the interiors. This makes it quite likely that they will mark out a pattern with bilateral symmetry something like the Stickman.

Mars is much more clement than Venus, although the *Viking* landers provided no compelling evidence for life. Its terrain is extremely heterogeneous and diverse. With 100,000 or so close-up photographs available, it is not surprising that claims have been made over the years about something unusual on Mars. There is, for example, a cheerful "happy face" sitting inside a Martian impact crater 8 kilometers (5 miles) across, with a set of radial splash marks outside, making it look like the conventional representation of a smiling Sun. But no one claims that this has been engineered by an advanced (and excessively genial) Martian civilization, perhaps to attract our attention. We recognize that, with objects of all sizes falling out of the sky, with the surface rebounding, slumping, and reconfiguring itself after each impact, and with ancient water and mudflows and modern windborne sand sculpting the surface, a wide variety of landforms must be generated. If we scrutinize 100,000 pictures, it's not surprising that occasionally we'll come upon something like a face. With our brains programmed for this from infancy, it would be amazing if we couldn't find one here and there.

A few small mountains on Mars resemble pyramids. In the Elysium high plateau, there is a cluster of them—the biggest a few kilometers across at the base—all oriented in the same direction. There is something a little eerie about these pyramids in the desert, so reminiscent of the Gizeh plateau in Egypt, and I would love to examine them more closely. Is it reasonable, though, to deduce Martian pharaohs?

Similar features are also known on Earth in miniature, especially in Antarctica. Some of them would come up to your knees. If we knew nothing else about them, would it be fair to conclude that they've been manufactured by scale-model Egyptians living in the Antarctic wasteland? (The hypothesis loosely fits the observations, but much else we know about the polar environment and the physiology of humans speaks against it.) They are, in fact, generated by wind erosion — the splatter of fine particles picked up by strong winds blowing mainly in the same direction and, over the years, sculpting what once were irregular hummocks into nicely symmetrical pyramids. They're called dreikanters, from a German word meaning three sides. This is order generated out of chaos by natural processes — something we see over and over again throughout the Universe (in rotating spiral galaxies, for instance). Each time it happens we're tempted to infer the direct intervention of a Maker.

On Mars, there is evidence of winds much fiercer than any ever experienced on Earth, ranging up to half the speed of sound. Planet-wide duststorms are common — carrying fine grains of sand. A steady pitter-patter of particles moving much faster than in the fiercest gales of Earth should, over ages of geological time, work profound changes in rock faces and landforms. It would not be too surprising if a few features — even very large ones — were sculpted by aeolian processes into the pyramidal forms we see.

—

There is a place on Mars called Cydonia, where a great stone face a kilometer across stares unblinkingly up at the sky. It is an unfriendly face, but one that seems recognizably human. In some representations, it could have been sculpted by Praxiteles. It lies in a landscape where many low hills have been molded into odd forms, perhaps by some mixture of ancient mudflows and subsequent wind erosion. From the number of impact craters, the surrounding terrain looks to be at least hundreds of millions of years old.

Intermittently, "The Face" has attracted attention, both in the United States and in the former Soviet Union. The headline in the November 20, 1984 *Weekly World News*, a supermarket tabloid not celebrated for its integrity, read:

SOVIET SCIENTIST'S AMAZING CLAIM:
RUINED TEMPLES FOUND ON MARS.
SPACE PROBE DISCOVERS REMAINS OF
50,000-YEAR-OLD CIVILIZATION.

The revelations are attributed to an anonymous Soviet source and breathlessly describe discoveries made by a nonexistent Soviet space vehicle.

But the story of "The Face" is almost entirely an American one. It was found by one of the *Viking* orbiters in 1976. There was an unfortunate dismissal of the feature by a project official as a trick of light and shadow, which prompted a later accusation that NASA was covering up the discovery of the Millennium. A few engineers, computer specialists, and others—some of them contract employees of NASA—worked on their own time to digitally enhance the image. Perhaps they hoped for stunning revelations. That's permissible in science, even encouraged—as long as your standards of evidence are high. Some of them were fairly cautious and deserve to be commended for advancing the subject. Others were less restrained, deducing not only that the Face was a genuine, monumental sculpture of a human being, but claiming to find a city nearby with temples and fortifications.* From spurious arguments, one writer announced that the monuments had a particular astronomical orientation—not now, though, but half a million years ago—from which it followed that the Cydonian wonders were erected in that remote epoch. But then how could the builders have been human? Half a million years ago, our ancestors were busy mastering stone tools and fire. They did not have spaceships.

The Martian Face is compared to "similar faces . . . constructed in civilizations on Earth. The faces are looking up at the sky because they're looking up to God." Or the Face was constructed by the survivors of an interplanetary war that left the surface of Mars (and the Moon) pockmarked and ravaged. What causes all those craters anyway? Is the Face a remnant of a long-extinct human civilization? Were the builders originally from Earth or Mars? Could the Face have been

---

* The general idea is quite old, going back at least a century to the Martian canal myth of Percival Lowell. As one of many examples, P. E. Cleator, in his 1936 book *Rockets Through Space: The Dawn of Interplanetary Travel*, speculated: "On Mars, the crumbling remains of ancient civilizations may be found, mutely testifying to the one-time glory of a dying world."

sculpted by interstellar visitors stopping briefly on Mars? Was it left for us to discover? Might they also have come to Earth and initiated life here? Or at least human life? Were they, whoever they were, gods? Much fervent speculation is evoked.

More recently, claims have been made for a connection between "monuments" on Mars and "crop circles" on Earth; of inexhaustible supplies of energy waiting to be extracted from ancient Martian machines; and of a massive NASA coverup to hide the truth from the American public. Such pronouncements go far beyond mere incautious speculation about enigmatic landforms.

When, in August 1993, the *Mars Observer* spacecraft failed within hailing distance of Mars, there were those who accused NASA of faking the mishap so it could study the Face in detail without having to release the images to the public. (If so, the charade is quite elaborate: All the experts on Martian geomorphology know nothing about it, and some of us have been working hard to design new missions to Mars less vulnerable to the malfunction that destroyed *Mars Observer*.) There was even a handful of pickets outside the gates of the Jet Propulsion Laboratory, worked up over this supposed abuse of power.

The tabloid *Weekly World News* for September 14, 1993 devoted its front page to the headline "New NASA Photo Proves Humans Lived on Mars!" A fake face, allegedly taken by *Mars Observer* in orbit about Mars (in fact, the spacecraft seems to have failed before achieving orbit), is said by a nonexistent "leading space scientist" to prove that Martians colonized Earth 200,000 years ago. The information is being suppressed, he is made to say, to prevent "world panic."

Put aside the improbability that such a revelation would actually lead to "world panic." For anyone who has witnessed a portentous scientific finding in the making—the July 1994 impact of Comet Shoemaker-Levy 9 with Jupiter comes to mind—it will be clear that scientists tend to be effervescent and uncontainable. They have an indomitable compulsion to share new data. Only through prior agreement, not *ex post facto*, do scientists abide military secrecy. I reject the notion that science is by its nature secretive. Its culture and ethos are, and for very good reason, collective, collaborative, and communicative.

If we restrict ourselves to what is actually known, and ignore the tabloid industry that manufactures epochal discoveries out of thin air, where are we? When we know only a little about the Face, it raises goosebumps. When we know a little more, the mystery quickly shallows.

Mars has a surface area of almost 150 million square kilometers, about the land area of the Earth. The area covered by the Martian "sphinx" is about one square kilometer. Is it so astonishing that one (comparatively) postage-stamp-sized patch in 150 million should look artificial—especially given our penchant, since infancy, for finding faces? When we examine the neighboring jumble of hillocks, mesas, and other complex surface forms, we recognize that the feature is akin to many that do not at all resemble a human face. Why this resemblance? Would the ancient Martian engineers rework only this mesa (well, maybe a few others) and leave all others unimproved by monumental sculpture? Or shall we conclude that other blocky mesas are also sculpted into the form of faces, but weirder faces, unfamiliar to us on Earth?

If we study the original image more carefully, we find that a strategically placed "nostril"—one that adds much to the impression of a face—is in fact a black dot corresponding to lost data in the radio transmission from Mars to Earth. The best picture of the Face shows one side lit by the Sun, the other in deep shadow. Using the original digital data, we can severely enhance the contrast in the shadows. When we do, we find something rather unfacelike there. The Face is at best half a face. Despite our shortness of breath and the beating of our hearts, the Martian sphinx looks natural—not artificial, not a dead ringer for a human face. It was probably sculpted by slow geological process over millions of years.

But I might be wrong. It's hard to be sure about a world we've seen so little of in extreme close-up. These features merit closer attention with higher resolution. Much more detailed photos of the "Face" would surely settle issues of symmetry and help resolve the debate between geology and monumental sculpture. Small impact craters found on or near the Face can settle the question of its age. In the case (most unlikely in my view) that the nearby structures were really once a city, that fact should also be obvious on closer examination. Are there broken streets? Crenelations in the "fort"? Ziggurats, towers, columned temples, monumental statuary, immense frescoes? Or just rocks?

Even if these claims are extremely improbable—as I think they are—they are worth examining. Unlike the UFO phenomenon, we have here the opportunity for a definitive experiment. This kind of hypothesis is falsifiable, a property that brings it well into the scientific arena. I hope that forthcoming American and Russian missions to Mars, especially orbiters with high-resolution television cameras, will make a special effort—among hundreds of other scientific questions—to look much more closely at the pyramids and what some people call the Face and the city.

—

Even if it becomes plain to everyone that these Martian features are geological and not artificial, monumental faces in space (and allied wonders) will not, I fear, go away. Already there are supermarket tabloids reporting nearly identical faces seen from Venus to Neptune (floating in the clouds?). The "findings" are typically attributed to fictitious Russian spacecraft and imaginary space scientists—which of course makes it marginally harder for a skeptic to check the story out.

One of the Mars face enthusiasts now announces

BREAKTHRU NEWS OF THE CENTURY
CENSORED BY NASA
FOR FEAR OF RELIGIOUS UPHEAVALS AND BREAKDOWNS.
THE DISCOVERY OF ANCIENT

ALIEN RUINS ON THE MOON.

A "giant city, size of Los Angeles basin, covered by immense glass dome, abandoned millions of years ago, and shattered by meteors with gigantic tower 5 miles tall, with giant one mile square cube on top" is breathlessly "CONFIRMED"—on the well-studied Moon. The evidence? Photos taken by NASA robotic and Apollo missions whose significance was suppressed by the government and overlooked by all those lunar scientists in many countries who don't work for the "government."

The August 18, 1992 issue of *Weekly World News* reports the discovery by "a secret NASA satellite" of "thousands maybe even millions of voices" emanating from the black hole at the center of the galaxy M51, all singing "'Glory, glory, glory to the Lord on high' over and over again." In English. There is even a tabloid report, fully although murkily illustrated, of a space probe that photographed God, or at least his eyes and the bridge of his nose, up there in the Orion Nebula.

The July 20, 1993 *WWN* sports a banner headline, "Clinton Meets With JFK!" along with a faked photo of a plausibly aged, slumped-over John Kennedy, having secretly survived the assassination attempt, in a wheelchair at Camp David. Many pages inside the tabloid, we are informed about another item of possible interest. In "Doomsday Asteroids," an alleged top-secret document quotes alleged "top" scientists about an alleged asteroid ("M-167") that will allegedly hit the Earth on November 11, 1993 and "could mean the end of life on Earth." President Clinton is described as being kept "constantly informed of the asteroid's position and speed." Perhaps it was one of the items he discussed in his meeting with President Kennedy. Somehow, the fact that the Earth escaped this catastrophe did not merit even a retrospective paragraph after November 11, 1993 uneventfully passed. At least the headline writer's judgment not to burden the front page with the news of the end of the world was vindicated.

Some see this as just a kind of fun. However, we live in a time when a real long-term statistical threat of an impact of an asteroid with the Earth has been identified. (This real science is of course the inspiration, if that's the word, of the *WWN* story.) Government agencies are studying what to do about it. Stories like this suffuse the subject with apocalyptic exaggeration and whimsy, make it difficult for the public to distinguish real perils from tabloid fiction, and conceivably can impede our ability to take precautionary steps to mitigate the danger.

The tabloids are often sued—sometimes by actors and actresses who stoutly deny they have performed loathsome acts—and large sums of money occasionally change hands. The tabloids must consider such suits as just one of the costs of doing a very profitable business. In their defense they often say that they are at the mercy of their writers and have no institutional responsibility to check out the truth of what they publish. Sal Ivone, the managing editor of *Weekly World News*, discussing the stories he publishes, says "For all I know, they could be the product of active imaginations. But because we're a tabloid, we don't have to question ourselves out of a story." Skepticism doesn't sell newspapers. Writers who have defected from the tabloids describe "creative" sessions in which writers and editors dream up stories and headlines out of whole cloth, the more outrageous the better.

Out of their immense readership, are there not many who take the stories at face value, who believe the tabloids "couldn't" print it if it wasn't so? Some readers I talk to insist they read them only for entertainment, just as they watch "wrestling" on television, that they're not in the least taken in, that the tabloids are understood by publisher and reader alike to be whimsies that explore the absurd. They merely exist outside any universe burdened by rules of evidence. But my mail suggests that large numbers of Americans take the tabloids very seriously indeed.

In the 1990s the tabloid universe is expanding, voraciously gobbling up other media. Newspapers, magazines, or television programs that labor under prissy restraints imposed by what is actually known are outsold by media outlets with less scrupulous standards. We can see this in the new generation of acknowledged tabloid television, and increasingly in what passes for news and information programs.

Such reports persist and proliferate because they sell. And they sell, I think, because there are so many of us who want so badly to be jolted out of our humdrum lives, to rekindle that sense of wonder we remember from childhood, and also, for a few of the stories, to be able, really and truly, to believe—in Someone older, smarter, and wiser who is looking out for us. Faith is clearly not enough for many people. They crave hard evidence, scientific proof. They long for the scientific seal of approval, but are unwilling to put up with the rigorous standards of evidence that impart credibility to that seal. What a relief it would be: doubt reliably abolished! Then, the irksome burden of looking after ourselves would be lifted. We're worried—and for good reason—about what it means for the human future if we have only ourselves to rely upon.

These are the modern miracles—shamelessly vouched for by those who make them up from scratch, bypassing any formal skeptical scrutiny, and available at low cost in every supermarket, grocery store and convenience outlet in the land. One of the pretenses of the tabloids is to make science, the very instrument of our disbelief, confirm our ancient faiths and effect a convergence of pseudoscience and pseudoreligion.

By and large, scientists' minds are open when exploring new

worlds. If we knew beforehand what we'd find, it would be unnecessary to go. In future missions to Mars or to the other fascinating worlds in our neck of the cosmic woods, surprises—even some of mythic proportions—are possible, maybe even likely. But we humans have a talent for deceiving ourselves. Skepticism must be a component of the explorer's toolkit, or we will lose our way. There are wonders enough out there without our inventing any.

Chapter 4

ALIENS

"Truly, that which makes me believe there
is no inhabitant on this sphere, is that it
seems to me that no sensible being would
be willing to live here."
"Well, then!" said Micromegas, "perhaps
the beings that inhabit it do not possess good sense."

One alien to another, on approaching
the Earth, in VOLTAIRE'S
Micromegas: A Philosophical History (1752)

It's still dark out. You're lying in bed, fully awake. You discover you're utterly paralyzed. You sense someone in the room. You try to cry out. You cannot. Several small gray beings, less than four feet tall, are standing at the foot of the bed. Their heads are pear-shaped, bald, and large for their bodies. Their eyes are enormous, their faces expressionless and identical. They wear tunics and boots. You hope this is only a dream. But as nearly as you can tell it's really happening. They lift you up and, eerily, they and you slip through the wall of your bedroom. You float out into the air. You rise high toward a metallic saucer-shaped spacecraft. Once inside, you are escorted into a medical examining room. A larger but similar being—evidently some kind of physician—takes over. What follows is even more terrifying.

Your body is probed with instruments and machines, especially your sexual parts. If you're a man, they may take sperm samples; if you're a woman, they may remove ova or fetuses, or implant semen. They may force you to have sex. Afterwards you may be ushered into a different room where hybrid babies or fetuses, partly human and partly like these creatures, stare back at you. You may be given an admonition about human misbehavior, especially in despoiling the environment or in allowing the AIDS pandemic; tableaus of future devastation are offered. Finally, these cheerless gray emissaries escort you out of the spacecraft and ooze you back through the walls into your bed. By the time you're able to move and talk . . . they're gone.

You may not remember the incident right away. Instead you might simply find some period of time unaccountably missing, and puzzle over it. Because all this seems so weird, you're a little concerned about your sanity. Naturally you're reluctant to talk about it. At the same time the experience is so disturbing that it's hard to keep it bottled up. It all pours out when you hear of similar accounts, or when you're under hypnosis with a sympathetic therapist, or even when you see a picture of an "alien" in one of the many popular magazines, books, and TV "specials" on UFOs. Some people say they can recall such experiences from early childhood. Their own children, they think, are now being abducted by aliens. It runs in families. It's a eugenics program, they say, to improve the human breeding stock. Maybe aliens have always done this. Maybe, some say, that's where humans came from in the first place.

As revealed by repeated polls over the years, most Americans believe that we're being visited by extraterrestrial beings in UFOs. In a 1992 Roper poll of nearly 6,000 American adults—especially commissioned by those who accept the alien abduction story at face value—18 percent reported sometimes waking up paralyzed, aware of one or more strange beings in the room. About 13 percent report odd episodes of missing time, and 10 percent claim to have flown through the air without mechanical assistance. From nothing more than these results, the poll's sponsors conclude that two percent of all Americans have been abducted, many repeatedly, by beings from other worlds. The question of whether respondents had been abducted by aliens was never actually put to them.

If we believed the conclusion drawn by those who bankrolled and interpreted the results of this poll, and if aliens are not partial to Americans, then the number for the whole planet would be more than a hundred million people. This means an abduction every few seconds over the past few decades. It's surprising more of the neighbors haven't noticed.

What's going on here? When you talk with self-described abductees, most seem very sincere, although caught in the grip of powerful emotions. Some psychiatrists who've examined them say they find no more evidence of psychopathology in them than in the rest of us. Why should anyone claim to have been abducted by alien creatures if it never happened? Could all these people be mistaken, or lying, or hallucinating the same (or a similar) story? Or is it arrogant and contemptuous even to question the good sense of so many?

On the other hand, could there really be a massive alien invasion; repugnant medical procedures performed on millions of innocent men, women, and children; humans apparently used as breeding stock over many decades—and all this not generally known and dealt with by responsible media, physicians, scientists, and the governments sworn to protect the lives and well-being of their citizens? Or, as many have suggested, is there a massive government conspiracy to keep the citizens from the truth?

Why should beings so advanced in physics and engineering—crossing vast interstellar distances, walking like ghosts through walls—be so backward when it comes to biology? Why, if the aliens are trying to do their business in secret, wouldn't they perfectly expunge all memories of the abductions? Too hard for them to do? Why are the examining instruments macroscopic and so reminiscent of what can be found at the neighborhood medical clinic? Why go to all the trouble of repeated sexual encounters between aliens and humans? Why not steal a few egg and sperm cells, read the full genetic code, and then manufacture as many copies as you like with whatever genetic variations happen to suit your fancy? Even we humans, who as yet cannot quickly cross interstellar space or slither through walls, are able to clone cells. How could humans be the result of an alien breeding program if we share 99.6 percent of our active genes with the chimpanzees? We're more closely related to chimps than rats are to mice. The preoccupation with reproduction in these accounts raises a warning flag—especially considering the uneasy balance between sexual impulse and societal repression that has always characterized the human condition, and the fact that we live in a time fraught with numerous ghastly accounts, both true and false, of childhood sexual abuse.

Contrary to many media reports,* the Roper pollsters and those who wrote the "official" report never asked whether their subjects had been abducted by aliens. They deduced it: Those who've ever awakened with strange presences around them, who've ever unaccountably seemed to fly through the air, and so on, have therefore been abducted. The pollsters didn't even check to see if sensing presences, flying, etc. were part of the same or separate incidents. Their conclusion—that millions of Americans have been so abducted—is spurious, based on careless experimental design.

Still, at least hundreds of people, perhaps thousands, claiming they have been abducted, have sought out sympathetic therapists or joined abductee support groups. Others may have similar complaints but,

---

* For example, the September 4, 1994 *Publishers Weekly:* "According to a Gallup [sic] poll, more than three million Americans believe they have been abducted by aliens."

fearing ridicule or the stigma of mental illness, have refrained from speaking up or getting help.

Some abductees are also said to be reluctant to talk for fear of hostility and rejection by hardline skeptics (although many willingly appear on radio and TV talk shows). Their diffidence supposedly extends even to audiences that already believe in alien abductions. But maybe there's another reason: Might the subjects themselves be unsure—at least at first, at least before many retellings of their story—whether it was an external event they are remembering or a state of mind?

—

"One unerring mark of the love of truth," wrote John Locke in 1690, "is not entertaining any proposition with greater assurance than the proofs it is built upon will warrant." On the matter of UFOs, how strong are the proofs?

The phrase "flying saucer" was coined when I was entering high school. The newspapers were full of stories about ships from beyond in the skies of Earth. It seemed pretty believable to me. There were lots of other stars, at least some of which probably had planetary systems like ours. Many stars were as old or older than the Sun, so there was plenty of time for intelligent life to evolve. Caltech's Jet Propulsion Laboratory had just flown a two-stage rocket high above the Earth. Clearly we were on our way to the Moon and the planets. Why shouldn't other, older, wiser beings be able to travel from their star to ours? Why not?

This was only a few years after the bombing of Hiroshima and Nagasaki. Maybe the UFO occupants were worried about us, and sought to help us. Or maybe they wanted to make sure that we and our nuclear weapons didn't come and bother *them*. Many people seemed to see flying saucers—sober pillars of the community, police officers, commercial airplane pilots, military personnel. And apart from some harumphs and giggles, I couldn't find any counterarguments. How could all these eyewitnesses be mistaken? What's more, the saucers had been picked up on radar, and pictures had been taken of them. You could see the photos in newspapers and glossy magazines. There were even reports about crashed flying saucers and little alien bodies with perfect teeth stiffly languishing in Air Force freezers in the Southwest.

The prevailing climate was summarized in *Life* magazine a few years later, in these words: "These objects cannot be explained by present science as natural phenomena—but solely as artificial devices, created and operated by a high intelligence." Nothing "known or projected on Earth could account for the performance of these devices."

And yet not a single adult I knew was preoccupied with UFOs. I couldn't figure out why not. Instead they were worried about Communist China, nuclear weapons, McCarthyism, and the rent. I wondered if they had their priorities straight.

In college, in the early 1950s, I began to learn a little about how science works, the secrets of its great success, how rigorous the standards of evidence must be if we are really to know something is true, how many false starts and dead ends have plagued human thinking, how our biases can color our interpretation of the evidence, and how often belief systems widely held and supported by the political, religious, and academic hierarchies turn out to be not just slightly in error, but grotesquely wrong.

I came upon a book called *Extraordinary Popular Delusions and the Madness of Crowds*, written by Charles Mackay in 1841, and still in print. In it could be found the histories of boom-and-bust economic crazes, including the Mississippi and South Sea "Bubbles" and the extravagant run on Dutch tulips, scams that bamboozled the wealthy and titled of many nations; a legion of alchemists, including the poignant tale of Mr. Kelly and Dr. Dee (and Dee's eight-year-old son Arthur, impressed by his desperate father into communicating with the spirit world by peering into a crystal); dolorous accounts of unfulfilled prophecy, divination, and fortune-telling; the persecution of witches; haunted houses; "popular admiration of great thieves"; and much else. Entertainingly portrayed was the Count of St. Germain, who dined out on the cheerful pretension that he was centuries old if not actually immortal. (When, at dinner, incredulity was expressed at his recounting of his conversations with Richard the Lion-Hearted, he turned to his man-servant for confirmation. "You forget, sir," was the reply, "I have been only five hundred years in your service." "Ah, true," said St. Germain, "it was a little before your time.")

A riveting chapter on the Crusades began

Every age has its peculiar folly; some scheme, project, or phantasy into which it plunges, spurred on either by the love of gain, the necessity of excitement, or the mere force of imitation. Failing in these, it has some madness, to which it is goaded by political or religious causes, or both combined.

The edition I first read was adorned by a quote from the financier and adviser of presidents, Bernard M. Baruch, attesting that reading Mackay had saved him millions.

There had been a long history of spurious claims that magnetism could cure disease. Paracelsus, for example, used a magnet to suck diseases out of the human body and dispose of them into the Earth. But the key figure was Franz Mesmer. I had vaguely understood the word "mesmerize" to mean something like hypnotize. But my first real knowledge of Mesmer came from Mackay. The Viennese physician had thought that the positions of the planets influenced human health, and was caught up in the wonders of electricity and magnetism. He catered to the declining French nobility on the eve of the Revolution. They crowded into a darkened room. Dressed in a gold-flowered silk robe and waving an ivory wand, Mesmer seated his marks around a vat of dilute sulfuric acid. The Magnetizer and his young male assistants peered deeply into the eyes of their patients, and rubbed their bodies. They grasped iron bars protruding into the solution or held each other's hands. In contagious frenzy, aristocrats—especially young women—were cured left and right.

Mesmer became a sensation. He called it "animal magnetism." For the more conventional medical practitioner, though, this was bad for business, so French physicians pressured King Louis XVI to crack down. Mesmer, they said, was a menace to public health. A commission was appointed by the French Academy of Sciences that included the pioneering chemist, Antoine Lavoisier, and the American diplomat and expert on electricity, Benjamin Franklin. They performed the obvious control experiment: When the magnetizing effects were performed without the patient's knowledge, no cures were effected. The cures, if any, the commission concluded, were all in the mind of the beholder. Mesmer and his followers were undeterred. One of them later urged the following attitude of mind for best results:

Forget for a while all of your knowledge of physics. . . Remove from your mind all objections that may occur. . . Never reason for six weeks. . . Be very credulous; be very persevering; reject all past experience, and do not listen to reason.

Oh, yes, a final piece of advice: "Never magnetize before inquisitive persons."

Another eye-opener was Martin Gardner's *Fads and Fallacies in the Name of Science*. Here was Wilhelm Reich uncovering the key to the structure of galaxies in the energy of the human orgasm; Andrew Crosse creating microscopic insects electrically from salts; Hans Hörbiger under Nazi aegis announcing that the Milky Way was made not of stars, but of snowballs; Charles Piazzi Smyth discovering in the dimensions of the Great Pyramid of Gizeh a world chronology from the Creation to the Second Coming; L. Ron Hubbard writing a manuscript able to drive its readers insane (was it ever proofed? I wondered); the Bridey Murphy case, which led millions into concluding that at last there was serious evidence of reincarnation; Joseph Rhine's "demonstrations" of ESP; appendicitis cured by cold water enemas, bacterial diseases by brass cylinders, and gonorrhea by green light — and amid all these accounts of self-deception and charlatanry, to my surprise a chapter on UFOs.

Of course, merely by writing books cataloging spurious beliefs, Mackay and Gardner came across, at least a little, as grumpy and superior. Was there nothing they accepted? Still, it was stunning how many passionately argued and defended claims to knowledge had amounted to nothing. It slowly dawned on me that, human fallibility being what it is, there might be other explanations for flying saucers.

I had been interested in the possibility of extraterrestrial life from childhood, from long before I ever heard of flying saucers. I've remained fascinated long after my early enthusiasm for UFOs waned — as I understood more about that remorseless taskmaster called the scientific method: Everything hinges on the matter of evidence. On so important a question, the evidence must be airtight. The more we want it to be true, the more careful we have to be. No witness's say-so is good enough. People make mistakes. People play practical jokes. People stretch the truth for money or attention or fame. People occasionally misunderstand what they're seeing. People sometimes even see things that aren't there.

Essentially all the UFO cases were anecdotes, something asserted.

UFOs were described variously as rapidly moving or hovering; disc-shaped, cigar-shaped, or ball-shaped; moving silently or noisily; with a fiery exhaust, or with no exhaust at all; accompanied by flashing lights, or uniformly glowing with a silvery cast, or self-luminous. The diversity of the observations hinted that they had no common origin, and that the use of such terms as UFOs or "flying saucers" served only to confuse the issue by grouping generically a set of unrelated phenomena.

There was something odd about the very invention of the phrase "flying saucer." As I write this chapter, I have before me a transcript of an April 7, 1950 interview between Edward R. Murrow, the celebrated CBS newsman, and Kenneth Arnold, a civilian pilot who saw something peculiar near Mount Rainier in the state of Washington on June 24, 1947 and who in a way coined the phrase. Arnold claims that the newspapers

> did not quote me properly. . . When I told the press they misquoted me, and in the excitement of it all, one newspaper and another one got it so ensnarled up that nobody knew just exactly what they were talking about. . . These objects more or less fluttered like they were, oh, I'd say, boats on very rough water. . . And when I described how they flew, I said that they flew like they take a saucer and throw it across the water. Most of the newspapers misunderstood and misquoted that, too. They said that I said that they were saucer-like; I said that they flew in a saucer-like fashion.

Arnold thought he saw a train of nine objects, one of which produced a "terrific blue flash." He concluded they were a new kind of winged aircraft. Murrow summed up: "That was an historic misquote. While Mr. Arnold's original explanation has been forgotten, the term 'flying saucer' has become a household word." Kenneth Arnold's flying saucers looked and behaved quite differently from what in only a few years would be rigidly particularized in the public understanding of the term: something like a very large and highly maneuverable frisbee.

Most people honestly reported what they saw, but what they saw were natural, if unfamiliar, phenomena. Some UFO sightings turned out to be unconventional aircraft, conventional aircraft with unusual lighting patterns, high-altitude balloons, luminescent insects, planets seen under unusual atmospheric conditions, optical mirages and looming, lenticular clouds, ball lightning, sundogs, meteors including green fireballs, and satellites, nosecones, and rocket boosters spectacularly reentering the atmosphere.* Just conceivably, a few might be small comets dissipating in the upper air. At least some radar reports were due to "anomalous propagation"—radio waves traveling curved paths due to atmospheric temperature inversions. Traditionally, they were also called radar "angels"—something that seems to be there but isn't. You could have simultaneous visual and radar sightings without there being any "there" there.

When we notice something strange in the sky, some of us become excitable and uncritical, bad witnesses. There was the suspicion that the field attracted rogues and charlatans. Many UFO photos turned out to be fakes—small models hanging by thin threads, often photographed in a double exposure. A UFO seen by thousands of people at a football game turned out to be a college fraternity prank—a piece of cardboard, some candles, and a thin plastic bag that dry cleaning comes in, all cobbled together to make a rudimentary hot air balloon.

The original crashed saucer account (with the little alien men and their perfect teeth) turned out to be a straight-out hoax. Frank Scully, columnist for *Variety*, passed on a story told by an oilman friend; it played a central dramatic role in Scully's best-selling 1950 book, *Behind the Flying Saucers*. Sixteen dead aliens from Venus, each three feet high, had been found in one of three crashed saucers. Booklets with alien pictograms had been recovered. The military was covering up. The implications were profound.

The hoaxers were Silas Newton, who said he used radio waves to prospect for gold and oil, and a mysterious "Dr. Gee" who turned out to be a Mr. GeBauer. Newton produced a gear from the UFO machinery and flashed close-up saucer photos. But he did not allow close inspection. When a prepared skeptic, through sleight of hand, switched gears and sent the alien artifact away for analysis, it turned out to be made of kitchen-pot aluminum.

The crashed saucer scam was a small interlude in a quarter-century of frauds by Newton and GeBauer—chiefly selling worthless oil leases

---

* There are so many artificial satellites up there that they're always making garish displays somewhere in the world. Two or three decay every day in the Earth's atmosphere, the flaming debris often visible to the naked eye.

and prospecting machines. In 1952 they were arrested by the FBI, and the following year found guilty of conducting a confidence game. Their exploits—chronicled by the historian Curtis Peebles—ought to have made UFO enthusiasts cautious forever about crashed saucer stories from the American Southwest around 1950. No such luck.

On October 4, 1957, *Sputnik 1*, the first Earth-orbiting artificial satellite, was launched. Of 1,178 recorded UFO sightings in America that year, 701, or 60 percent—rather than the 25 percent you'd expect—occurred between October and December. The clear implication is that Sputnik and its attendant publicity somehow generated UFO reports. Perhaps people were looking at the night sky more, and saw more natural phenomena they didn't understand. Or could it be they looked up more and saw more of the alien spacecraft that are there all the time?

The idea of flying saucers had dubious antecedents, tracing back to a conscious hoax entitled *I Remember Lemuria!*, written by Richard Shaver, and published in the March 1945 number of the pulp fiction periodical *Amazing Stories*. It was exactly the sort of stuff I devoured as a child. Lost continents were settled by space aliens 150,000 years ago, I was informed, leading to the creation of a race of demonic underground beings responsible for human tribulations and the existence of evil. The editor of the magazine, Ray Palmer—who was, like the subterranean beings he warned about, roughly four feet high—promoted the notion, well before Arnold's sighting, that the Earth is being visited by disc-shaped alien spacecraft and that the government is covering up its knowledge and complicity. Merely from the newsstand covers of such magazines, millions of Americans were exposed to the idea of flying saucers well before the term was coined.

All in all, the alleged evidence seemed thin—most often devolving into gullibility, hoax, hallucination, misunderstanding of the natural world, hopes and fears disguised as evidence, and a craving for attention, fame, and fortune. Too bad, I remember thinking.

Since then, I've been lucky enough to be involved in sending spacecraft to other planets to look for life, and in listening for possible radio signals from alien civilizations, if any, on planets of distant stars. We've had a few tantalizing moments. But if the suspected signal isn't available for every grumpy skeptic to pick over, we cannot call it evidence of extraterrestrial life—no matter how appealing we find the notion.

We'll just have to wait until, if such a time ever comes, better data are available. We've not yet found compelling evidence for life beyond the Earth. We're only at the very beginning of the search, though. New and better information might emerge, for all we know, tomorrow.

I don't think anyone could be more interested than I am in whether we're being visited. It would save me so much time and effort to be able to study extraterrestrial life directly and nearby, rather than at best indirectly and at a great distance. Even if the aliens are short, dour, and sexually obsessed—if they're here, I want to know about them.

—

How modest our expectations are about "aliens," and how shoddy the standards of evidence that many of us are willing to accept, can be found in the saga of the crop circles. Originating in Great Britain and spreading throughout the world was something surpassing strange.

Farmers or passersby would discover circles (and, in later years, much more complex pictograms) impressed upon fields of wheat, oats, barley, and rapeseed. Beginning with simple circles in the middle 1970s, the phenomenon progressed year by year, until by the late 1980s and early 1990s the countryside, especially in southern England, was graced by immense geometrical figures, some the size of football fields, imprinted on cereal grain before the harvest—circles tangent to circles, or connected by axes, parallel lines drooping off, "insectoids." Some of the patterns showed a central circle surrounded by four symmetrically-placed smaller circles—clearly, it was concluded, caused by a flying saucer and its four landing pods.

A hoax? Impossible, almost everyone said. There were hundreds of cases. It was done sometimes in only an hour or two in the dead of night, and on *such* a large scale. No footprints of pranksters leading towards or away from the pictograms could be found. And besides, what possible motive could there be for a hoax?

Many less conventional conjectures were offered. People with some scientific training examined sites, spun arguments, instituted whole journals devoted to the subject. Were the figures caused by strange whirlwinds called "columnar vortices," or even stranger ones called "ring vortices"? What about ball lightning? Japanese investigators tried to simulate, in the laboratory and on a small scale, the plasma physics they thought was working its way on far-off Wiltshire.

But especially as the crop figures became more complex, meteorological or electrical explanations became more strained. Plainly, it was due to UFOs, the aliens communicating to us in a geometrical language. Or perhaps it was the devil, or the long-suffering Earth complaining about the depredations visited upon it by the hand of Man. New Age tourists came in droves. All-night vigils were undertaken by enthusiasts equipped with audio recorders and infrared vision scopes. Print and electronic media from all over the world tracked the intrepid cerealogists. Best-selling books on extraterrestrial crop distorters were purchased by a breathless and admiring public. True, no saucer was actually seen settling down on the wheat, no geometrical figure was filmed in the course of being generated. But dowsers authenticated their alien origin, and channelers made contact with the entities responsible. "Orgone energy" was detected within the circles.

Questions were asked in Parliament. The royal family called in for special consultation Lord Solly Zuckerman, former principal scientific adviser to the Ministry of Defence. Ghosts were said to be involved; also, the Knights Templar of Malta and other secret societies. Satanists were implicated. The Defence Ministry was covering the matter up. A few inept and inelegant circles were judged attempts by the military to throw the public off the track. The tabloid press had a field day. The *Daily Mirror* hired a farmer and his son to make five circles in hope of tempting a rival tabloid, the *Daily Express*, into reporting the story. The *Express* was, in this case at least, not taken in.

"Cerealogical" organizations grew and splintered. Competing groups sent each other intimidating doggerel. Accusations were made of incompetence or worse. The number of crop "circles" rose into the thousands. The phenomenon spread to the United States, Canada, Bulgaria, Hungary, Japan, the Netherlands. The pictograms—especially the more complex of them—began to be quoted increasingly in arguments for alien visitation. Strained connections were drawn to the "Face" on Mars. One scientist of my acquaintance wrote to me that extremely sophisticated mathematics was hidden in these figures; they could only be the result of a superior intelligence. In fact, one matter on which almost all of the contending cerealogists agreed is that the later crop figures were much too complex and elegant to be due to mere human intervention, much less to some ragged and irresponsible hoaxers. Extraterrestrial intelligence was apparent at a glance. . .

In 1991, Doug Bower and Dave Chorley, two blokes from Southampton, announced they had been making crop figures for 15 years. They dreamed it up over stout one evening in their regular pub, The Percy Hobbes. They had been amused by UFO reports and thought it might be fun to spoof the UFO gullibles. At first they flattened the wheat with the heavy steel bar that Bower used as a security device on the back door of his picture framing shop. Later on they used planks and ropes. Their first efforts took only a few minutes. But, being inveterate pranksters as well as serious artists, the challenge began to grow on them. Gradually, they designed and executed more and more demanding figures.

At first no one seemed to notice. There were no media reports. Their artforms were neglected by the tribe of UFOlogists. They were on the verge of abandoning crop circles to move on to some other, more emotionally rewarding hoax.

Suddenly crop circles caught on. UFOlogists fell for it hook, line, and sinker. Bower and Chorley were delighted—especially when scientists and others began to announce their considered judgment that no merely human intelligence could be responsible.

Carefully they planned each nocturnal excursion—sometimes following meticulous diagrams they had prepared in watercolors. They closely tracked their interpreters. When a local meteorologist deduced a kind of whirlwind because all of the crops were deflected downward in a clockwise circle, they confounded him by making a new figure with an exterior ring flattened counterclockwise.

Soon other crop figures appeared in southern England and elsewhere. Copycat hoaxsters had appeared. Bower and Chorley carved out a responsive message in wheat: "WEARENOTALONE." Even this some took to be a genuine extraterrestrial message (although it would have been better had it read "YOUARENOTALONE"). Doug and Dave began signing their artworks with two Ds; even this was attributed to a mysterious alien purpose. Bower's nocturnal disappearances aroused the suspicions of his wife Ilene. Only with great difficulty—Ilene accompanying Dave and Doug one night, and then joining the credulous in admiring their handiwork next day—was she convinced that his absences were, in this sense, innocent.

Eventually Bower and Chorley tired of the increasingly elaborate prank. While in excellent physical condition, they were both in their sixties now and a little old for nocturnal commando operations in the fields of unknown and often unsympathetic farmers. They may have been annoyed at the fame and fortune accrued by those who merely photographed their art and announced aliens to be the artists. And they became worried that if they delayed much longer, no statement of theirs would be believed.

So they confessed. They demonstrated to reporters how they made even the most elaborate insectoid patterns. You might think that never again would it be argued that a sustained hoax over many years is impossible, and never again would we hear that no one could possibly be motivated to deceive the gullible into thinking that aliens exist. But the media paid brief attention. Cerealogists urged them to go easy; after all, they were depriving many of the pleasure of imagining wondrous happenings.

Since then, other crop circle hoaxers have kept at it, but mostly in a more desultory and less inspired manner. As always, the confession of the hoax is greatly overshadowed by the sustained initial excitement. Many have heard of the pictograms in cereal grains and their alleged UFO connection, but draw a blank when the names of Bower and Chorley or the very idea that the whole business may be a hoax are raised. An informative exposé by the journalist Jim Schnabel (*Round in Circles*; Penguin Books, 1994) — from which much of my account is taken — is in print. Schnabel joined the cerealogists early and in the end made a few successful pictograms himself. (He prefers a garden roller to a wooden plank, and found that simply stomping grain with one's feet does an acceptable job.) But Schnabel's work, which one reviewer called "the funniest book I've read in ages," had only modest success. Demons sell; hoaxers are boring and in bad taste.

—

The tenets of skepticism do not require an advanced degree to master, as most successful used car buyers demonstrate. The whole idea of a democratic application of skepticism is that everyone should have the essential tools to effectively and constructively evaluate claims to knowledge. All science asks is to employ the same levels of skepticism we use in buying a used car or in judging the quality of analgesics or beer from their television commercials.

But the tools of skepticism are generally unavailable to the citizens of our society. They're hardly ever mentioned in the schools, even in the presentation of science, its most ardent practitioner, although skepticism repeatedly sprouts spontaneously out of the disappointments of everyday life. Our politics, economics, advertising, and religions (New Age and Old) are awash in credulity. Those who have something to sell, those who wish to influence public opinion, those in power, a skeptic might suggest, have a vested interest in discouraging skepticism.

Chapter 5

SPOOFING
AND
SECRECY

Trust a witness in all matters in which neither his self-interest, his passions, his prejudices, nor the love of the marvelous is strongly concerned. When they are involved, require corroborative evidence in exact proportion to the contravention of probability by the thing testified.

THOMAS HENRY HUXLEY
(1825–1895)

When the mother of celebrity abductee Travis Walton was informed that a UFO had zapped her son with a bolt of lightning and then carried him off into space, she replied incuriously, "Well, that's the way these things happen." Is it?

To agree that UFOs are in our skies is not committing to very much: "UFO" is an abbreviation for "Unidentified Flying Object." It is a more inclusive term than "Flying Saucer." That there are things seen which the ordinary observer, or even an occasional expert, does not understand is inevitable. But why, if we see something we don't recognize, should we conclude it's a ship from the stars? A wide variety of more prosaic possibilities present themselves.

After misapprehended natural events and hoaxes and psychological aberrations are removed from the data set, is there any residue of very credible but extremely bizarre cases, especially ones supported by physical evidence? Is there a "signal" hiding in all that noise? In my view, no signal has been detected. There are reliably reported cases that are unexotic, and exotic cases that are unreliable. There are no cases—despite well over a million UFO reports since 1947—in which something so strange that it could only be an extraterrestrial spacecraft is reported so reliably that misapprehension, hoax, or hallucination can be reliably excluded. There's still a part of me that says, "Too bad."

We're regularly bombarded with extravagant UFO claims vended in bite-sized packages, but only rarely do we get to hear about their comeuppance. This isn't hard to understand: Which sells more newspapers and books, which garners higher ratings, which is more fun to believe, which is more resonant with the torments of our time—real crashed alien ships, or experienced con men preying on the gullible; extraterrestrials of immense powers toying with the human species, or such claims deriving from human weakness and imperfection?

Over the years I've continued to spend time on the UFO problem.

I receive many letters about it, frequently with detailed first-hand accounts. Sometimes momentous revelations are promised if only I will call the letter writer. After I give lectures—on almost any subject—I often am asked, "Do you believe in UFOs?" I'm always struck by how the question is phrased, the suggestion that this is a matter of belief and not of evidence. I'm almost never asked, "How good is the evidence that UFOs are alien spaceships?"

I've found that the going-in attitude of many people is highly predetermined. Some are convinced that eyewitness testimony is reliable, that people do not make things up, that hallucinations or hoaxes on such a scale are impossible, and that there must be a long-standing, high-level government conspiracy to keep the truth from the rest of us. Gullibility about UFOs thrives on widespread mistrust of government, arising naturally enough from all those circumstances where—in the tension between public well-being and "national security"—the government lies. As government deceit and conspiracies of silence have been exposed on so many other matters, it's hard to argue that a cover-up on this odd subject is impossible, that the government would never hide important information from its citizens. A common explanation on why there would be a cover-up is to prevent worldwide panic or erosion of confidence in the government.

I was a member of the U.S. Air Force Scientific Advisory Board committee that investigated the Air Force's UFO study—called "Project Bluebook," but earlier and revealingly called "Project Grudge." We found the on-going effort to be lackadaisical and dismissive. In the middle 1960s, "Project Bluebook" was headquartered at Wright-Patterson Air Force Base in Ohio—where "Foreign Technical Intelligence" (chiefly, understanding what new weapons the Soviets had) was also based. They had state-of-the-art technology in file retrieval. You asked about a given UFO incident and, somewhat like sweaters and suits at the dry cleaner's today, reams of files made their way past you, until the engine stopped when the file you wanted arrived before you.

But what was *in* those files wasn't worth much. For example, senior citizens reported lights hovering over their small New Hampshire town for more than an hour, and the case is explained as a wing of strategic bombers from a nearby Air Force base on a training exercise. Could the bombers take an hour to pass over the town? No. Did the bombers fly over at the time the UFOs were reported? No. Can you explain to us, Colonel, how strategic bombers can be described as "hovering"? No. The slipshod Bluebook investigations played little scientific role, but they did serve the important bureaucratic purpose of convincing much of the public that the Air Force was on the job; and that maybe there was nothing to UFO reports.

Of course, this doesn't preclude the possibility that another, more serious, more scientific study of UFOs was going on somewhere else—headed, say, by a brigadier general rather than a lieutenant colonel. I think something like this is even likely, not because I believe we're being visited by aliens, but because hiding in the UFO phenomena must be data once considered of significant military interest. Certainly if UFOs are as reported—very fast, very maneuverable craft—there is a military duty to find out how they work. If UFOs were built by the Soviet Union it was the Air Force's responsibility to protect us. Considering the remarkable performance characteristics reported, the strategic implications of Soviet UFOs flagrantly overflying American military and nuclear facilities were worrisome. If on the other hand the UFOs were built by extraterrestrials, we might copy the technology (if we could get our hands on just one saucer) and secure a huge advantage in the Cold War. And even if the military believed that UFOs were manufactured neither by Soviets nor by extraterrestrials, there was a good reason to follow the reports closely:

In the 1950s balloons were being extensively used by the Air Force—not just as weather measurement platforms, as prominently advertised, and radar reflectors, as acknowledged, but also, secretly, as robotic espionage craft, with high-resolution cameras and signal intelligence devices. While the balloons themselves were not very secret, the reconnaissance packages they carried were. High-altitude balloons can seem saucer-shaped when seen from the ground. If you misestimate how far away they are, you can easily imagine them going absurdly fast. Occasionally, propelled by a gust of wind, they make abrupt changes in direction, uncharacteristic of aircraft and in seeming defiance of the conservation of momentum—if you don't realize that they're hollow and weigh almost nothing.

The most famous of these military balloon systems, widely tested over the United States in the early 1950s, was called "Skyhook." Other balloon systems and projects were designated "Mogul," "Moby Dick," "Grandson," and "Genetrix." Urner Lidell, who had some responsibility for these missions at the Naval Research Laboratory, and who was later a NASA official, once told me he thought all UFO reports were due to military balloons. While "all" is going too far, their role has, I think, been insufficiently appreciated. So far as I know there has never been a systematic and intentional control experiment—in which high-altitude balloons were secretly released and tracked, and UFO reports from visual and radar observers noted.

In 1956, overflights of the Soviet Union by U.S. reconnaissance balloons began. At their peak there were dozens of balloon launches a day. Balloon overflights were then replaced by high-altitude aircraft, such as the U-2, which in turn were largely replaced by reconnaissance satellites. Many UFOs dating from this period were clearly scientific balloons, as are some since. High-altitude balloons are still being launched—including platforms carrying cosmic ray sensors, optical and infrared telescopes, radio receivers probing the cosmic background radiation, and other instruments above most of the Earth's atmosphere.

A great to-do has been made of one or more alleged crashed flying saucers near Roswell, New Mexico, in 1947. Some initial reports and newspaper photographs of the incident are entirely consistent with the idea that the debris was a crashed high-altitude balloon. But other residents of the region—especially decades later—remember more exotic materials, enigmatic hieroglyphics, threats by military personnel to witnesses if they didn't keep what they knew to themselves, and the canonical story that alien machinery and body parts were packed into an airplane and flown to the Air Materiel Command at Wright-Patterson Air Force Base. Some, but not all, of the recovered alien body stories are associated with this incident.

Philip Klass, a long-time and dedicated UFO skeptic, has uncovered a subsequently declassified letter dated July 27, 1948, a year after the Roswell "incident," from Major General C. B. Cabell—then Director of Intelligence for the U.S. Air Force (and later, as a CIA official, a major figure in the abortive U.S. invasion of Cuba at the Bay of Pigs). Cabell was inquiring of those who reported to him on what UFOs might be. He hadn't a clue. In an October 11, 1948 summary response, explicitly including information in the possession of the Air Materiel Command, we find the Director of Intelligence being told that nobody else in the Air Force had a clue either. This makes it unlikely that UFO fragments and occupants had made their way to Wright-Patterson the year before.

What the Air Force was mostly worried about was that UFOs were Russian. Why Russians would be testing flying saucers over the United States was a puzzle to which the following four answers were proposed: "(1) To negate U.S. confidence in the atom bomb as the most advanced and decisive weapon in warfare. (2) To perform photographic reconnaissance missions. (3) To test U.S. air defenses. (4) To conduct familiarization flights [for strategic bombers] over U.S. territory." We now know that UFOs neither were nor are Russian, and however dedicated the Soviet interest may have been to objectives (1) through (4), flying saucers weren't how they pursued these objectives.

Much of the evidence regarding the Roswell "incident" seems to point to a cluster of high-altitude classified balloons, perhaps launched from nearby Alamogordo Army Air Field or White Sands Proving Ground, that crashed near Roswell, the debris of secret instruments hurriedly collected by earnest military personnel, early press reports announcing that it was a spaceship from another planet ("RAAF Captures Flying Saucer on Ranch in Roswell Region"), diverse recollections simmering over the years, and memories refreshed by the opportunity for a little fame and fortune. (Two UFO museums in Roswell are leading tourist stops.)

A 1994 report ordered by the Secretary of the Air Force and the Department of Defense in response to prodding from a New Mexico Congressman identifies the Roswell debris as remnants of a long-range, highly secret, balloon-borne low-frequency acoustic detection system called "Project Mogul"—an attempt to sense Soviet nuclear weapons explosions at tropopause altitudes. The Air Force investigators, rummaging comprehensively through the secret files of 1947, found no evidence of heightened message traffic:

There were no indications and warnings, notice of alerts, or a higher tempo of operational activity reported that would be logically generated if an alien craft, whose intentions were unknown, entered U.S. territory. . . The records indicate that none of this happened (or if it did, it was controlled by a security system so efficient and tight that no one, U.S. or otherwise, has been able to duplicate it since. If such a system had been in effect at the time, it would have also been used to protect our atomic secrets from the Soviets, which history has shown obviously was not the case.).

The radar targets carried by the balloons were partly manufactured by novelty and toy companies in New York, whose inventory of decorative icons seems to have been remembered many years later as alien hieroglyphics.

The heyday of UFOs corresponds to the time when the main delivery vehicle for nuclear weapons was being switched from aircraft to missiles. An early and important technical problem concerned re-entry—returning a nuclear-armed nosecone through the bulk of the Earth's atmosphere without burning it up in the process (as small asteroids and comets are destroyed in their passage through the upper air). Certain materials, nosecone geometries, and angles of entry are better than others. Observations of re-entry (or the more spectacular launches) could very well reveal U.S. progress in this vital strategic technology or, worse, inefficiencies in the design; such observations might suggest what defensive measures an adversary should take. Understandably, the subject was considered highly sensitive.

Inevitably there must have been cases in which military personnel were told not to talk about what they had seen, or where seemingly innocuous sightings were suddenly classified top secret with severely constrained need-to-know criteria. Air Force officers and civilian scientists thinking back about it in later years might very well conclude that the government had engineered a UFO cover-up. If nosecones are judged UFOs, the charge is a fair one.

Consider spoofing. In the strategic confrontation between the United States and the Soviet Union, the adequacy of air defenses was a vital issue. It was item (3) on General Cabell's list. If you could find a weakness, it might be the key to "victory" in an all-out nuclear war. The only sure way to test your adversary's defenses is to fly an aircraft over their borders and see how long it takes for them to notice. The United States did this routinely to test Soviet air defenses.

In the 1950s and '60s, the United States had state-of-the-art radar defense systems covering its west and east coasts, and especially its northern approaches (over which a Soviet bomber or missile attack would most likely come). But there was a soft underbelly—no significant early warning system to detect the geographically much more taxing southern approach. This is of course information vital for a potential adversary. It immediately suggests a spoof: One or more of the adversary's high-performance aircraft zoom out of the Caribbean, let's say, into U.S. airspace, penetrating, let's say, a few hundred miles up the Mississippi River until a U.S. air defense radar locks on. Then the intruders hightail it out of there. (Or, as a control experiment, a unit of U.S. high-performance aircraft is sequestered and sent in unannounced sorties to determine how porous American air defenses are.) In such a case, there may be combined visual and radar sightings by military and civilian observers and large numbers of independent reports. What is reported corresponds to no known aircraft. The Air Force and civilian aviation authorities truthfully state that none of their aircraft was responsible. Even if they've been urging Congress to fund a southern Early Warning System, the Air Force is unlikely to admit that Soviet or Cuban aircraft got to New Orleans, much less Memphis, before anybody caught on.

Here again, we have every reason to expect a high-level technical investigating team, Air Force and civilian observers told to keep their mouths shut, and not just the appearance but the reality of suppression of the data. Again, this conspiracy of silence need have nothing to do with alien spacecraft. Even decades later, there are bureaucratic reasons for the Department of Defense to be close-mouthed about such embarrassments. There is a potential conflict of interest between parochial concerns of the Department of Defense and the solution of the UFO enigma.

In addition, something that both the Central Intelligence Agency and the U.S. Air Force worried about then was UFOs as a means of clogging communication channels in a national crisis, and confusing visual and radar sightings of enemy aircraft—a signal-to-noise problem that in a way is the flip side of spoofing.

In view of all this, I'm perfectly prepared to believe that at least some UFO reports and analyses, and perhaps voluminous files, have been made inaccessible to the public which pays the bills. The Cold War is over, the missile and balloon technology is largely obsolete or widely available, and those who would be embarrassed are no longer on active duty. The worst that would happen, from the military's point of view, is that there would be one more acknowledged instance of the American public being misled or lied to in the interest of national security. It's time for the files to be declassified and made generally available.

Another instructive intersection of the conspiracy temperament and the secrecy culture concerns the National Security Agency. This organization monitors the telephone, radio, and other communications of both friends and adversaries of the United States. Surreptitiously, it reads the world's mail. Its daily intercept traffic is huge. In times of tension, vast arrays of NSA personnel fluent in the relevant languages are sitting with earphones, monitoring in real time everything from encrypted commands from the target nation's General Staff to pillow talk. For other material there are key words by which computers cull out for human attention specific messages or conversations of current urgent concern. Everything is stored, so that retrospectively it is possible to go back to the magnetic tapes—to trace the first appearance of a codeword, say, or command responsibility in a crisis. Some of the intercepts are made from listening posts in nearby countries (Turkey for Russia, India for China), from aircraft and ships patrolling nearby, or from ferret satellites in Earth orbit. There is a continuing dance of measures and countermeasures between the NSA and the security services of other nations, who understandably do not wish to be listened in on.

Now add to this already heady mix the Freedom of Information Act (FOIA). A request is made to the NSA for all information it has available on UFOs. It is required by law to be responsive, but of course without revealing "methods and sources." NSA also feels a deep obligation not to alert other nations, friends or foes, in an obtrusive and politically embarrassing way, of its activities. So a more or less typical intercept released by NSA in response to an FOIA request will be a third of a page blacked out, a fragment of a line saying "reported a UFO at low altitude," followed by two-thirds of a page blacked out. The NSA's position is that releasing the rest of the page would potentially compromise sources and methods, or at least alert the nation in question to how readily its aviation radio traffic is being intercepted. (If NSA released surrounding, seemingly bland aircraft-to-tower transmissions, it would then be possible for the nation in question to recognize that its military air traffic control dialogues are being monitored and to switch to communications means—frequency hopping, for example—that make NSA intercepts more difficult.) But UFO conspiracy theorists receiving, in response to their FOIA requests, dozens of pages of material, almost all of it blacked out, understandably deduce that the NSA possesses extensive information on UFOs and is part of a conspiracy of silence.

In talking not for attribution with NSA officials, I am told the following story: Typical intercepts are of military and civilian aircraft radioing that they see a UFO, by which they mean an unidentified object in the surrounding airspace. It may even be U.S. aircraft on reconnaissance or spoofing missions. In most cases it is something much more ordinary, and the clarification is also reported on later NSA intercepts.

Similar logic can be used to make NSA seem a part of any conspiracy. For example, they say, a response was required to an FOIA request on what the NSA knew about the singer Elvis Presley. (Apparitions of Mr. Presley and resulting miraculous cures have been reported.) Well, the NSA knew a few things. For example, a report on the economic health of a certain nation reported how many Elvis Presley tapes and CDs were sold there. This information also was supplied as a few lines of clear in a vast ocean of censorship black. Was NSA engaged in an Elvis Presley cover-up? While of course I have not personally investigated NSA's UFO-related traffic, their story seems to me very plausible.

If we are convinced that the government is keeping visits of aliens from us, then we should take on the secrecy culture of the military and intelligence establishments. At the very least we can push for declassification of relevant information from decades ago—of which the July 1994 Air Force report on the "Roswell Incident" is a good example.

You can catch a flavor of the paranoid style of many UFOlogists, as well as a naïveté about the secrecy culture, in a book by a former *New York Times* reporter, Howard Blum (*Out There*; Simon and Schuster, 1990):

> I could not, no matter how inventively I tried, avoid slamming into sudden dead ends. The whole story was always lingering, deliberately, I came to believe, just out of my grasp.
> Why?
> This was the single, practical, impossible question that was balanced ominously on the tall peak of my mounting suspicions.

Why were all these official spokesmen and institutions doing their collusive best to hinder and obstruct my efforts? Why were stories true one day, and false the next? Why all the tense, unyielding secretiveness? Why were military intelligence agents spreading disinformation, driving UFO believers mad? What had the government found out there? What was it trying to hide?

Of course there's resistance. Some information is classified legitimately; as with military hardware, secrecy sometimes really is in the national interest. Further, military, political, and intelligence communities tend to value secrecy for its own sake. It's a way of silencing critics and evading responsibility—for incompetence or worse. It generates an elite, a band of brothers in whom the national confidence can be reliably vested, unlike the great mass of citizenry on whose behalf the information is presumably made secret in the first place. With a few exceptions, secrecy is deeply incompatible with democracy and with science.

One of the most provocative purported intersections of UFOs and secrecy are the so-called MJ-12 documents. In late 1984, so the story goes, an envelope containing a·canister of exposed but undeveloped film was thrust into the home mail slot of a film producer, Jaime Shandera, interested in UFOs and government cover-up—remarkably, just as he was about to go out and have lunch with the author of a book on the alleged events in Roswell, New Mexico. When developed, it "proved to be" page after page of a highly-classified "eyes only" executive order dated 24 September, 1947 in which President Harry S Truman seemingly established a committee of twelve scientists and government officials to examine a set of crashed flying saucers and little alien bodies. The membership of the MJ-12 committee is remarkable because these are just the military, intelligence, science and engineering people who might have been called to investigate such crashes if they had occurred. In the MJ-12 documents there are tantalizing references to appendices about the nature of the aliens, the technology of their ships and so on, but the appendices were not included in the mysterious film.

The Air Force says that the document is bogus. The UFO expert Philip J. Klass and others find lexicographic and typographic inconsistencies that suggest that the whole thing is a hoax. Those who purchase fine art are concerned about the provenance of their painting—that is, who owned it most recently and who before that . . . and so on all the way back to the original artist. If there are breaks in the chain—if a 300-year-old painting can be tracked back only 60 years and then we have no idea in what home or museum it was hanging—then the forgery warning flags go up. Because the rewards of forgery in fine art are high, collectors must be very cautious. Where the MJ-12 documents are most vulnerable and suspect is exactly on this question of provenance—the evidence miraculously dropped on a doorstep like something out of a fairy story, perhaps "The Shoemaker and the Elves."

There are many cases in human history of a similar character—where a document of dubious provenance suddenly appears carrying information of great import which strongly supports the case of those who have made the discovery. After careful and in some cases courageous investigation the document is proved to be a hoax. There is no difficulty in understanding the motivation of the hoaxers. A more or less typical example is the book of Deuteronomy—discovered hidden in the Temple in Jerusalem by King Josiah, who, miraculously, in the midst of a major reformation struggle, found in Deuteronomy confirmation of all his views.

Another case is what is called the Donation of Constantine. Constantine the Great is the Emperor who made Christianity the official religion of the Roman Empire. The city of Constantinople (now Istanbul), for over a thousand years the capital of the Eastern Roman Empire, was named after him. He died in the year 337. In the ninth century, references to a Donation of Constantine suddenly appeared in Christian writings; in it Constantine wills to his contemporary Pope Sylvester I the entire Western Roman Empire, including Rome. This little gift, so the story went, was partly in gratitude for Sylvester's cure of Constantine's leprosy. By the eleventh century, popes were regularly referring to the Donation of Constantine to justify their claims to be not only the ecclesiastical but also the secular rulers of central Italy. Through the Middle Ages the Donation was judged genuine both by those who supported and by those who opposed the temporal claims of the Church.

Lorenzo of Valla was one of the polymaths of the Italian Renaissance. A controversialist, crusty, critical, arrogant, a pedant, he was attacked by his contemporaries for sacrilege, impudence, temerity and presumption—among other imperfections. After he concluded that the Apostles' Creed could not on grammatical grounds have actually been written by the Twelve Apostles, the Inquisition declared him a heretic, and only the intervention of his patron, Alfonso, King of Naples, prevented his immolation. Undeterred, in 1440, he published a treatise demonstrating that the Donation of Constantine is a crude forgery. The language in which it was written was to fourth century court Latin as Cockney was to the King's English. Because of Lorenzo of Valla, the Roman Catholic Church no longer presses its claim to rule European nations because of the Donation of Constantine. This work, whose provenance has a five-century hole in it, is generally understood to have been forged by a cleric attached to the Church's curia around the time of Charlemagne, when the papacy (and especially Pope Adrian I) was arguing for unification of church and state.

Assuming they both belong to the same category, the MJ-12 documents are a more clever hoax than the Donation of Constantine. But on matters of provenance, vested interest, and lexicographic inconsistencies, they have much in common.

A cover-up to keep knowledge of extraterrestrial life or alien abductions almost wholly secret for 45 years, with hundreds if not thousands of government employees privy to it, is a remarkable notion. Certainly, government secrets are routinely kept, even secrets of substantial general interest. But the ostensible point of such secrecy is to protect the country and its citizens. Here, though, it's different. The alleged conspiracy of those with security clearances is to keep from the citizens knowledge of a continuing alien assault on the human species. If extraterrestrials really were abducting millions of us, it would be much more than a matter of national security. It would impact the security of all human beings everywhere on Earth. Given such stakes, is it plausible that no one with real knowledge and evidence, in nearly 200 nations, would blow the whistle, speak out and side with the humans rather than the aliens?

Since the end of the Cold War NASA has been flailing about, trying to find missions that justify its existence—particularly a good reason for humans in space. If the Earth were being visited daily by hostile aliens, wouldn't NASA leap on this opportunity to augment its funding? And if an alien invasion were in progress, why would the Air

Force, traditionally led by pilots, step back from manned spaceflight and launch all its payloads on unmanned boosters?

Consider the former Strategic Defense Initiative Organization, in charge of "Star Wars." It's fallen on hard times now, particularly its objective of basing defenses in space. Its name and perspective have been demoted. It's the Ballistic Missile Defense Organization these days. It no longer even reports directly to the Secretary of Defense. The inability of such technology to protect the United States against a massive attack by nuclear-armed missiles is manifest. But wouldn't we want at least to attempt deployment of defenses in space if we were facing an alien invasion?

The Department of Defense, like similar ministries in every nation, thrives on enemies, real or imagined. It is implausible in the extreme that the existence of such an adversary would be suppressed by the very organization that would most benefit from its presence. The entire post-Cold War posture of the military and civilian space programs of the United States (and other nations) speaks powerfully against the idea that there are aliens among us—unless, of course, the news is also being kept from those who plan the national defense.

—

Just as there are those who accept every UFO report at face value, there are also those who dismiss the idea of alien visitation out of hand and with great passion. It is, they say, unnecessary to examine the evidence, and "unscientific" even to contemplate the issue. I once helped to organize a public debate at the annual meeting of the American Association for the Advancement of Science between proponent and opponent scientists of the proposition that some UFOs were spaceships; whereupon a distinguished physicist, whose judgment in many other matters I respected, threatened to sic the Vice President of the United States on me if I persisted in this madness. (Nevertheless, the debate was held and published, the issues were a little better clarified, and I did not hear from Spiro T. Agnew.)

A 1969 study by the National Academy of Sciences, while recognizing that there are reports "not easily explained," concluded that "the least likely explanation of UFOs is the hypothesis of extraterrestrial visitations by intelligent beings." Think of how many other "explanations" there might be: time travelers; demons from witchland;

tourists from another dimension—like Mr. Mxyztplk (or was it Mxyzptlk?; I always forget) from the land of Zrfff in the Fifth Dimension in the old *Superman* comic books; the souls of the dead; or a "noncartesian" phenomenon that doesn't obey the rules of science or even of logic. Each of these "explanations" has in fact been seriously proffered. "Least likely" is really saying something. This rhetorical excess is an index of how distasteful the whole subject has become to many scientists.

It's telling that emotions can run so high on a matter about which we really know so little. This is especially true of the more recent flurry of alien abduction reports. After all, if true, either hypothesis—invasion by sexually manipulative extraterrestrials or an epidemic of hallucinations—teaches us something we certainly ought to know about. Maybe the reason for strong feelings is that both alternatives have such unpleasant implications.

Aurora

> The number of reports and their consistency
> suggest that there may be some basis for these sightings
> other than hallucinogenic drugs.
>
> Mystery Aircraft, *report*,
> Federation of American Scientists,
> *August 20, 1992*

Aurora is a high-altitude, extremely secret American reconnaissance aircraft—a successor to the U-2 and the SR-71 *Blackbird*. It either exists or it doesn't. By 1993, there were reports by observers near California's Edwards Air Force Base and Groom Lake, Nevada, and particularly a region of Groom Lake called Area 51 where experimental aircraft for the Department of Defense are tested, that seemed by and large mutually consistent. Confirming reports were filed from all over the world. Unlike its predecessors, the aircraft is said to be hypersonic, to travel much faster, perhaps 6 to 8 times faster, than the speed of sound. It leaves an odd contrail described as "donuts-on-a-rope." Perhaps it is also a means of launching small secret satellites into orbit, developed, it is speculated, after the *Challenger* disaster indicated the shuttle's episodic unreliability for defense payloads. But the CIA "swears up and down there's no such program," says U.S. Senator and former astronaut John Glenn. The principal designer of some of the most secret U.S. aircraft says the same thing. A Secretary of the Air Force has vehemently denied the existence of such an airplane, or any program to build one, in the U.S. Air Force or anywhere else. Would he lie? "We have looked into all such sightings, as we have for UFO reports," says an Air Force spokesman, in perhaps carefully chosen words, "and we cannot explain them." Meanwhile, in April 1995 the

Air Force seized 4,000 more acres near Area 51. The area to which public access is denied is growing.

Consider then the two possibilities: that *Aurora* exists, and that it does not. If it exists, it's striking that an official cover-up of its very existence has been attempted, that secrecy could be so effective, and that the aircraft could be tested or refueled all over the world without a single photograph of it or any other hard evidence being published. On the other hand, if *Aurora* does not exist, it's striking that a myth has been propagated so vigorously, and gone so far. Why should insistent official denials have carried so little weight? Could the very existence of a designation—*Aurora* in this case—serve to pin a common label on a range of diverse phenomena? Either way, *Aurora* seems relevant to UFOs.

Chapter 6

HALLUCINATIONS

[A]s children tremble and fear everything
in the blind darkness, so we in the light
sometimes fear what is no more to be
feared than the things children in the dark
hold in terror. . .

LUCRETIUS,
On the Nature of Things
(ca. 60 B.C.)

Advertisers must know their audiences. It's a simple matter of product and corporate survival. So we can learn how commercial, free enterprise America views UFO buffs by examining the advertisements in magazines devoted to UFOs. Here are some (entirely typical) ad headlines from an issue of *UFO Universe*:

- Senior Research Scientist Discovers 2,000-Year-Old Secret to Wealth, Power, and Romantic Love.
- Classified! Above Top Secret. The Most Sensational Government Conspiracy of Our Time Is Finally Revealed to the World by a Retired Military Officer.
- What Is Your "Special Mission" While on Earth? The Cosmic Awakening of Light Workers, Walk-Ins, & All Star-Born Representatives Has Begun!
- This Is What You Have Been Waiting For. 24 Superb, Incredible Life-Improving UFO Seals of the Spirits.
- I Got a Girl. Do You? Stop Missing Out! Get Girls Now!
- Subscribe Today to the Most Amazing Magazine in the Universe.
- Bring Miraculous Good Luck, Love, and Money into Your Life! These Powers Have Worked for Centuries! They Can Work for You.
- Amazing Psychic Research Breakthrough. It Takes Only 5 Minutes to Prove that Psychic Magic Powers Really Work!
- Have You the Courage to Be Lucky, Loved and Rich? Guaranteed Good Fortune Will Come Your Way! Get Everything You Want with the Most Powerful Talismans in the World.
- Men in Black: Government Agents or Aliens?
- Increase the Power of Gemstones, Charms, Seals and Symbols. Improve the Effectiveness of Everything You Do. Magnify Your Mind Power and Abilities with the Mind Power MAGNIFIER.
- The Famous Money Magnet: Would You Like More Money?
- Testament of Lael, Sacred Scriptures of a Lost Civilization.

- A New Book by "Commander X" from Inner Light: The Controllers, the Hidden Rulers of Earth Identified. We Are the Property of an Alien Intelligence!

What is the common thread that binds these ads together? Not UFOs. Surely it's the expectation of unlimited audience gullibility. That's why they're placed in UFO magazines—because by and large the very act of buying such a magazine so categorizes the reader. Doubtless, there are moderately skeptical and fully rational purchasers of these periodicals who are demeaned by such expectations of advertisers and editors. But if they're right even about the bulk of their readers, what might it mean for the alien abduction paradigm?

Occasionally, I get a letter from someone who is in "contact" with extraterrestrials. I am invited to "ask them anything." And so over the years I've prepared a little list of questions. The extraterrestrials are very advanced, remember. So I ask things like, "Please provide a short proof of Fermat's Last Theorem." Or the Goldbach Conjecture. And then I have to explain what these are, because extraterrestrials will not call it Fermat's Last Theorem. So I write out the simple equation with the exponents. I never get an answer. On the other hand, if I ask something like "Should we be good?" I almost always get an answer. Anything vague, especially involving conventional moral judgments, these aliens are extremely happy to respond to. But on anything specific, where there is a chance to find out if they actually know anything beyond what most humans know, there is only silence.* Something can be deduced from this differential ability to answer questions.

In the good old days before the alien abduction paradigm, people taken aboard UFOs were offered, so they reported, edifying lectures on the dangers of nuclear war. Nowadays, when such instruction is given, the extraterrestrials seem fixated on environmental degradation and AIDS. How is it, I ask myself, that UFO occupants are so bound to fashionable or urgent concerns on this planet? Why not even an incidental warning about CFCs and ozone depletion in the 1950s, or

---

* It's a stimulating exercise to think of questions to which no human today knows the answers, but where a correct answer would immediately be recognized as such. It's even more challenging to formulate such questions in fields other than mathematics. Perhaps we should hold a contest and collect the best responses in "Ten Questions to Ask an Alien."

about the HIV virus in the 1970s, when it might really have done some good? Why not alert us now to some public health or environmental threat we haven't yet figured out? Can it be that aliens know only as much as those who report their presence? And if one of the chief purposes of alien visitations is admonitions about global dangers, why tell it only to a few people whose accounts are suspect anyway? Why not take over the television networks for a night, or appear with vivid cautionary audiovisuals before the United Nations Security Council? Surely this is not too difficult for those who wing across the light-years.

———

The earliest commercially successful UFO "contactee" was George Adamski. He operated a tiny restaurant at the foot of California's Mount Palomar, and set up a small telescope out in back. At the summit of the mountain was the largest telescope on Earth, the 200-inch reflector of the Carnegie Institution of Washington and the California Institute of Technology. Adamski styled himself *Professor* Adamski of Mount Palomar *Observatory*. He published a book—it caused quite a sensation, I recall—in which he described how in the desert nearby he had encountered nice-looking aliens with long blond hair and, if I remember correctly, white robes, who warned Adamski about the dangers of nuclear war. They hailed from the planet Venus (whose 900° Fahrenheit surface temperature we can now recognize as a barrier to Adamski's credibility). In person, he was utterly convincing. The Air Force officer nominally in charge of UFO investigations at the time described Adamski in these words:

> To look at the man and to listen to his story you had an immediate
> urge to believe him. Maybe it was his appearance. He was dressed
> in well worn, but neat, overalls. He had slightly graying hair and
> the most honest pair of eyes I've ever seen.

Adamski's star slowly faded as he aged, but he self-published other books and was a long-standing fixture at conventions of flying saucer "believers."

The first alien abduction story in the modern genre was that of Betty and Barney Hill, a New Hampshire couple—she a social worker and he a Post Office employee. During a late-night drive in 1961 through the White Mountains, Betty spotted a bright, initially starlike UFO that seemed to follow them. Because Barney feared it might harm them, they left the main highway for narrow mountain roads, arriving home two hours later than they'd expected. The experience prompted Betty to read a book that described UFOs as spaceships from other worlds; their occupants were little men who sometimes abducted humans.

Soon after, she experienced a terrifying, repetitive nightmare in which she and Barney were abducted and taken aboard the UFO. Barney overheard her describing this dream to friends, coworkers, and volunteer UFO investigators. (It's curious that Betty didn't discuss it with her husband directly.) By a week or so after the experience, they were describing a "pancake"-like UFO with uniformed figures seen through the craft's transparent windows.

Several years later, Barney's psychiatrist referred him to a Boston hypnotherapist, Benjamin Simon, M.D. Betty came to be hypnotized as well. Under hypnosis they separately filled in details of what had happened during the "missing" two hours: They watched the UFO land on the highway, and were taken, partly immobilized, inside the craft—where short, gray, humanoid creatures with long noses (a detail discordant with the current paradigm) subjected them to unconventional medical examinations, including a needle in her navel (before amniocentesis had been invented on Earth). There are those who now believe that eggs were taken from Betty's ovaries and sperm from Barney, although that isn't part of the original story.* The captain showed Betty a map of interstellar space with the ship's routes marked.

Martin S. Kottmeyer has shown that many of the motifs in the Hills' account can be found in a 1953 motion picture, *Invaders from Mars*. And Barney's story of what the aliens looked like, especially their enormous eyes, emerged in a hypnosis session just twelve days after the airing of an episode of the television series *The Outer Limits* in which such an alien was portrayed.

The Hill case was widely discussed. It was made into a 1975 TV movie that introduced the idea that short, gray alien abductors are

---

* In more recent times, Ms. Hill has written that in real alien abductions, "no sexual interest is shown. However, frequently they help themselves to some of [the abductee's] belongings, such as fishing rods, jewelry of different types, eyeglasses or a cup of laundry soap."

among us into the psyches of millions of people. But even the few scientists of the time who thought that some UFOs might in fact be alien
spaceships were wary. The alleged encounter was conspicuous by its
absence from the list of suggestive UFO cases compiled by James E.
McDonald, a University of Arizona atmospheric physicist. In general,
those scientists who have taken UFOs seriously have tended to keep
the alien abduction accounts at arm's length—while those who take
alien abductions at face value see little reason to analyze mere lights
in the sky.

McDonald's view on UFOs was based, he said, not on irrefutable
evidence, but was a conclusion of last resort: All the alternative explanations seemed to him even less credible. In the middle 1960s I
arranged for McDonald to present his best cases in a private meeting
with leading physicists and astronomers who had not before staked a
claim on the UFO issue. Not only did he fail to convince them that we
were being visited by extraterrestrials; he failed even to excite their interest. And this was a group with a very high wonder quotient. It was
simply that where McDonald saw aliens, they saw much more prosaic
explanations.

I was glad to have an opportunity to spend several hours with Mr.
and Mrs. Hill and with Dr. Simon. There was no mistaking the
earnestness and sincerity of Betty and Barney, and their mixed feelings
about becoming public figures under such odd and awkward circumstances. With the Hills' permission, Simon played for me (and, at my
invitation, McDonald) some of the audiotapes of their sessions under
hypnosis. By far my most striking impression was the absolute terror in
Barney's voice as he described—"re-lived" would be a better word—
the encounter.

Simon, while a leading proponent of the virtues of hypnosis in war
and peace, had not been caught up in the public frenzy about UFOs.
He shared handsomely in the royalties of John Fuller's best-seller, *The
Interrupted Journey*, about the Hills' experience. If Simon had pronounced their account authentic, the sales of the book might have
gone through the roof and his own financial reward been considerably
augmented. But he didn't. He also instantly rejected the notion that
they were lying, or, as suggested by another psychiatrist, that this was a
folie à deux—a shared delusion in which, generally, the submissive
partner goes along with the delusion of the dominant partner. So

what's left? The Hills, said their psychotherapist, had experienced a species of "dream." Together.

—

There may very well be more than one source of alien abduction accounts, just as there are for UFO sightings. Let's run through some of the possibilities:

In 1894 *The International Census of Waking Hallucinations* was published in London. From that time to this, repeated surveys have shown that 10 to 25 percent of ordinary, functioning people have experienced, at least once in their lifetimes, a vivid hallucination—hearing a voice, usually, or seeing a form when there's no one there. More rarely, people sense a haunting aroma, or hear music, or receive a revelation that arrives independent of the senses. In some cases these become transforming personal events or profound religious experiences. Hallucinations may be a neglected low door in the wall to a scientific understanding of the sacred.

Probably a dozen times since their deaths I've heard my mother or father, in a conversational tone of voice, call my name. Of course they called to me often during my life with them—to do a chore, to remind me of a responsibility, to come to dinner, to engage in conversation, to hear about an event of the day. I still miss them so much that it doesn't seem at all strange that my brain will occasionally retrieve a lucid recollection of their voices.

Such hallucinations may occur to perfectly normal people under perfectly ordinary circumstances. Hallucinations can also be elicited: by a campfire at night, or under emotional stress, or during epileptic seizures or migraine headaches or high fever, or by prolonged fasting or sleeplessness* or sensory deprivation (for example, in solitary con-

---

* Dreams are associated with a state called REM sleep, the abbreviation standing for rapid eye movement. (Under the closed eyelids the eyes move, perhaps following the action in the dream, or perhaps randomly.) The REM state is strongly correlated with sexual arousal. Experiments have been performed in which sleeping subjects are awakened whenever the REM state emerges, while members of a control group are awakened just as often each night but not when they're dreaming. After some days, the control group is a little groggy, but the experimental group—the ones who are prevented from dreaming—is hallucinating in daytime. It's not that a few people with a particular abnormality can be made to hallucinate in this way; anyone is capable of hallucinations.

finement), or through hallucinogens such as LSD, psilocybin, mescaline, or hashish. (Delirium tremens, the dreaded alcohol-induced "DTs," is one well-known manifestation of a withdrawal syndrome from alcoholism.) There are also molecules, such as the phenothiazines (Thorazine, for example), that make hallucinations go away. It is very likely that the normal human body generates substances — perhaps including the morphinelike small brain proteins called endorphins — that cause hallucinations, and others that suppress them. Such celebrated (and unhysterical) explorers as Admiral Richard Byrd, Captain Joshua Slocum, and Sir Ernest Shackleton all experienced vivid hallucinations when coping with unusual isolation and loneliness.

Whatever their neurological and molecular antecedents, hallucinations feel real. They are sought out in many cultures, and considered a sign of spiritual enlightenment. Among the Native Americans of the Western Plains, for example, or many indigenous Siberian cultures, a young man's future was foreshadowed by the nature of the hallucination he experienced after a successful "vision quest"; its meaning was discussed with great seriousness among the elders and shamans of the tribe. There are countless instances in the world's religions where patriarchs, prophets, or saviors repair themselves to desert or mountain and, assisted by hunger and sensory deprivation, encounter gods or demons. Psychedelic-induced religious experiences were a hallmark of the Western youth culture of the 1960s. The experience, however brought about, is often described respectfully by words such as "transcendent," "numinous," "sacred," and "holy."

Hallucinations are common. If you have one, it doesn't mean you're crazy. The anthropological literature is replete with hallucination ethnopsychiatry, REM dreams, and possession trances, which have many common elements transculturally and across the ages. The hallucinations are routinely interpreted as possession by good or evil spirits. The Yale anthropologist Weston La Barre goes so far as to argue that "a surprisingly good case could be made that much of culture is hallucination," and that "the whole intent and function of ritual appears to be . . . [a] group wish to hallucinate reality."

Here is a description of hallucinations as a signal-to-noise problem by Louis J. West, former medical director of the Neuropsychiatric Clinic at the University of California, Los Angeles. It is taken from the 15th edition of the *Encyclopaedia Britannica:*

[I]magine a man standing at a closed glass window opposite his fireplace, looking out at his garden in the sunset. He is so absorbed by the view of the outside world that he fails to visualize the interior of the room at all. As it becomes darker outside, however, images of the objects in the room behind him can be seen reflected dimly in the window glass. For a time he may see either the garden (if he gazes into the distance) or the reflection of the room's interior (if he focusses on the glass a few inches from his face). Night falls, but the fire still burns brightly in the fireplace and illuminates the room. The watcher now sees in the glass a vivid reflection of the interior of the room behind him, which appears to be outside the window. This illusion becomes dimmer as the fire dies down, and, finally, when it is dark both outside and within, nothing more is seen. If the fire flares up from time to time, the visions in the glass reappear.

In an analogous way, hallucinatory experiences such as those of normal dreams occur when the "daylight" (sensory input) is reduced while the "interior illumination" (general level of brain arousal) remains "bright," and images originating within the "rooms" of our brains may be perceived (hallucinated) as though they came from outside the "windows" of our senses.

Another analogy might be that dreams, like the stars, are shining all the time. Though the stars are not often seen by day, since the sun shines too brightly, if, during the day, there is an eclipse of the sun, or if a viewer chooses to be watchful awhile after sunset or awhile before sunrise, or if he is awakened from time to time on a clear night to look at the sky, then the stars, like dreams, though often forgotten, may always be seen.

A more brain-related concept is that of a continuous information-processing activity (a kind of "preconscious stream") that is influenced continually by both conscious and unconscious forces and that constitutes the potential supply of dream content. The dream is an experience during which, for a few minutes, the individual has some awareness of the stream of data being processed. Hallucinations in the waking state also would involve the same phenomenon, produced by a somewhat different set of psychological or physiological circumstances . . .

It appears that all human behaviour and experience (normal as well as abnormal) is well attended by illusory and hallucinatory phenomena. While the relationship of these phenomena to mental illness has been well documented, their role in everyday life has perhaps not been considered enough. Greater understanding of illusions and hallucinations among normal people may provide explanations for experiences otherwise relegated to the uncanny, "extrasensory," or supernatural.

We would surely be missing something important about our own nature if we refused to face up to the fact that hallucinations are part of being human. However, none of this makes hallucinations part of an external rather than an internal reality. Five to ten percent of us are extremely suggestible, able to move at a command into a deep hypnotic trance. Roughly ten percent of Americans report having seen one or more ghosts. This is more than the number who allegedly remember being abducted by aliens, about the same as the number who've reported seeing one or more UFOs, and less than the number who in the last week of Richard Nixon's Presidency—before he resigned to avoid impeachment—thought he was doing a good-to-excellent job as President. At least 1 percent of all of us is schizophrenic. This amounts to over 50 million schizophrenics on the planet, more than the population of, say, England.

In his 1970 book on nightmares, the psychiatrist John Mack—about whom I will have more to say—writes:

> There is a period in early childhood in which dreams are regarded as real and in which the events, transformations, gratifications, and threats of which they are composed are regarded by the child as if they were as much a part of his actual daily life as his daytime experiences. The capacity to establish and maintain clear distinctions between the life of dreams and life in the outside world is hard-won and requires several years to accomplish, not being completed even in normal children before ages eight to ten. Nightmares, because of their vividness and compelling affective intensity, are particularly difficult for the child to judge realistically.

When a child tells a fabulous story—a witch was grimacing in the darkened room; a tiger is lurking under the bed; the vase was broken by a multicolored bird that flew in the window and not because, contrary to family rules, a soccer ball was being kicked inside the house—is he or she consciously lying? Surely parents often act as if the child cannot fully distinguish between fantasy and reality. Some children have active imaginations; others are less well endowed in this department. Some families may respect the ability to fantasize and encourage the child, while at the same time saying something like "Oh, that's not real; that's just your imagination." Other families may be impatient about confabulating—it makes running the household and adjudicating disputes at least marginally more difficult—and discourage their children from fantasizing, perhaps even teaching them to think it's something shameful. A few parents may be unclear about the distinction between reality and fantasy themselves, or may even seriously enter into the fantasy. Out of all these contending propensities and child-rearing practices, some people emerge with an intact ability to fantasize, and a history, extending well into adulthood, of confabulation. Others grow up believing that anyone who doesn't know the difference between reality and fantasy is crazy. Most of us are somewhere in between.

Abductees frequently report having seen "aliens" in their childhood—coming in through the window or from under the bed or out of the closet. But everywhere in the world children report similar stories—with fairies, elves, brownies, ghosts, goblins, witches, imps, and a rich variety of imaginary "friends." Are we to imagine two different groups of children—one that sees imaginary earthly beings and the other that sees genuine extraterrestrials? Isn't it more reasonable that both groups are seeing, or hallucinating, the same thing?

Most of us recall being frightened at the age of two and older by real-seeming but wholly imaginary "monsters," especially at night or in the dark. I can still remember occasions when I was absolutely terrified, hiding under the bedclothes until I could stand it no longer, and then bolting for the safety of my parents' bedroom—if only I could get there before falling into the clutches of . . . The Presence. The American cartoonist Gary Larson, who draws in the horror genre, dedicates one of his books as follows:

> When I was a boy, our house was filled with monsters. They lived
> in the closets, under the beds, in the attic, in the basement, and—

when it was dark—just about everywhere. This book is dedicated to my father, who kept me safe from all of them.

Maybe the abduction therapists should be doing more of that.

Part of the reason that children are afraid of the dark may be that, in our entire evolutionary history up until just a moment ago, they never slept alone. Instead, they nestled safely, protected by an adult—usually Mom. In the enlightened West we stick them alone in a dark room, say goodnight, and have difficulty understanding why they're sometimes upset. It makes good evolutionary sense for children to have fantasies of scary monsters. In a world stalked by lions and hyenas, such fantasies help prevent defenseless toddlers from wandering too far from their guardians. How can this safety machinery be effective for a vigorous, curious young animal unless it delivers industrial-strength terror? Those who are not afraid of monsters tend not to leave descendants. Eventually, I imagine, over the course of human evolution, almost all children become afraid of monsters. But if we're capable of conjuring up terrifying monsters in childhood, why shouldn't some of us, at least on occasion, be able to fantasize something similar, something truly horrifying, a shared delusion, as adults?

It is telling that alien abductions occur mainly on falling asleep or when waking up, or on long automobile drives where there is a well-known danger of falling into some autohypnotic reverie. Abduction therapists are puzzled when their patients describe crying out in terror while their spouses sleep leadenly beside them. But isn't this typical of dreams—our shouts for help unheard? Might these stories have something to do with sleep and, as Benjamin Simon proposed for the Hills, a kind of dream?

A common, although insufficiently well-known, psychological syndrome rather like alien abduction is called sleep paralysis. Many people experience it. It happens in that twilight world between being fully awake and fully asleep. For a few minutes, maybe longer, you're immobile and acutely anxious. You feel a weight on your chest as if some being is sitting or lying there. Your heartbeat is quick, your breathing labored. You may experience auditory or visual hallucinations—of people, demons, ghosts, animals, or birds. In the right setting, the experience can have "the full force and impact of reality," according to Robert Baker, a psychologist at the University of Kentucky. Sometimes there's a marked sexual component to the hallucination. Baker argues that these common sleep disturbances are behind many if not most of the alien abduction accounts. (He and others suggest that there are other classes of abduction claims as well, made by fantasy-prone individuals, say, or hoaxers.)

Similarly, the *Harvard Mental Health Letter* (September 1994) comments,

> Sleep paralysis may last for several minutes, and is sometimes accompanied by vivid dreamlike hallucinations that give rise to stories about visitations from gods, spirits, and extraterrestrial creatures.

We know from early work of the Canadian neurophysiologist Wilder Penfield that electrical stimulation of certain regions of the brain elicits full-blown hallucinations. People with temporal lobe epilepsy—involving a cascade of naturally generated electrical impulses in the part of the brain beneath the forehead—experience a range of hallucinations almost indistinguishable from reality: including the presence of one or more strange beings, anxiety, floating through the air, sexual experiences, and a sense of missing time. There is also what feels like profound insight into the deepest questions and a need to spread the word. A continuum of spontaneous temporal lobe stimulation seems to stretch from people with serious epilepsy to the most average among us. In at least one case reported by another Canadian neuroscientist, Michael Persinger, administration of the antiepileptic drug, carbamazepine, eliminated a woman's recurring sense of experiencing the standard alien abduction scenario. So such hallucinations, generated spontaneously, or with chemical or experiential assists, may play a role—perhaps a central role—in the UFO accounts.

But such a view is easy to burlesque: UFOs explained away as "mass hallucinations." Everyone knows there's no such thing as a shared hallucination. Right?

——

As the possibility of extraterrestrial life began to be widely popularized—especially around the turn of the last century by Percival Lowell with his Martian canals—people began to report contact with aliens, mainly Martians. The psychologist Theodore Flournoy's 1901 book, *From India to the Planet Mars*, describes a French-speaking medium who in a trance state drew pictures of the Martians (they look just like us) and presented their alphabet and language (remarkably like French). The psychiatrist Carl Jung in his 1902 doctoral dissertation described a young Swiss woman who was agitated to discover, sitting across from her on the train, a "star-dweller" from Mars. Martians are innocent of science, philosophy, and souls, she was told, but have advanced technology. "Flying machines have long been in existence on Mars; the whole of Mars is covered with canals," and so on. Charles Fort, a collector of anomalous reports who died in 1932, wrote, "Perhaps there are inhabitants of Mars, who are secretly sending reports upon the ways of this world to their governments." In the 1950s there was a book by Gerald Heard that revealed the saucer occupants to be intelligent Martian bees. Who else could survive the fantastic right angle turns reported for UFOs?

But after the canals were shown to be illusory by *Mariner* 9 in 1971, and after no compelling evidence even for microbes was found on Mars by *Vikings* 1 and 2 in 1976, popular enthusiasm for the Lowellian Mars waned and we heard little about visiting Martians. Aliens were then reported to come from somewhere else. Why? Why no more Martians? And after the surface of Venus was found to be hot enough to melt lead, there were no more visiting Venusians. Does some part of these stories adjust to the current canons of belief? What does that imply about their origin?

There's no doubt that humans commonly hallucinate. There's considerable doubt about whether extraterrestrials exist, frequent our planet, or abduct and molest us. We might argue about details, but the one category of explanation is surely much better supported than the other. The main reservation you might then have is: Why do so many people today report this particular set of hallucinations? Why somber little beings, and flying saucers, and sexual experimentation?

Chapter 7

———

THE
DEMON-
HAUNTED
WORLD

There are demon-haunted worlds,
regions of utter darkness.

THE ISA UPANISHAD
(India, ca. 600 B.C.)

Fear of things invisible is the natural
seed of that which every one in himself
calleth religion.

THOMAS HOBBES,
Leviathan
(1651)

The gods watch over us and guide our destinies, many human cultures teach; other entities, more malevolent, are responsible for the existence of evil. Both classes of beings, whether considered natural or supernatural, real or imaginary, serve human needs. Even if they're wholly fanciful, people feel better believing in them. So in an age when traditional religions have been under withering fire from science, is it not natural to wrap up the old gods and demons in scientific raiment and call them aliens?

———

Belief in demons was widespread in the ancient world. They were thought of as natural rather than supernatural beings. Hesiod casually mentions them. Socrates described his philosophical inspiration as the work of a personal, benign demon. His teacher, Diotima of Mantineia, tells him (in Plato's *Symposium*) that "Everything demonic is intermediate between God and mortal. God has no contact with man," she continues; "only through the demonic is there intercourse and conversation between man and gods, whether in the waking state or during sleep."

Plato, Socrates' most celebrated student, assigned a high role to demons: "No human nature invested with supreme power is able to order human affairs," he said, "and not overflow with insolence and wrong. . ."

> We do not appoint oxen to be the lords of oxen, or goats of goats, but we ourselves are a superior race and rule over them. In like manner God, in his love of mankind, placed over us the demons, who are a superior race, and they with great ease and pleasure to themselves, and no less to us, taking care of us and giving us peace and reverence and order and justice never failing, made the tribes of men happy and united.

He stoutly denied that demons were a source of evil, and represented Eros, the keeper of sexual passions, as a demon, not a god, "neither mortal nor immortal," "neither good nor bad." But all later Platonists, including the Neo-Platonists who powerfully influenced Christian philosophy, held that some demons were good and others evil. The pendulum was swinging. Aristotle, Plato's famous student, seriously considered the contention that dreams are scripted by demons. Plutarch and Porphyry proposed that the demons, who filled the upper air, came from the Moon.

The early Church Fathers, despite having imbibed Neo-Platonism from the culture they swam in, were anxious to separate themselves from "pagan" belief-systems. They taught that all of pagan religion consisted of the worship of demons and men, both misconstrued as gods. When St. Paul complained (Ephesians 6:14) about wickedness in high places, he was referring not to government corruption, but to demons, who lived in high places:

> For we wrestle not against flesh and blood, but against principalities, against powers, against the rulers of the darkness of this world, against spiritual wickedness in high places.

From the beginning, much more was intended than demons as a mere poetic metaphor for the evil in the hearts of men.

St. Augustine was much vexed with demons. He quotes the pagan thinking prevalent in his time: "The gods occupy the loftiest regions, men the lowest, the demons the middle region. . . They have immortality of body, but passions of the mind in common with men." In Book VIII of *The City of God* (begun in 413), Augustine assimilates this ancient tradition, replaces gods by God, and demonizes the demons— arguing that they are, without exception, malign. They have no redeeming virtues. They are the fount of all spiritual and material evil. He calls them "aerial animals. . . most eager to inflict harm, utterly alien from righteousness, swollen with pride, pale with envy, subtle in deceit." They may profess to carry messages between God and man, disguising themselves as angels of the Lord, but this pose is a snare to lure us to our destruction. They can assume any form, and know many things—"demon" *means* "knowledge" in Greek*—especially about the material world. However intelligent, they are deficient in charity.

---

* "Science" means "knowledge" in Latin. A jurisdictional dispute is exposed, even if we look no further.

They prey on "the captive and outwitted minds of men," wrote Tertullian. "They have their abode in the air, the stars are their neighbors, their commerce is with the clouds."

In the eleventh century, the influential Byzantine theologian, philosopher, and shady politician, Michael Psellus, described demons in these words:

> These animals exist in our own life, which is full of passions, for they are present abundantly in the passions, and their dwelling-place is that of matter, as is their rank and degree. For this reason they are also subject to passions and fettered to them.

One Richalmus, abbot of Schönthal, around 1270 penned an entire treatise on demons, rich in first-hand experience: He sees (but only when his eyes are shut) countless malevolent demons, like motes of dust, buzzing around his head—and everyone else's. Despite successive waves of rationalist, Persian, Jewish, Christian, and Moslem world views, despite revolutionary social, political, and philosophical ferment, the existence, much of the character, and even the name of demons remained unchanged from Hesiod through the Crusades.

Demons, the "powers of the air," come down from the skies and have unlawful sexual congress with women. Augustine believed that witches were the offspring of these forbidden unions. In the Middle Ages, as in classical antiquity, nearly everyone believed such stories. The demons were also called devils, or fallen angels. The demonic seducers of women were labeled incubi; of men, succubi. There are cases in which nuns reported, in some befuddlement, a striking resemblance between the incubus and the priest-confessor, or the bishop, and awoke the next morning, as one fifteenth-century chronicler put it, to "find themselves polluted just as if they had commingled with a man." There are similar accounts, but in harems not convents, in ancient China. So many women reported incubi, argued the Presbyterian religious writer Richard Baxter (in his *Certainty of the World of Spirits*, 1691), "that 'tis impudence to deny it."*

---

* Likewise, in the same work, "The raising of storms by witches is attested by so many, that I think it needless to recite them." The theologian Meric Casaubon argued—in his 1668 book, *Of Credulity and Incredulity*—that witches must exist because, after all, everyone believes in them. Anything that a large number of people believe must be true.

As they seduced, the incubi and succubi were perceived as a weight bearing down on the chest of the dreamer. *Mare*, despite its Latin meaning, is the Old English word for incubus, and *nightmare* meant originally the demon that sits on the chests of sleepers, tormenting them with dreams. In Athanasius' *Life of St. Anthony* (written around 360) demons are described as coming and going at will in locked rooms; 1400 years later, in his work *De Daemonialitae*, the Franciscan scholar Ludovico Sinistrari assures us that demons pass through walls.

The external reality of demons was almost entirely unquestioned from antiquity through late medieval times. Maimonides denied their reality, but the overwhelming majority of rabbis believed in *dybbuks*. One of the few cases I can find where it is even hinted that demons might be *internal*, generated in our minds, is when Abba Poemen—one of the desert fathers of the early Church—was asked,

"How do the demons fight against me?"

"The demons fight against you?" Father Poemen asked in turn. "Our own wills become the demons, and it is these which attack us."

The medieval attitudes on incubi and succubi were influenced by Macrobius' fourth-century *Commentary on the Dream of Scipio*, which went through dozens of editions before the European Enlightenment. Macrobius described phantoms (*phantasma*) seen "in the moment between wakefulness and slumber." The dreamer "imagines" the phantoms as predatory. Macrobius had a skeptical side which his medieval readers tended to ignore.

Obsession with demons began to reach a crescendo when, in his famous Bull of 1484, Pope Innocent VIII declared,

> It has come to Our ears that members of both sexes do not avoid to have intercourse with evil angels, incubi, and succubi, and that by their sorceries, and by their incantations, charms, and conjurations, they suffocate, extinguish, and cause to perish the births of women

as well as generate numerous other calamities. With this Bull, Innocent initiated the systematic accusation, torture, and execution of countless "witches" all over Europe. They were guilty of what Augustine had described as "a criminal tampering with the unseen world."

Despite the evenhanded "members of both sexes" in the language of the Bull, unsurprisingly it was mainly girls and women who were so persecuted.

Many leading Protestants of the following centuries, their differences with the Catholic Church notwithstanding, adopted nearly identical views. Even humanists such as Desiderius Erasmus and Thomas More believed in witches. "The giving up of witchcraft," said John Wesley, the founder of Methodism, "is in effect the giving up of the Bible." William Blackstone, the celebrated jurist, in his *Commentaries on the Laws of England* (1765), asserted:

> To deny the possibility, nay, actual existence of witchcraft and sorcery is at once flatly to contradict the revealed word of God in various passages of both the Old and New Testament.

Innocent commended "Our dear sons Henry Kramer and James Sprenger," who "have been by Letters Apostolic delegated as Inquisitors of these heretical [de]pravities." If "the abominations and enormities in question remain unpunished," the souls of multitudes face eternal damnation.

The Pope appointed Kramer and Sprenger to write a comprehensive analysis, using the full academic armory of the late fifteenth century. With exhaustive citations of Scripture and of ancient and modern scholars, they produced the *Malleus Maleficarum*, the "Hammer of Witches"—aptly described as one of the most terrifying documents in human history. Thomas Ady, in *A Candle in the Dark*, condemned it as "villainous Doctrines & Inventions," "horrible lyes and impossibilities," serving to hide "their unparalleled cruelty from the ears of the world." What the *Malleus* comes down to, pretty much, is that if you're accused of witchcraft, you're a witch. Torture is an unfailing means to demonstrate the validity of the accusation. There are no rights of the defendant. There is no opportunity to confront the accusers. Little attention is given to the possibility that accusations might be made for impious purposes—jealousy, say, or revenge, or the greed of the inquisitors who routinely confiscated for their own private benefit the property of the accused. This technical manual for torturers also includes methods of punishment tailored to release demons from the victim's body before the process kills her. The *Malleus* in hand, the

Pope's encouragement guaranteed, inquisitors began springing up all over Europe.

It quickly became an expense account scam. All costs of investigation, trial, and execution were borne by the accused or her relatives — down to per diems for the private detectives hired to spy on her, wine for her guards, banquets for her judges, the travel expenses of a messenger sent to fetch a more experienced torturer from another city, and the faggots, tar and hangman's rope. Then there was a bonus to the members of the tribunal for each witch burned. The convicted witch's remaining property, if any, was divided between Church and State. As this legally and morally sanctioned mass murder and theft became institutionalized, as a vast bureaucracy arose to serve it, attention was turned from poor hags and crones to the middle class and well-to-do of both sexes.

The more who, under torture, confessed to witchcraft, the harder it was to maintain that the whole business was mere fantasy. Since each "witch" was made to implicate others, the numbers grew exponentially. These constituted "frightful proofs that the Devil is still alive," as it was later put in America in the Salem witch trials. In a credulous age, the most fantastic testimony was soberly accepted — that tens of thousands of witches had gathered for a Sabbath in public squares in France, or that 12,000 of them darkened the skies as they flew to Newfoundland. The Bible had counseled, "Thou shalt not suffer a witch to live." Legions of women were burnt to death.* And the most horrendous tortures were routinely applied to every defendant, young or old, after the instruments of torture were first blessed by the priests. Innocent himself died in 1492, following unsuccessful attempts to keep him alive by transfusion (which resulted in the deaths of three boys) and by suckling at the breast of a nursing mother. He was mourned by his mistress and their children.

In Britain witch-finders, also called "prickers," were employed, receiving a handsome bounty for each girl or woman they turned over for execution. They had no incentive to be cautious in their accusations. Typically they looked for "devil's marks" — scars or birthmarks or nevi — that when pricked with a pin neither hurt nor bled. A simple

---

* This mode of execution was adopted by the Holy Inquisition apparently to guarantee literal accord with a well-intentioned sentence of canon law (Council of Tours, 1163): "The Church abhors bloodshed."

sleight of hand often gave the appearance that the pin penetrated deep into the witch's flesh. When no visible marks were apparent, "invisible marks" sufficed. Upon the gallows, one mid-seventeenth-century pricker "confessed he had been the death of above 220 women in England and Scotland, for the gain of twenty shillings apiece."*

In the witch trials, mitigating evidence or defense witnesses were inadmissible. In any case, it was nearly impossible to provide compelling alibis for accused witches: The rules of evidence had a special character. For example, in more than one case a husband attested that his wife was asleep in his arms at the very moment she was accused of frolicking with the devil at a witch's Sabbath; but the archbishop patiently explained that a demon had taken the place of the wife. The husbands were not to imagine that their powers of perception could exceed Satan's powers of deception. The beautiful young women were perforce consigned to the flames.

There were strong erotic and misogynistic elements—as might be expected in a sexually repressed, male-dominated society with inquisitors drawn from the class of nominally celibate priests. The trials paid close attention to the quality and quantity of orgasm in the supposed copulations of defendants with demons or the Devil (although Augustine had been certain "we cannot call the Devil a fornicator"), and to the nature of the Devil's "member" (cold, by all reports). "Devil's marks" were found "generally on the breasts or private parts" according to Ludovico Sinistrari's 1700 book. As a result pubic hair was shaved, and the genitalia were carefully inspected by the exclusively male inquisitors. In the immolation of the 20-year-old Joan of Arc, after her dress had caught fire the Hangman of Rouen slaked the flames so onlookers could view "all the secrets which can or should be in a woman."

The chronicle of those who were consumed by fire in the single German city of Würzburg in the single year 1598 penetrates the statistics and lets us confront a little of the human reality:

---

* In the murky territory of bounty hunters and paid informers, vile corruption is often the rule—worldwide and through all of human history. To take an example almost at random, in 1994, for a fee, a group of postal inspectors from Cleveland agreed to go underground and ferret out wrongdoers; they then contrived criminal cases against 32 innocent postal workers.

The steward of the senate, named Gering; old Mrs. Kanzler; the tailor's fat wife; the woman cook of Mr. Mengerdorf; a stranger; a strange woman; Baunach, a senator, the fattest citizen in Würtzburg; the old smith of the court; an old woman; a little girl, nine or ten years old; a younger girl, her little sister; the mother of the two little aforementioned girls; Liebler's daughter; Goebel's child, the most beautiful girl in Würtzburg; a student who knew many languages; two boys from the Minster, each twelve years old; Stepper's little daughter; the woman who kept the bridge gate; an old woman; the little son of the town council bailiff; the wife of Knertz, the butcher; the infant daughter of Dr. Schultz; a blind girl; Schwartz, canon at Hach. . .

On and on it goes. Some were given special humane attention: "The little daughter of Valkenberger was privately executed and burnt." There were 28 public immolations, each with 4 to 6 victims on average, in that small city in a single year. This was a microcosm of what was happening all across Europe. No one knows how many were killed altogether—perhaps hundreds of thousands, perhaps millions. Those responsible for prosecuting, torturing, judging, burning, and justifying were selfless. Just ask them.

They could not be mistaken. The confessions of witchcraft could not be based on hallucinations, say, or desperate attempts to satisfy the inquisitors and stop the torture. In such a case, explained the witch judge Pierre de Lancre (in his 1612 book, *Description of the Inconstancy of Evil Angels*), the Catholic Church would be committing a great crime by burning witches. Those who raise such possibilities are thus attacking the Church and *ipso facto* committing a mortal sin. Critics of witch-burning were punished and, in some cases, themselves burnt. The inquisitors and torturers were doing God's work. They were saving souls. They were foiling demons.

Witchcraft of course was not the only offense that merited torture and burning at the stake. Heresy was a still more serious crime, and both Catholics and Protestants punished it ruthlessly. In the sixteenth century the scholar William Tyndale had the temerity to contemplate translating the New Testament into English. But if people could actually read the Bible in their own language instead of arcane Latin, they could form their own, independent religious views. They might conceive of their own private unintermediated line to God. This was a challenge to the job security of Roman Catholic priests. When Tyndale tried to publish his translation, he was hounded and pursued all over Europe. Eventually he was captured, garroted, and then, for good measure, burned at the stake. His copies of the New Testament (which a century later became the basis of the exquisite King James translation) were then hunted down house-to-house by armed posses—Christians piously defending Christianity by preventing other Christians from knowing the words of Christ. Such a cast of mind, such a climate of absolute confidence that knowledge should be rewarded by torture and death were unlikely to help those accused of witchcraft.

Burning witches is a feature of Western civilization that has, with occasional political exceptions, declined since the sixteenth century. In the last judicial execution of witches in England, a woman and her nine-year-old daughter were hanged. Their crime was raising a rain storm by taking their stockings off. In our time, witches and djinns are found as regular fare in children's entertainment, exorcism of demons is still practiced by the Roman Catholic and other churches, and the proponents of one cult still denounce as sorcery the cultic practices of another. We still use the word "pandemonium" (literally, all demons). A crazed and violent person is still said to be demonic. (Not until the eighteenth century was mental illness no longer generally ascribed to supernatural causes; even insomnia had been considered a punishment inflicted by demons.) More than half of Americans tell pollsters they "believe" in the Devil's existence, and 10 percent have communicated with him, as Martin Luther reported he did regularly. In a 1992 "spiritual warfare manual" called *Prepare for War*, Rebecca Brown informs us that abortion and sex outside of marriage "will almost always result in demonic infestation"; that meditation, yoga and martial arts are designed so unsuspecting Christians will be seduced into worshiping demons; and that "rock music didn't 'just happen,' it was a carefully masterminded plan by none other than Satan himself." Sometimes "your loved ones are demonically bound and blinded." Demonology is today still part and parcel of many earnest faiths.

And what is it that demons do? In the *Malleus*, Kramer and Sprenger reveal that "devils . . . busy themselves by interfering with the process of normal copulation and conception, by obtaining human semen, and themselves transferring it." Demonic artificial insemination in the Middle Ages goes back at least to St. Thomas Aquinas, who tells us in *On the Trinity* that "demons can transfer the semen which they have collected and inject it into the bodies of others." His contemporary, St. Bonaventura, spells it out in a little more detail: Succubi "yield to males and receive their semen; by cunning skill, the demons preserve its potency, and afterwards, with the permission of God, they become incubi and pour it out into female repositories." The products of these demon-mediated unions are also, when they grow up, visited by demons. A multigenerational transspecies sexual bond is forged. And these creatures, we recall, are well known to fly; indeed they inhabit the upper air.

There is no spaceship in these stories. But most of the central elements of the alien abduction account are present, including sexually obsessive non-humans who live in the sky, walk through walls, communicate telepathically, and perform breeding experiments on the human species. Unless *we* believe that demons really exist, how can we understand so strange a belief system, embraced by the whole Western world (including those considered the wisest among us), reinforced by personal experience in every generation, and taught by Church and State? Is there any real alternative besides a shared delusion based on common brain wiring and chemistry?

—

In Genesis we read of angels who couple with "the daughters of men." The culture myths of ancient Greece and Rome told of gods appearing to women as bulls or swans or showers of gold and impregnating them. In one early Christian tradition, philosophy derived not from human ingenuity but out of demonic pillow talk—the fallen angels betraying the secrets of Heaven to their human consorts. Accounts with similar elements appear in cultures around the world. Parallels to incubi include Arabian djinn, Greek satyrs, Hindu bhuts, Samoan hotua poro, Celtic dusii, and many others. In an epoch of demon hysteria, it was easy enough to demonize those we feared or hated. So Merlin was said to have been fathered by an incubus. So were Plato, Alexander the Great, Augustus, and Martin Luther. Occasionally an entire people—for example the Huns or the inhabitants of Cyprus— were accused by their enemies of having been sired by demons.

In Talmudic tradition the archetypical succubus was Lilith, whom

God made from the dust along with Adam. She was expelled from Eden for insubordination—not to God, but to Adam. Ever since, she spends her nights seducing Adam's descendants. In ancient Iranian and many other cultures, nocturnal seminal emissions were believed to be elicited by succubi. St. Teresa of Avila reported a vivid sexual encounter with an angel—an angel of light, not of darkness, she was sure—as did other women later sanctified by the Catholic Church. Cagliostro, the eighteenth-century magician and con man, let it be understood that he, like Jesus of Nazareth, was a product of the union "between the children of heaven and earth."

In 1645 a Cornish teenager, Anne Jefferies, was found groggy, crumpled on the floor. Much later, she recalled being attacked by half-a-dozen little men, carried paralyzed to a castle in the air, seduced, and returned home. She called the little men fairies. (For many pious Christians, as for the inquisitors of Joan of Arc, this was a distinction without a difference. Fairies were demons, plain and simple.) They returned to terrify and torment her. The next year she was arrested for witchcraft. Fairies traditionally have magical powers, and can cause paralysis by the merest touch. The ordinary passage of time is slowed in fairyland. Fairies are reproductively impaired, so they have sex with humans and carry off babies from their cradles—sometimes leaving a fairy substitute, a "changeling." Now it seems a fair question: If Anne Jefferies had grown up in a culture touting aliens rather than fairies, and UFOs rather than castles in the air, would her story have been distinguishable in any significant respect from the ones "abductees" tell?

In his 1982 book *The Terror That Comes in the Night: An Experience-Centered Study of Supernatural Assault Traditions*, David Hufford describes an executive, university-educated, in his mid-thirties, who recalled a summer spent as a teenager in his aunt's house. One night, he saw mysterious lights moving in the harbor. Afterwards, he fell asleep. From his bed he then witnessed a white, glowing figure climbing the stairs. She entered his room, paused, and then said—anticlimactically, it seems to me—"That is the linoleum." Some nights the figure was an old woman; in others, an elephant. Sometimes the young man was convinced the entire business was a dream; other times he was certain he was awake. He was pressed down into his bed, paralyzed, unable to move or cry out. His heart was pounding. He was short of breath. Similar events transpired on many consecutive nights. What is happening here? These events took place before alien abductions were widely described. If the young man had known about alien abductions, would his old woman have had a larger head and bigger eyes?

In several famous passages in *The Decline and Fall of the Roman Empire*, Edward Gibbon described the balance between credulity and skepticism in late classical antiquity:

> Credulity performed the office of faith; fanaticism was permitted to assume the language of inspiration, and the effects of accident or contrivance were ascribed to supernatural causes. . .
>
> In modern times [Gibbon is writing in the middle eighteenth century], a latent and even involuntary scepticism adheres to the most pious dispositions. Their admission of supernatural truths is much less an active consent than a cold and passive acquiescence. Accustomed long since to observe and to respect the invariable order of Nature, our reason, or at least our imagination, is not sufficiently prepared to sustain the visible action of the Deity. But in the first ages of Christianity the situation of mankind was extremely different. The most curious, or the most credulous, among the Pagans were often persuaded to enter into a society which asserted an actual claim of miraculous powers. The primitive Christians perpetually trod on mystic ground, and their minds were exercised by the habits of believing the most extraordinary events. They felt, or they fancied, that on every side they were incessantly assaulted by dæmons, comforted by visions, instructed by prophecy, and surprisingly delivered from danger, sickness, and from death itself, by the supplications of the church. . .
>
> It was their firm persuasion that the air which they breathed was peopled with invisible enemies; with innumerable dæmons, who watched every occasion, and assumed every form, to terrify, and above all to tempt, their unguarded virtue. The imagination, and even the senses, were deceived by the illusions of distempered fanaticism; and the hermit, whose midnight prayer was oppressed by involuntary slumber, might easily confound the phantoms of horror or delight which had occupied his sleeping and his waking dreams. . .
>
> [T]he practice of superstition is so congenial to the multitude that, if they are forcibly awakened, they still regret the loss of their pleasing vision. Their love of the marvellous and supernatural,

their curiosity with regard to future events, and their strong propensity to extend their hopes and fears beyond the limits of the visible world, were the principal causes which favoured the establishment of Polytheism. So urgent on the vulgar is the necessity of believing, that the fall of any system of mythology will most probably be succeeded by the introduction of some other mode of superstition. . .

Put aside Gibbon's social snobbery: The devil tormented the upper classes too, and even a king of England—James I, the first Stuart monarch—wrote a credulous and superstitious book on demons (*Daemonologie*, 1597). He also was the patron of the great translation of the Bible into English that still bears his name. It was King James' opinion that tobacco is the "devil's weed," and a number of witches were exposed through their addiction to this drug. But by 1618, James had become a thoroughgoing skeptic—mainly because adolescents had been found faking demonic possession, in which state they had accused innocent people of witchcraft. If we reckon the skepticism that Gibbon says characterized his time to have declined in ours, and if even a little of the rampant gullibility he attributes to late classical times is left over in ours, should we not expect something like demons to find a niche in the popular culture of the present?

Of course, as enthusiasts for extraterrestrial visitations are quick to remind me, there's another interpretation of these historical parallels: Aliens, they say, have *always* been visiting us, poking at us, stealing our sperms and eggs, impregnating us. In earlier times we recognized them as gods, demons, fairies, or spirits; only now do we understand that it's aliens who've been diddling us all these millennia. Jacques Vallee has made such arguments. But then why are there virtually no reports of flying saucers prior to 1947? Why is it that none of the world's major religions uses saucers as icons of the divine? Why no warnings about the dangers of high technology then? Why isn't this genetic experiment, whatever its objective, completed by now—thousands of years or more after its initiation by beings supposedly of vastly superior technological attainments? Why are we in such trouble if the breeding program is designed to improve our lot?

Following this line of argument, we might anticipate present adherents of the old beliefs to understand "aliens" to be fairies, gods, or demons. In fact, there are several contemporary sects—the "Raelians," for example—that hold gods or God to come to Earth in UFOs. Some abductees describe the aliens, however repulsive, as "angels," or "emissaries of God." And there are those who still think it's demons:

In Whitley Strieber's *Communion*, a first-hand account of "alien abduction," the author relates

> Whatever was there seemed so monstrously ugly, so filthy and dark and sinister. Of course they were demons. They had to be. . .
> I still remember that thing crouching there, so terribly ugly, its arms and legs like the limbs of a great insect, its eyes glaring at me.

Reportedly, Strieber is now open to the possibility that these nighttime terrors were dreams or hallucinations.

Articles on UFOs in *The Christian News Encyclopedia*, a fundamentalist compilation, include "Unchristian Fanatic Obsession," and "Scientist Believes UFOs Work of Devil." The Spiritual Counterfeits Project of Berkeley, California, teaches that UFOs are of demonic origin; the Aquarian Church of Universal Service of McMinnville, Oregon, that all aliens are hostile. A 1993 newsletter of "Cosmic Awareness Communications" informs us that UFO occupants think of humans as laboratory animals, wish us to worship them, but tend to be deterred by the Lord's Prayer. Some abductees have been cast out of their evangelical religious congregations; their stories sound too close to Satanism. A 1980 fundamentalist tract, *The Cult Explosion*, by Dave Hunt, reveals that

> UFOs . . . are clearly not physical and seem to be demonic manifestations from another dimension calculated to alter man's way of thinking. . . . [T]he alleged UFO entities that have presumably communicated psychically with humans have always preached the same four lies that the serpent introduced to Eve. . . . [T]hese beings are demons and they are preparing for the Antichrist.

A number of sects hold UFOs and alien abductions to be premonitions of "end-times."

If UFOs come from another planet or another dimension, were they sent by the same God who has been revealed to us in any of the major religions? Nothing in the UFO phenomena, the fundamentalist complaint goes, requires belief in the one, true God, while much in it contradicts the God portrayed in the Bible and Christian tradition. *The New Age: A Christian Critique* by Ralph Rath (1990) discusses UFOs—typically for such literature, with extreme credulity. It serves their purpose to accept UFOs as real and revile them as instruments of Satan and the Antichrist, rather than to use the blade of scientific skepticism. That tool, once honed, might accomplish more than just a limited heresiotomy.

The Christian fundamentalist author Hal Lindsey, in his 1994 religious best-seller *Planet Earth—2000 A.D.*, writes,

> I have become thoroughly convinced that UFOs are real. . . They are operated by alien beings of great intelligence and power. . . I believe these beings are not only extraterrestrial but supernatural in origin. To be blunt, I think they are demons. . . part of a Satanic plot.

And what is the evidence for this conclusion? Chiefly, it is the 11th and 12th verses of Luke, Chapter 21, in which Jesus talks about "great signs from Heaven"—nothing like a UFO is described—in the last days. Typically, Lindsey ignores verse 32, in which Jesus makes it very clear he is talking about events in the first, not the twentieth, century.

There is also a Christian tradition according to which extraterrestrial life cannot exist. In *Christian News* for May 23, 1994, for example, W. Gary Crampton, Doctor of Theology, tells us why:

> The Bible, either explicitly or implicitly, speaks to every area of life; it never leaves us without an answer. The Bible nowhere explicitly affirms or negates intelligent extraterrestrial life. Implicitly, however, Scripture does deny the existence of such beings, thus also negating the possibility of flying saucers. . . Scripture views earth as the center of the universe. . . According to Peter, a "planet hopping" Savior is out of the question. Here is an answer to intelligent life on other planets. If there were such, who would redeem them? Certainly not Christ. . . Experiences which are out of line with the teachings of Scripture must always be renounced as fallacious. The Bible has a monopoly on the truth.

But many other Christian sects—Roman Catholics, for example—are completely open-minded, with no *a priori* objections to and no insistence on the reality of aliens and UFOs.

In the early 1960s, I argued that the UFO stories were crafted chiefly to satisfy religious longings. At a time when science has complicated uncritical adherence to the old-time religions, an alternative is proffered to the God hypothesis: Dressed in scientific jargon, their immense powers "explained" by superficially scientific terminology, the gods and demons of old come down from heaven to haunt us, to offer prophetic visions, and to tantalize us with visions of a more hopeful future: a space-age mystery religion aborning.

The folklorist Thomas E. Bullard wrote in 1989 that

abduction reports sound like rewrites of older supernatural encounter traditions with aliens serving the functional roles of divine beings.

He concludes:

Science may have evicted ghosts and witches from our beliefs, but it just as quickly filled the vacancy with aliens having the same functions. Only the extraterrestrial outer trappings are new. All the fear and the psychological dramas for dealing with it seem simply to have found their way home again, where it is business as usual in the legend realm where things go bump in the night.

Is it possible that people in all times and places occasionally experience vivid, realistic hallucinations, often with sexual content, about abduction by strange, telepathic, aerial creatures who ooze through walls—with the details filled in by the prevailing cultural idioms, sucked out of the *Zeitgeist*? Others, who have not personally had the experience, find it stirring and in a way familiar. They pass the story on. Soon it takes on a life of its own, inspires others trying to understand their own visions and hallucinations, and enters the realm of folklore, myth, and legend. The connection between the content of spontaneous temporal lobe hallucinations and the alien abduction paradigm is consistent with such a hypothesis.

Perhaps when everyone knows that gods come down to Earth, we hallucinate gods; when all of us are familiar with demons, it's incubi and succubi; when fairies are widely accepted, we see fairies; in an age of spiritualism, we encounter spirits; and when the old myths fade and we begin thinking that extraterrestrial beings are plausible, then that's where our hypnogogic imagery tends.

Snatches of song or foreign languages, images, events that we witnessed, stories that we overheard in childhood can be accurately recalled decades later without any conscious memory of how they got into our heads. "[I]n violent fevers, men, all ignorance, have talked in ancient tongues," says Herman Melville in *Moby-Dick*; "and . . . when the mystery is probed, it turns out always that in their wholly forgotten childhood those ancient tongues had been really spoken in their hearing." In our everyday life, we effortlessly and unconsciously incorporate cultural norms and make them our own.

A similar inhaling of motifs is present in schizophrenic "command hallucinations." Here people feel they are being told what to do by an imposing or mythic figure. They are ordered to assassinate a political leader or a folk hero, or defeat the British invaders, or harm themselves, because it is the wish of God, or Jesus, or the Devil, or demons, or angels, or—lately—aliens. The schizophrenic is transfixed by a clear and powerful command from a voice that no one else can hear, and that the subject must somehow identify. Who *would* issue such a command? Who *could* speak inside our heads? The culture in which we've been raised offers up an answer.

Think of the power of repetitive imagery in advertising, especially to suggestible viewers and readers. It can make us believe almost anything—even that smoking cigarettes is cool. In our time, putative aliens are the subject of innumerable science fiction stories, novels, TV dramas, and films. UFOs are a regular feature of the weekly tabloids devoted to falsification and mystification. One of the highest-grossing motion pictures of all time is about aliens very like those described by abductees. Alien abduction accounts were comparatively rare until 1975, when a credulous television dramatization of the Hill case was aired; another leap into public prominence occurred after 1987, when Strieber's purported first-hand account with a haunting cover painting of a large-eyed "alien" became a best-seller. In contrast, we hear very little lately about incubi, elves, and fairies. Where have they all gone?

Far from being global, such alien abduction stories are disappointingly local. The vast majority emanate from North America. They hardly transcend American culture. In other countries, bird-headed, insect-headed, reptilian, robot, and blond and blue-eyed aliens are reported (the last, predictably, from northern Europe). Each group of aliens is said to behave differently. Clearly cultural factors are playing an important role.

Long before the terms "flying saucer" or "UFOs" were invented, science fiction was replete with "little green men" and "bug-eyed monsters." Somehow small hairless beings with big heads (and eyes) have been our staple aliens for a long time. You could see them routinely in the science fiction pulp magazines of the '20s and '30s (and, for example, in an illustration of a Martian sending radio messages to Earth in the December 1937 issue of the magazine *Short Wave and Television*). It goes back perhaps to our remote descendants as depicted by the British science fiction pioneer, H. G. Wells. Wells argued that humans evolved from smaller-brained but hairier primates with an athleticism far exceeding that of Victorian academics; extrapolating this trend into the far future, he suggested that our descendants should be nearly hairless, with immense heads, although barely able to walk around on their own. Advanced beings from other worlds might be similarly endowed.

The typical modern extraterrestrial reported in America in the '80s and early '90s is small, with disproportionately large head and eyes, undeveloped facial features, no visible eyebrows or genitals, and smooth gray skin. It looks to me eerily like a fetus in roughly the twelfth week of pregnancy, or a starving child. Why so many of us might be obsessing on fetuses or malnourished children, and imagining them attacking and sexually manipulating us, is an interesting question.

In recent years in America, aliens different from the short gray motif have been on the rise. One psychotherapist, Richard Boylan of Sacramento, says:

You've got three-and-a-half-foot to four-foot types; you've got five- to six-foot types; you've got seven- to eight-foot types; you've got three-, four-, and five-finger types, pads on the ends of fingers or suction cups; you've got webbed or non-webbed fingers; you've got large almond-shape eyes slanted upward, outward, or horizontally; in some cases large ovoid eyes without the almond slant; you've got extraterrestrials with slit pupils; you've got other different body types—the so-called Praying Mantis type, the reptoid types . . . These are the ones that I keep getting recurrently. There are a few exotic and single case reports that I tend to be a little cautious about until I get a lot more corroborative.

Despite this apparent variety of extraterrestrials, the UFO abduction syndrome portrays, it seems to me, a banal Universe. The form of the supposed aliens is marked by a failure of the imagination and a preoccupation with human concerns. Not a single being presented in all these accounts is as astonishing as a cockatoo would be if you had never before beheld a bird. Any protozoology or bacteriology or mycology textbook is filled with wonders that far outshine the most exotic descriptions of the alien abductionists. The believers take the common elements in their stories as tokens of verisimilitude, rather than as evidence that they have contrived their stories out of a shared culture and biology.

Chapter 8

———

ON THE
DISTINCTION
BETWEEN
TRUE AND
FALSE VISIONS

A credulous mind . . . finds most
delight in believing strange things,
and the stranger they are the easier
they pass with him; but never regards
those that are plain and feasible,
for every man can believe such.

SAMUEL BUTLER,
Characters
(1667–1669)

For just an instant I sense an apparition in the darkened room—could it be a ghost? Or there's a flicker of motion; I see it out of the corner of my eye, but when I turn my head there's nothing there. Is that a telephone ringing, or is it just my "imagination"? In astonishment, I seem to be smelling the salt air of the Coney Island summer seashore of my childhood. I turn a corner in the foreign city I'm visiting for the first time, and before me is a street so familiar I feel I've known it all my life.

In these commonplace experiences, we're generally unsure what to do next. Were my eyes (or ears, or nose, or memory) playing "tricks" on me? Or did I, really and truly, witness something out of the ordinary course of Nature? Shall I keep quiet about it, or shall I tell?

The answer depends very much on my environment, friends, loved ones, and culture. In an obsessively rigid, practically oriented society, perhaps I would be cautious about admitting to such experiences. They might mark me as flighty, unsound, unreliable. But in a society that readily believes in ghosts, say, or "apporting," accounts of such experiences might gain approval, even prestige. In the former, I would be sorely tempted to suppress the thing altogether; in the latter, maybe even to exaggerate or elaborate just a little to make it even more miraculous than it seemed.

Charles Dickens, who lived in a flourishing rational culture in which, however, spiritualism was also thriving, described the dilemma in these words (from his short story, "To Be Taken with a Grain of Salt"):

> I have always noticed a prevalent want of courage, even among persons of superior intelligence and culture, as to imparting their own psychological experiences when those have been of a strange sort. Almost all men are afraid that what they could relate in such wise would find no parallel or response in a listener's internal life,

and might be suspected or laughed at. A truthful traveller who should have seen some extraordinary creature in the likeness of a sea-serpent, would have no fear of mentioning it; but the same traveller having had some singular presentiment, impulse, vagary of thought, vision (so-called), dream, or other remarkable mental impression, would hesitate considerably before he would own to it. To this reticence I attribute much of the obscurity in which such subjects are involved.

In our time, there is still much dismissive chortling and ridicule. But the reticence and obscurity is more readily overcome—for example, in a "supportive" setting provided by a therapist or hypnotist. Unfortunately—and, for some people, unbelievably—the distinction between imagination and memory is often blurred.

Some "abductees" say they remember the experience without hypnosis; many do not. But hypnosis is an unreliable way to refresh memory. It often elicits imagination, fantasy, and play as well as true recollections, with neither patient nor therapist able to distinguish the one from the other. Hypnosis seems to involve, in a central way, a state of heightened suggestibility. Courts have banned its use as evidence or even as a tool of criminal investigation. The American Medical Association calls memories surfacing under hypnosis less reliable than those recalled without it. A standard medical school text (Harold I. Kaplan, *Comprehensive Textbook of Psychiatry*, 1989) warns of "a high likelihood that the beliefs of the hypnotist will be communicated to the patient and incorporated into what the patient believes to be memories, often with strong conviction." So the fact that, when hypnotized, people sometimes relate alien abduction stories carries little weight. There's a danger that subjects are—at least on some matters—so eager to please the hypnotist that they sometimes respond to subtle cues of which even the hypnotist is unaware.

In a study by Alvin Lawson of California State University, Long Beach, eight subjects, pre-screened to eliminate UFO buffs, were hypnotized by a physician and informed that they had been abducted, brought to a spaceship, and examined. With no further prompting, they were asked to describe the experience. Their accounts, most of which were easily elicited, were almost indistinguishable from the accounts that self-described abductees present. True, Lawson had cued his subjects briefly and directly; but in many cases the therapists who routinely deal with alien abductions cue *their* subjects—some in great detail, others more subtly and indirectly.

The psychiatrist George Ganaway (as related by Lawrence Wright) once proposed to a highly suggestible patient under hypnosis that five hours were missing from her memory of a certain day. When he mentioned a bright light overhead, she promptly told him about UFOs and aliens. When he insisted she had been experimented on, a detailed abduction story emerged. But when she came out of the trance, and examined a video of the session, she recognized that something like a dream had been caught surfacing. Over the next year, though, she repeatedly flashed back to the dream material.

The University of Washington psychologist Elizabeth Loftus has found that unhypnotized subjects can easily be made to believe they saw something they didn't. In a typical experiment, subjects will view a film of a car accident. In the course of being questioned about what they saw, they're casually given false information. For example, a stop sign is off-handedly referred to, although there wasn't one in the film. Many subjects then dutifully recall seeing a stop sign. When the deception is revealed, some vehemently protest, stressing how vividly they remember the sign. The greater the time lag between viewing the film and being given the false information, the more people allow their memories to be tampered with. Loftus argues that "memories of an event more closely resemble a story undergoing constant revision than a packet of pristine information."

There are many other examples, some—a spurious memory of being lost as a child in a shopping mall, for instance—of greater emotional impact. Once the key idea is suggested, the patient often plausibly fleshes out the supporting details. Lucid but wholly false recollections can easily be induced by a few cues and questions, especially in the therapeutic setting. Memory can be contaminated. False memories can be implanted even in minds that do not consider themselves vulnerable and uncritical.

Stephen Ceci of Cornell University, Loftus, and their colleagues have found, unsurprisingly, that preschoolers are exceptionally vulnerable to suggestion. The child who, when first asked, correctly denies having caught his hand in a mousetrap later remembers the event in vivid, self-generated detail. When more directly told about "some things that happened to you when you were little," over time they easily enough assent to the implanted memories. Professionals watching videotapes of the children can do no better than chance in distinguishing false memories from true ones. Is there any reason to think that adults are wholly immune to the fallibilities exhibited by children?

President Ronald Reagan, who spent World War II in Hollywood, vividly described his own role in liberating Nazi concentration camp victims. Living in the film world, he apparently confused a movie he had seen with a reality he had not. On many occasions in his Presidential campaigns, Mr. Reagan told an epic story of World War II courage and sacrifice, an inspiration for all of us. Only it never happened; it was the plot of the movie A *Wing and a Prayer*—that made quite an impression on me, too, when I saw it at age 9. Many other instances of this sort can be found in Reagan's public statements. It is not hard to imagine serious public dangers emerging out of instances in which political, military, scientific or religious leaders are unable to distinguish fact from vivid fiction.

In preparing for courtroom testimony, witnesses are coached by their lawyers. Often, they are made to repeat the story over and over again, until they get it "right." Then, on the stand what they remember is the story they've been telling in the lawyer's office. The nuances have been shaded. Or it may no longer correspond, even in its major features, to what really happened. Conveniently, the witnesses may have forgotten that their memories were reprocessed.

These facts are relevant in evaluating the societal effects of advertising and of national propaganda. But here they suggest that on alien abduction matters—where interviews typically take place years after the alleged event—therapists must be very careful that they do not accidentally implant or select the stories they elicit.

Perhaps what we actually remember is a set of memory fragments stitched onto a fabric of our own devising. If we sew cleverly enough, we have made ourselves a memorable story easy to recall. Fragments by themselves, unencumbered by association, are harder to retrieve. The situation is rather like the method of science itself—where many isolated data points can be remembered, summarized, and explained in the framework of a theory. We then much more easily recall the theory and not the data.

In science the theories are always being reassessed and confronted with new facts; if the facts are seriously discordant—beyond the error bars—the theory may have to be revised. But in everyday life it is very rare that we are confronted with new facts about events of long ago. Our memories are almost never challenged. They can, instead, be frozen in place, no matter how flawed they are, or become a work in continual artistic revision.

—

More than gods and demons, the best-attested apparitions are those of saints—especially the Virgin Mary in Western Europe from late medieval to modern times. While alien abduction stories have much more the flavor of profane, demonic apparitions, insight into the UFO myth can also be gained from visions described as sacred. Perhaps best-known are those of Jeanne d'Arc in France, St. Bridget in Sweden, and Girolamo Savonarola in Italy. But more appropriate for our purpose are the apparitions seen by shepherds and peasants and children. In a world plagued by uncertainty and horror, these people longed for contact with the divine. A detailed record of such events in Castile and Catalonia is provided by William A. Christian, Jr., in his book *Apparitions in Late Medieval and Renaissance Spain* (Princeton University Press, 1981):

In a typical case, a rural woman or child reports encountering a girl or an oddly tiny woman—perhaps three or four feet tall—who reveals herself to be the Virgin Mary, the Mother of God. She requests the awestruck witness to go to the village fathers or the local Church authorities and order them to say prayers for the dead, or obey the Commandments, or build a shrine at this very spot in the countryside. If they do not comply, dire penalties are threatened, perhaps the plague. Alternatively, in plague-infested times, Mary promises to cure the disease but only if her request is satisfied.

The witness tries to do as she is told. But when she informs her father or husband or priest, she is ordered to repeat the story to no one; it is mere female foolishness or frivolity or demonic hallucination. So she keeps quiet. Days later she is confronted again by Mary, a little put out that her request has not been honored.

"They will not believe me," the witness complains. "Give me a sign." *Evidence* is needed.

So Mary—who seems to have had no foreknowledge that evidence would have to be provided—provides a sign. The villagers and priests are promptly convinced. The shrine is built. Miraculous cures occur in its vicinity. Pilgrims come from far and wide. Priests are busy. The economy of the region booms. The original witness is appointed keeper of the sacred shrine.

In most of the cases we know of, there was a commission of inquiry, comprised of leaders civic and ecclesiastic, who attested to the genuineness of the apparition—despite initial, almost exclusively male, skepticism. But the standards of evidence were not generally high. In one case the testimony of a delirious eight-year-old boy, taken two days before his death from plague, was soberly accepted. Some of these commissions deliberated decades or even a century after the event.

In *On the Distinction Between True and False Visions*, an expert on the subject, Jean Gerson, in around 1400, summarized the criteria for recognizing a credible witness of an apparition: One was the willingness to accept advice from the political and religious hierarchy. Thus anyone seeing a vision disturbing to those in power was *ipso facto* an unreliable witness, and saints and virgins could be made to say whatever the authorities wanted to hear.

The "signs" allegedly provided by Mary, the evidence offered and considered compelling, included an ordinary candle, a piece of silk, and a magnetic stone; a piece of colored tile; footprints; the witness's unusually quick gathering of thistles; a simple wooden cross inserted in the ground; welts and wounds on the witness; and a variety of contortions—a 12-year-old with her hand held funny, or legs folded back, or a closed mouth making her temporarily mute—that are "cured" the moment her story is accepted.

In some cases accounts may have been compared and coordinated before testimony was given. For example, multiple witnesses in a small town might tell of a tall, glowing woman dressed all in white carrying an infant son and surrounded by a radiance that lit up the street the previous night. But in other cases, people standing directly beside the witness could see nothing, as in this report of a 1617 apparition from Castile:

'Ay, Bartolomé, the lady who came to me these past days is coming through the meadow, and she is kneeling and embracing the cross there—look at her, look at her!' The youth though he looked as hard as he could saw nothing except some small birds flying around above the cross.

Possible motives for inventing and accepting such stories are not hard to find: jobs for priests, notaries, carpenters and merchants, and other boosts to the regional economy in a time of depression; augmented social status of the witness and her family; prayers once again offered for relatives buried in graveyards later abandoned because of plague, drought, and war; rousing public spirit against enemies, especially Moors; improving civility and obedience to canon law; and confirming the faith of the pious. The fervor of pilgrims in such shrines was impressive; it was not uncommon for rock scrapings or dirt from the shrine to be mixed with water and drunk as medicine. But I'm not suggesting that most witnesses made the whole business up. Something else was going on.

Almost all the urgent requests by Mary were remarkable for their prosaicness—for example, in this 1483 apparition from Catalonia:

> I charge you by your soul to charge the souls of the men of the parishes of El Torn, Milleras, El Salent, and Sant Miquel de Campmaior to charge the souls of the priests to ask the people to pay up the tithes and all the duties of the church and restore other things that they hold covertly or openly which are not theirs to their rightful owners within thirty days, for it will be necessary, and observe well the holy Sunday.
>
> And second that they should cease and desist from blaspheming and they should pay the usual *charitas* mandated by their dead ancestors.

Often the apparition is seen just after the witness awakes. Francisca la Brava testified in 1523 that she had gotten out of bed "without knowing if she was in control of her senses," although in later testimony she claimed to be fully awake. (This was in response to a question which allowed a gradation of possibilities: fully awake, dozing, in a trance, asleep.) Sometimes details are wholly missing, such as what the accompanying angels looked like; or Mary is described as both tall and short, both mother and child—characteristics that unmistakably suggest themselves as dream material. In the *Dialogue on Miracles* written around 1223 by Caesarius of Heisterbach, clerical visions of the Virgin Mary often occurred during *matins*, which took place at the sleepy midnight hour.

It is natural to suspect that many, perhaps all, of these apparitions were a species of dream, waking or sleeping, compounded by hoaxes (and by forgeries; there was a thriving business in contrived miracles: religious paintings and statues dug up by accident or divine command). The matter was addressed in the *Siete Partidas*, the codex of canon and civil law compiled under the direction of Alfonso the Wise, king of Castile, around 1248. In it we can read the following:

> Some men fraudulently discover or build altars in fields or in towns, saying that there are relics of certain saints in those places and pretending that they perform miracles, and, for this reason, people from many places are induced to go there as on a pilgrimage, in order to take something away from them; and there are others who influenced by dreams or empty phantoms which appear to them, erect altars and pretend to discover them in the above named localities.

In listing the reason for erroneous beliefs, Alfonso lays out a continuum from sect, opinion, fantasy, and dream to hallucination. A kind of fantasy named *antoiança* is defined as follows:

> Antoiança is something that stops before the eyes and then disappears, as one sees or hears it in a trance, and so is without substance.

A 1517 papal bull distinguishes between apparitions that appear "in dreams or divinely." Clearly, the secular and ecclesiastical authorities, even in times of extreme credulity, were alert to the possibilities of hoax and delusion.

Nevertheless, in most of medieval Europe, such apparitions were greeted warmly by the Roman Catholic clergy—especially because the Marian admonitions were so congenial to the priesthood. A pathetic few "signs" of evidence—a stone or a footprint and never anything unfakeable—sufficed. But beginning in the fifteenth century, around the time of the Protestant Reformation, the attitude of the Church changed. Those who reported an independent channel to Heaven were outflanking the Church's chain of command up to God. Moreover, a few of the apparitions—Jeanne d'Arc's, for example—had awkward political or moral implications. The perils represented by Jeanne d'Arc's visions were described by her inquisitors in 1431 in these words:

> The great danger was shown to her that comes of someone so presumptuous to believe they have such apparitions and revelations, and therefore lie about matters concerning God, giving out false prophecies and divinations not known from God, but invented. From which could follow the seduction of peoples, the inception of new sects, and many other impieties that subvert the Church and Catholics.

Both Jeanne d'Arc and Girolamo Savonarola were burnt at the stake for their visions.

In 1516 the Fifth Lateran Council reserved to "the Apostolic seat" the right to examine the authenticity of apparitions. For poor peasants whose visions had no political content, the punishments fell short of the ultimate severity. The Marian apparition seen by Francisca la Brava, a young mother, was described by Licenciado Mariana, the Lord Inquisitor, as "to the detriment of our holy Catholic faith and the diminution of its authority." Her apparition "was all vanity and frivolity." "By rights we could have treated her more rigorously," the Inquisitor continued.

> But in deference to certain just reasons that move us to mitigate the rigor of the sentences we decree as a punishment to Francisca la Brava and an example to others not to attempt similar things that we condemn her to be put on an ass and given one hundred lashes in public through the accustomed streets of Belmonte naked from the waist up, and the same number in the town of El Quintanar in the same manner. And that from now on she not say or affirm in public or secretly by word or insinuation the things she said in her confessions or else she will be prosecuted as an impenitent and one who does not believe in or agree with what is in our holy Catholic faith.

Despite the penalties, it is striking how often the witness stuck to her guns and—ignoring the encouragements offered her to confess that she was lying or dreaming or confused—insisted that she really and truly had seen the vision.

In a time when nearly everyone was illiterate, before newspapers, radio, and television, how could the religious and iconographic detail of these apparitions have been so similar? William Christian believes there is a ready answer in cathedral dramaturgy (especially Christmas plays), in itinerant preachers and pilgrims, and in church sermons. Legends about nearby shrines spread quickly. People sometimes came from a hundred miles or more so that, say, their sick child could be cured by a pebble that had been trodden on by the Mother of God. Legends influenced apparitions and vice versa. In a time haunted by drought, plague, and war, with no social or medical services available to the average person, with public literacy and the scientific method unheard of, skeptical thinking was rare.

Why are the admonitions so prosaic? Why is a vision of so illustrious a personage as the Mother of God necessary so, in a tiny county populated by a few thousand souls, a shrine will be repaired or the populace will refrain from cursing? Why not important and prophetic messages whose significance could be recognized in later years as something that could have emanated only from God or the saints? Wouldn't this have greatly enhanced the Catholic cause in its mortal struggle with Protestantism and the Enlightenment? But we have no apparitions cautioning the Church against, say, accepting the delusion of an Earth-centered Universe, or warning it of complicity with Nazi Germany—two matters of considerable moral as well as historical import, on which Pope John Paul II, to his credit, has admitted that the Church has erred.

Not a single saint criticized the practice of torturing and burning "witches" and heretics. Why not? Were they unaware of what was going on? Could they not grasp its evil? And why is Mary always ordering the poor peasant to inform the authorities? Why doesn't she admonish the authorities herself? Or the King? Or the Pope? In the nineteenth and twentieth centuries, it is true, some of the apparitions have taken on greater import—at Fatima, Portugal, in 1917, where the Virgin was incensed that a secular government had replaced a government run by the Church, and at Garabandal, Spain, in 1961–1965, where the end of the world was threatened unless conservative political and religious doctrines were adopted forthwith.

I think I can see many parallels between Marian apparitions and alien abductions—even though the witnesses in the former cases are not promptly taken to Heaven and don't have their reproductive organs meddled with. The beings reported are diminutive, most often about two-and-a-half to four feet high. They come from the sky. The content of the communication is, despite its purported celestial origin, mundane. There seems to be a clear connection with sleep and dreams. The witnesses, often females, are troubled about speaking out, especially after encountering ridicule from males in positions of authority. Nevertheless they persist: They really saw such a thing, they insist. Means of conveying the stories exist; they are eagerly discussed, permitting details to be coordinated even among witnesses who have never met one another. Others present at the time and place of the apparition see nothing unusual. The purported "signs" or evidence are, without exception, nothing that humans couldn't acquire or fabricate on their own. Indeed, Mary seems unsympathetic to the need for evidence, and occasionally is willing to cure only those who had believed the account of her apparition *before* she supplied "signs." And while there are no therapists, per se, the society is suffused by a network of influential parish priests and their hierarchical superiors who have a vested interest in the reality of the visions.

In our time, there are still apparitions of Mary and other angels, but also—as summarized by G. Scott Sparrow, a psychotherapist and hypnotist—of Jesus. In *I Am With You Always: True Stories of Encounters with Jesus* (Bantam, 1995), first-hand accounts, some moving, some banal, of such encounters are laid out. Oddly, most of them are straightforward dreams, acknowledged as such, and the ones called visions are said to differ from dreams "only because we experience them while we are awake." But, for Sparrow, judging something "only a dream" does not compromise its external reality. For Sparrow, any being you dream of, and any incident, really exists in the world outside your head. He specifically denies that dreams are "purely subjective." Evidence doesn't enter into it. If you dreamt it, if it felt good, if it elicited wonder, why then it really happened. There's not a skeptical bone in Sparrow's body. When Jesus tells a conflicted woman in an "intolerable" marriage to throw the bum out, Sparrow admits that this poses problems for "advocates of a scripturally consistent position." In that case, "[u]ltimately, perhaps, one could say that virtually all presumed guidance is generated from within." What if someone reported a dream in which Jesus counseled, say, abortion—or vengeance? And if indeed somewhere, somehow we must eventually draw the line and conclude that *some* dreams are invented by the dreamer, why not all?

—

Why would people invent abduction stories? Why, for that matter, would people appear on TV audience participation programs devoted to sexual humiliation of the "guests"—the current rage in America's video wasteland? Discovering that you're an alien abductee is at least a break from the routine of everyday life. You gain the attention of peers, therapists, maybe even the media. There is a sense of discovery, exhilaration, awe. What will you remember next? You begin to believe that you may be the harbinger or even the instrument of momentous events now rolling towards us. And you don't want to disappoint your therapist. You crave his or her approval. I think there can very well be psychic rewards in becoming an abductee.

For comparison, consider product tampering cases, which convey very little of the sense of wonder that surrounds UFOs and alien abductions: Someone claims to find a hypodermic syringe in a popular soft drink can. Understandably, this is upsetting. It's reported in newspapers and especially on television news. Soon there's a spate, a virtual epidemic of similar reports from all over the country. But it's very hard to see how a hypodermic syringe could get into a can at the factory, and in none of the cases are witnesses present when an intact can is opened and a syringe discovered inside.

Slowly the evidence accumulates that this is a "copycat" crime. People have only been pretending to find syringes in soft drink cans. Why would anyone do it? What possible motives could they have? Some psychiatrists say that the primary motives are greed (they'll sue the manufacturer for damages), a craving for attention, and a wish to be portrayed as a victim. Note there are no therapists touting the reality of needles in cans and urging their patients—subtly or directly—to go public with the news. Also, serious penalties are levied for product tampering, and even for falsely alleging that products have been tampered with. In contrast, there *are* therapists who encourage abductees to tell their stories to mass audiences, and no legal penalties are exacted for falsely claiming you've been abducted by a UFO. Whatever your reason for going down this road, how much more satisfying it must be to convince others that you've been chosen by higher beings for their own enigmatic purpose than that by mere happenstance you've found a hypodermic syringe in your cola.

Chapter 9

———

THERAPY

It is a capital mistake to theorize before
one has data. Insensibly one begins to
twist facts to suit theories, instead of
theories to suit facts.

SHERLOCK HOLMES,
in Arthur Conan Doyle's *A Scandal in Bohemia*
(1891)

True memories seemed like phantoms,
while false memories were so convincing
that they replaced reality.

GABRIEL GARCÍA MÁRQUEZ,
Strange Pilgrims
(1992)

John Mack is a Harvard University psychiatrist whom I've known for many years.

Is there anything to this UFO business? he asked me long ago.

Not much, I replied. Except of course on the psychiatric side.

He looked into it, interviewed abductees, and was converted. He now accepts the accounts of abductees at face value. Why? "I wasn't looking for this," he says. "There's nothing in my background that prepared me" for the alien abduction story. "It's completely persuasive because of the emotional power of these experiences." In his book *Abductions*, Mack explicitly proposes the very dangerous doctrine that "the power or intensity with which something is felt" is a guide to whether it's true.

I can personally attest to the emotional power. But aren't powerful emotions a routine component of our dreams? Don't we sometimes awake in stark terror? Doesn't Mack, himself the author of a book on nightmares, know about the emotional power of hallucinations? Some of Mack's patients describe themselves as having hallucinated since childhood. Have the hypnotists and psychotherapists working with "abductees" made conscientious attempts to steep themselves in the body of knowledge on hallucinations and perceptual malfunctions? Why do they believe *these* witnesses but not those who reported, with comparable conviction, encounters with gods, demons, saints, angels, and fairies? And what about those who hear irresistible commands from a voice within? Are all deeply felt stories true?

A scientist of my acquaintance says "If the aliens would only keep all the folks they abduct, our world would be a little saner." But her judgment is too harsh. It doesn't seem to be a matter of sanity. It's something else. The Canadian psychologist Nicholas Spanos and his colleagues concluded that there are no obvious pathologies in those who report being abducted by UFOs. However, intense UFO experiences are more likely to occur in individuals who are disposed to esoteric beliefs in general and alien beliefs in particular and who interpret unusual sensory and imaginal experiences in terms of the alien hypothesis. Among UFO believers, those with stronger propensities toward fantasy production were particularly likely to generate such experiences. Moreover, such experiences were likely to be generated and interpreted as real events rather than imaginings when they were associated with restricted sensory environments . . . (e.g., experiences that occurred at night and in association with sleep).

What a more critical mind might recognize as a hallucination or a dream, a more credulous mind interprets as a glimpse of an elusive but profound external reality.

———

Some alien abduction accounts may conceivably be disguised memories of rape and childhood sexual abuse, with the father, stepfather, uncle, or mother's boyfriend represented as an alien. Surely it's more comforting to believe that an alien abused you than that it was done by someone you trusted and loved. Therapists who take the alien abduction stories at face value deny this, saying they would know if their patients were sexually abused. Some estimates from opinion surveys range as high as one in four American women and one in six American men having been sexually abused in childhood (although these estimates are probably too high). It would be astonishing if a significant number of patients who present themselves to alien abduction therapists had *not* been so abused, perhaps even a larger proportion than in the general population.

Both sexual abuse therapists and alien abduction therapists spend months, sometimes years, encouraging their subjects to remember being abused. Their methods are similar, and their goals are in a way the same—to recover painful memories, often of long ago. In both cases the therapist believes the patient to be suffering from trauma attendant to an event so terrible that it is repressed. I find it striking that alien abduction therapists find so few cases of sexual abuse, and vice versa.

Those who have in fact been subjected to childhood sexual abuse or incest are, for very understandable reasons, sensitive about anything that seems to minimize or deny their experience. They are angry, and they have every right to be. In the U.S., at least one in ten women have been raped, almost two-thirds before the age of eighteen. A recent survey reports that one-sixth of all rape victims reported to police are under the age of 12. (And this is the category of rape least likely to be reported.) One-fifth of these girls were raped by their fathers. They have been betrayed. I want to be very clear about this: There are many real cases of ghoulish sexual predation by parents, or those acting in the role of parents. Compelling physical evidence—photos, for example, or diaries, or gonorrhea or chlamydia in the child—have in some cases come to light. Abuse of children has been implicated as a major probable cause of social problems. According to one survey, 85 percent of all violent prison inmates were abused in childhood. Two-thirds of all teenage mothers were raped or sexually abused as children or teenagers. Rape victims are ten times more likely than other women to use alcohol and other drugs to excess. The problem is real and urgent. Most of these tragic and incontestable cases of childhood sexual abuse, however, have been continuously remembered into adulthood. There is no hidden memory to be retrieved.

While there is better reporting today than in the past, there does seem to be a significant increase in cases of child abuse reported each year by hospitals and law enforcement authorities, rising in the United States tenfold (to 1.7 million cases) between 1967 and 1985. Alcohol and other drugs, as well as economic stresses, are pointed to as the "reasons" adults are more prone to abuse children today than in the past. Perhaps increasing publicity given to contemporary cases of child abuse emboldens adults to remember and focus on the abuse they once suffered.

A century ago, Sigmund Freud introduced the concept of repression, the forgetting of events in order to avoid intense psychic pain, as a coping mechanism essential for mental health. It seemed to emerge especially in patients diagnosed with "hysteria," the symptoms of which included hallucinations and paralysis. At first Freud believed that behind every case of hysteria was a repressed instance of childhood sexual abuse. Eventually Freud changed his explanation to hysteria being caused by *fantasies*—not all of them unpleasant—of having been sexually abused as a child. The burden of guilt was shifted from parent to child. Something like this debate rages today. (The reason for Freud's change of heart is still being disputed—the explanations ranging from his provoking outrage among his Viennese middle-aged male peers, to his recognition that he was taking the stories of hysterics seriously.)

Instances in which the "memory" suddenly surfaces, especially at the ministrations of a psychotherapist or hypnotist, and where the first "recollections" have a ghost- or dreamlike quality are highly questionable. Many such claims of sexual abuse appear to be invented. The Emory University psychologist Ulric Neisser says:

> There is child abuse, and there are such things as repressed memories. But there are also such things as false memories and confabulations, and they are not rare at all. Misrememberings are the rule, not the exception. They occur all the time. They occur even in cases where the subject is absolutely confident—even when the memory is a seemingly unforgettable flashbulb, one of those metaphorical mental photographs. They are still more likely to occur in cases where suggestion is a lively possibility, where memories can be shaped and re-shaped to meet the strong interpersonal demands of a therapy session. And once a memory has been reconfigured in this way, it is very, very hard to change.
>
> These general principles cannot help us to decide with certainty where the truth lies in any individual case or claim. But on the average, across a large number of such claims, it is pretty obvious where we should place our bets. Misremembering and retrospective reworking of the past are a part of human nature; they go with the territory and they happen all the time.

Survivors of the Nazi death camps provide the clearest imaginable demonstration that even the most monstrous abuse can be carried continuously in human memory. Indeed, the problem for many Holocaust survivors has been to put some emotional distance between themselves and the death camps, to forget. But if in some alternative world of inexpressible evil they were forced to *live* in Nazi Germany—let's say a thriving post-Hitler nation with its ideology intact, except that it's changed its mind about anti-Semitism—imagine the psychological burden on Holocaust survivors then. Then perhaps they *would* be able to forget, because remembering would make their current lives unbearable. If there is such a thing as the repression and subsequent recall of ghastly memories, then perhaps it requires two conditions: (1) that the abuse actually happened, and (2) that the victim was required to pretend for long periods of time that it never happened.

The University of California social psychologist Richard Ofshe explains:

> When patients are asked to explain how the memories returned, they report assembling fragments of images, ideas, feelings, and sensations into marginally coherent stories. As the so-called memory work stretches out for months, feelings become vague images, images become figures, and figures become known persons. Vague discomfort in certain parts of the body is reinterpreted as childhood rape ... The original physical sensations, sometimes augmented by hypnosis, are then labeled "body memories." There is no conceivable mechanism by which the muscles of the body could store memories. If these methods fail to persuade, the therapist may resort to still more heavy-handed practices. Some patients are recruited into survivor groups in which peer pressure is brought to bear, and they are asked to demonstrate politically correct solidarity by establishing themselves as members of a survivor subculture.

A cautious 1993 statement by the American Psychiatric Association accepts the possibility that some of us forget childhood abuse as a means of coping, but warns,

> It is not known how to distinguish, with complete accuracy, memories based on true events from those derived from other sources ... Repeated questioning may lead individuals to report "memories" of events that never occurred. It is not known what proportion of adults who report memories of sexual abuse were actually abused ... A strong prior belief by the psychiatrist that sexual abuse, or other factors, are or are not the cause of the patient's problems is likely to interfere with appropriate assessment and treatment.

On the one hand, callously to dismiss charges of horrifying sexual abuse can be heartless injustice. On the other hand, to tamper with people's memories, to infuse false stories of childhood abuse, to break up intact families, and even to send innocent parents to prison is also heartless injustice. Skepticism is essential on both sides. Picking our way between these two extremes can be very tricky.

Early editions of the influential book by Ellen Bass and Laura Davis (*The Courage to Heal: A Guide for Women Survivors of Child Sexual Abuse*; Perennial Library, 1988) give illuminating advice to therapists:

> **Believe the survivor.** You must believe your client was sexually abused, even if she doubts it herself. . . Your client needs you to stay steady in the belief that she was abused. Joining a client in doubt would be like joining a suicidal client in her belief that suicide is the best way out. If a client is unsure that she was abused but thinks she might have been, work as though she was. So far, among the hundreds of women we've talked to and the hundreds more we've heard about, not one has suspected that she might have been abused, explored it, and determined that she wasn't.

But Kenneth V. Lanning, Supervisory Special Agent at the Behavioral Science Instruction and Research Unit of the FBI Academy in Quantico, Virginia, a leading expert on the sexual victimization of children, wonders: "Are we making up for centuries of denial by now blindly accepting *any* allegation of child abuse, no matter how absurd or unlikely?" "I don't care if it's true," replies one California therapist reported by *The Washington Post*. "What actually happened is irrelevant to me. . . We all live in a delusion."

The existence of *any* false accusation of childhood sexual abuse — especially those created under the ministrations of an authority figure — has, it seems to me, relevance to the alien abduction issue. If some people can with great passion and conviction be led to falsely remember being abused by their own parents, might not others, with comparable passion and conviction, be led to falsely remember being abused by aliens?

The more I look into claims of alien abduction, the more similar they seem to reports of "recovered memories" of childhood sexual abuse. And there's a third class of related claims, repressed "memories" of satanic ritual cults—in which sexual torture, coprophilia, infanticide, and cannibalism are said to be prominently featured. In a survey of 2,700 members of the American Psychological Association, 12 percent replied that they had treated cases of satanic ritual abuse (while 30 percent reported cases of abuse done in the name of religion). Something like 10,000 cases are reported annually in the United States in recent years. A significant number of those touting the peril of rampant Satanism in America, including law enforcement officers who organize seminars on the subject, turn out to be Christian fundamentalists; their sects explicitly require a literal devil to be meddling in everyday human life. The connection is neatly drawn in the saying "No Satan, no God."

Apparently, there is a pervasive police gullibility problem on this matter. Here are some excerpts from FBI expert Lanning's analysis of "Satanic, Occult and Ritualistic Crime," based on bitter experience, and published in the October 1989 issue of the professional journal, *The Police Chief*:

> Almost any discussion of Satanism and witchcraft is interpreted in the light of the religious beliefs of those in the audience. Faith, not logic and reason, governs the religious beliefs of most people. As a result, some normally skeptical law enforcement officers accept the information disseminated at these conferences without critically evaluating it or questioning the sources. . . For some people Satanism is any religious belief system other than their own.

Lanning then offers a long list of belief systems he has personally heard described as Satanism at such conferences. It includes Roman Catholicism, the Orthodox Churches, Islam, Buddhism, Hinduism, Mormonism, rock and roll music, channeling, astrology and New Age beliefs in general. Is there not a hint here about how witch hunts and pogroms get started?

"Within the personal religious belief system of a law enforcement officer," he continues,

> Christianity may be good and Satanism evil. Under the Constitution, however, both are neutral. This is an important, but difficult,

concept for many law enforcement officers to accept. They are paid to uphold the penal code, not the Ten Commandments. . . The fact is that far more crime and child abuse has been committed by zealots in the name of God, Jesus and Mohammed than has ever been committed in the name of Satan. Many people don't like that statement, but few can argue with it.

Many of those alleging satanic abuse describe grotesque orgiastic rituals in which infants are murdered and eaten. Such claims have been made about reviled groups by their detractors throughout European history—including the Cataline conspirators in Rome, the Passover "blood libel" against the Jews, and the Knights Templar as they were being dismantled in fourteenth century France. Ironically, reports of cannibalistic infanticide and incestuous orgies were among the particulars used by Roman authorities to persecute the early Christians. After all, Jesus himself is quoted as saying (John 6:53) "Except ye eat the flesh of the Son of man, and drink his blood, ye have no life in you." Although the next line makes it clear Jesus is talking about eating his own flesh and drinking his own blood, unsympathetic critics might have misunderstood the Greek "Son of man" to mean "child" or "infant." Tertullian and other early Church fathers defended themselves against these grotesque accusations as best they could.

Today, the lack of corresponding numbers of lost infants and young children in police files is explained by the claim that all over the world babies are being bred for this purpose—surely reminiscent of abductee claims that alien/human breeding experimen*: are rampant. Also similar to the alien abduction paradigm, satanic cult abuse is said to pass down from generation to generation in certain families. To the best of my knowledge, as in the alien abduction paradigm, no physical evidence has ever been offered in a court of law to support such claims. Their emotional power, though, is evident. The mere possibility that such things are going on rouses us mammals to action. When we give credence to satanic ritual, we also raise the social status of those who warn us of the supposed danger.

Consider these five cases: (1) Myra Obasi, a Louisiana schoolteacher, was—she and her sisters believed after consultation with a hoodoo practitioner—possessed by demons. Her nephew's nightmares were part of the evidence. So they left for Dallas, abandoned their five children, and the sisters then gouged out Ms. Obasi's eyes. At the trial, she defended her sisters. They were trying to help her, she said. But hoodoo is not devil-worship; it is a cross between Catholicism and African-Haitian nativist religion. (2) Parents beat their child to death because she would not embrace their brand of Christianity. (3) A child molester justifies his acts by reading the Bible to his victims. (4) A 14-year-old boy has his eyeball plucked out of his head in an exorcism ceremony. His assailant is not a satanist, but a Protestant fundamentalist minister engaged in religious pursuits. (5) A woman thinks her 12-year-old son is possessed by the devil. After an incestuous relationship with him, she decapitates him. But there is no satanic ritual content to the "possession."

The second and third cases come from FBI files. The last two come from a 1994 study by Dr. Gail Goodman, a psychologist at the University of California, Davis, and her colleagues, done for the National Center on Child Abuse and Neglect. They examined over 12,000 claims of sexual abuse involving satanic ritual cults, and could not find a single one that held up to scrutiny. Therapists reported satanic abuse based only on, for instance, "patient's disclosure via hypnotherapy" or children's "fear of satanic symbols." In some cases diagnosis was made on the basis of behavior common to many children. "In only a few cases was physical evidence mentioned—usually, 'scars.' " But in most cases the "scars" were very faint or nonexistent. "Even when there were scars, it was not determined whether the victims themselves had caused them." This also is very similar to alien abduction cases, as described below. George K. Ganaway, Professor of Psychiatry at Emory University, proposes that "the most common likely cause of cult-related memories may very well turn out to be a mutual deception between the patient and the therapist."

One of the most troublesome cases of "recovered memory" of satanic ritual abuse has been chronicled by Lawrence Wright in a remarkable book, *Remembering Satan* (Knopf, 1994). It concerns Paul Ingram, a man who may have had his life ruined because he was too gullible, too suggestible, too unpracticed in skepticism. Ingram was, in 1988, Chairman of the Republican Party in Olympia, Washington, the chief civil deputy in the local sheriff's department, well-regarded, highly religious, and responsible for warning children in school assemblies of the dangers of drugs. Then came the nightmare moment when one of his daughters—after a highly emotional session at a fundamentalist religious retreat—leveled the first of many charges, each more ghastly than the previous, that Ingram had sexually abused her, impregnated her, tortured her, made her available to other sheriff's deputies, introduced her to satanic rites, dismembered and eaten babies . . . This had gone on since her childhood, she said, almost to the day she began to "remember" it all.

Ingram could not see why his daughter should lie about this—although he himself had no recollection of it. But police investigators, a consulting psychotherapist, and his minister at the Church of Living Water all explained that sex offenders often repressed memories of their crimes. Strangely detached but at the same time eager to cooperate, Ingram tried to recall. After a psychologist employed a closed-eye hypnotic technique to induce trance, Ingram began to visualize something similar to what the police were describing. What came to mind were not like real memories, but something like snatches of images in a fog. Every time he produced one—the more so the more odious the content—he was encouraged and reinforced. His pastor assured him that God would permit only genuine memories to surface in his reveries.

"Boy, it's almost like I'm making it up," Ingram said, "but I'm not." He suggested that a demon might be responsible. Under the same sort of influences, with the Church grapevine circulating the latest horrors that Ingram was confessing, and the police pressuring them, his other children and his wife also began "remembering." Prominent citizens were accused of participating in the orgiastic rites. Law enforcement officers elsewhere in America began paying attention. This was only the tip of the iceberg, some said.

When Berkeley's Richard Ofshe was called in by the prosecution, he performed a control experiment. It was a breath of fresh air. Merely suggesting to Ingram that he had forced his son and daughter to commit incest and asking him to use the "memory recovery" technique he had learned, promptly elicited just such a "memory." It required no pressure, no intimidation—just the suggestion and the technique were enough. But the alleged participants, who had "remembered" so much else, denied it ever happened. Confronted with this evidence, Ingram vehemently denied he was making anything up or was influenced by others. His memory of this incident was as clear and "real" as all his other recollections.

One of the daughters described the terrible scars on her body from torture and forced abortions. But when she finally received a medical examination, there were no corresponding scars to be seen. The prosecution never tried Ingram on charges of satanic abuse. Ingram hired a lawyer who had never tried a criminal case. On his pastor's advice he did not even read Ofshe's report: it would only confuse him, he was told. He pled guilty to six counts of rape, and ultimately was sent to prison. In jail, while awaiting sentencing, away from his daughters, his police colleagues and his pastor, he reconsidered. He asked to withdraw his guilty plea. His memories had been coerced. He had not distinguished real memories from a kind of fantasy. His plea was rejected. He is serving a twenty-year sentence. If it was the sixteenth century instead of the twentieth, perhaps the whole family would have been burned at the stake—along with a good fraction of the leading citizens of Olympia, Washington.

The existence of a highly skeptical FBI report on the general subject of satanic abuse (Kenneth V. Lanning, "Investigator's Guide to Allegations of 'Ritual' Child Abuse," January 1992) is widely ignored by enthusiasts. Likewise, a 1994 study by the British Department of Health into claims of satanic abuse there concluded that, of 84 alleged instances, not one stood up to scrutiny. What then is all the furor about? The study explains,

> The Evangelical Christian campaign against new religious movements has been a powerful influence encouraging the identification of satanic abuse. Equally, if not more, important in spreading the idea of satanic abuse in Britain are the "specialists," American and British. They may have few or even no qualifications as professionals, but attribute their expertise to "experience of cases."

Those convinced that devil cults represent a serious danger to our society tend to be impatient with skeptics. Consider this analysis by Corydon Hammond, Ph.D., past President of the American Society for Clinical Hypnosis:

> I will suggest to you that these people [skeptics] are either, one, naïve and of limited clinical experience; two, have a kind of naïveté that people have of the Holocaust, or they're just such in-

tellectualizers and skeptics that they'll doubt everything; or, three, they're cult people themselves. And I can assure that there are people who are in that position. . . There are people who are physicians, who are mental health professionals, who are in the cults, who are raising trans-generational cults. . . I think the research is real clear: We got three studies, one found 25 percent, one found 20 percent of out-patient multiples [multiple personality disorders] appear to be cult-abuse victims, and another on a specialized in-patient unit found 50 percent.

In some of his statements, he seems to believe that satanic Nazi mind control experiments have been performed by the CIA on tens of thousands of unsuspecting American citizens. The overarching motive, Hammond believes, is to "create a satanic order that will rule the world."

In all three classes of "recovered memories," there are specialists — alien abduction specialists, satanic cult specialists, and specialists in recalling repressed memories of childhood sexual abuse. As is common in mental health practice, patients select or are referred to a therapist whose specialty seems relevant to their complaint. In all three classes, the therapist helps to draw forth images of events alleged to have occurred long ago (in some cases from decades past); in all three, therapists are profoundly moved by the unmistakably genuine agony of their patients; in all three, at least some therapists are known to ask leading questions — which are virtually orders by authority figures to suggestible patients insisting that they remember (I almost wrote "confess"); in all three, there are networks of therapists who trade client histories and therapeutic methods; in all three, practitioners feel the necessity of defending their practice against more skeptical colleagues; in all three, the iatrogenic hypothesis is given short shrift; in all three, the majority of those who report abuse are women. And in all three classes — with the exceptions mentioned — there is no physical evidence. So it's hard not to wonder whether alien abductions might be part of some larger picture.

What could this larger picture be? I posed this question to Dr. Fred H. Frankel, professor of psychiatry at Harvard Medical School, Chief of Psychiatry at Beth Israel Hospital in Boston, and a leading expert on hypnosis. His answer:

If alien abductions are a part of a larger picture, what indeed is the larger picture? I fear to rush in where angels fear to tread; however, the factors you outline all feed what was described at the turn of the century as "hysteria." The term, sadly, became so widely used that our contemporaries in their dubious wisdom . . . not only dropped it, but also lost sight of the phenomena it represented: high levels of suggestibility, imaginal capacity, sensitivity to contextual cues and expectations, and the element of contagion . . . Little of all of this seems to be appreciated by a large number of practicing clinicians.

In exact parallel to regressing people so they supposedly retrieve forgotten memories of "past lives," Frankel notes that therapists can as readily *progress* people under hypnosis so they can "remember" their futures. This elicits the same emotive intensity as in regression or in Mack's abductee hypnosis. "These people are not out to deceive the therapist. They deceive themselves," Frankel says. "They cannot distinguish their confabulations from their experiences."

If we fail to cope, if we're saddled with a burden of guilt for not having made more of ourselves, wouldn't we welcome the professional opinion of a therapist with a diploma on the wall that it's not our fault, that we're off the hook, that satanists, or sexual abusers, or aliens from another planet are the responsible parties? Wouldn't we be willing to pay good money for this reassurance? And wouldn't we resist smart-ass skeptics telling us that it's all in our heads, or that it's implanted by the very therapists who have made us happier about ourselves?

How much training in scientific method and skeptical scrutiny, in statistics, or even in human fallibility have these therapists received? Psychoanalysis is not a very self-critical profession, but at least many of its practitioners have M.D. degrees. Most medical curricula include significant exposure to scientific results and methods. But many of those dealing with abuse cases seem to have at best a casual acquaintance with science. Mental health providers in America are more likely by about two-to-one to be social workers than either psychiatrists or Ph.D. psychologists.

Most of these therapists contend that their responsibility is to support their patients, not to question, to be skeptical, or to raise doubts.

Whatever is presented, no matter how bizarre, is accepted. Sometimes the prompting by therapists is not at all subtle. Here [from the False Memory Syndrome Foundation's *FMS Newsletter*, vol. 4, no. 4, p. 3, 1995] is a hardly atypical report:

> My former therapist has testified that he still believes that my mother is a Satanist, [and] that my father molested me. . . It was my therapist's delusional belief system and techniques involving suggestion and persuasion that led me to believe the lies were memories. When I doubted the reality of the memories he insisted they were true. Not only did he insist they were true, he informed me that in order to get well I must not only accept them as real, but remember them all.

In a 1991 case in Allegheny County, Pennsylvania, a teenager, Nicole Althaus, encouraged by a teacher and a social worker, accused her father of having sexually abused her, resulting in his arrest. Nicole also reported that she had given birth to three children, whom her relatives had killed, that she had been raped in a crowded restaurant, and that her grandmother flew about on a broom. Nicole recanted her allegations the following year, and all charges against her father were dropped. Nicole and her parents brought a civil suit against the therapist and psychiatric clinic to whom Nicole had been referred shortly after she began making the accusations. The jury found that the doctor and the clinic had been negligent and awarded almost a quarter million dollars to Nicole and her parents. There are increasing numbers of cases of this sort.

Might the competition among therapists for patients, and the obvious financial interest of therapists in prolonged therapy, make them less likely to offend patients by evincing some skepticism about their stories? How aware are they of the dilemma of a naïve patient walking into a professional office and being told that the insomnia or obesity is due (in increasing order of bizarreness) to wholly forgotten parental abuse, satanic ritual, or alien abduction? While there are ethical and other constraints, we need something like a control experiment: perhaps the same patient sent to specialists in all three fields. Does any of them say, "No, your problem isn't due to forgotten childhood abuse" (or forgotten satanic ritual, or alien abduction, as appropriate)? How many of them say, "There's a much more prosaic explanation"? Instead, Mack goes so far as to tell one of his patients admiringly and reassuringly that he is on a "hero's journey." One group of "abductees" — each having a separate but similar experience — writes

> [S]everal of us had finally summoned enough courage to present our experiences to professional counselors, only to have them nervously avoid the subject, raise an eyebrow in silence or interpret the experience as a dream or waking hallucination and patronizingly "reassure" us that such things happen to people, "but don't worry, you're basically mentally sound." Great! We're not crazy, but if we take our experiences seriously, then we might become crazy!

With enormous relief, they found a sympathetic therapist who not only accepted their stories at face value, but was full of stories of alien bodies and high-level government coverup of UFOs.

A typical UFO therapist finds his subjects in three ways: They write letters to him at an address given in the back of his books; they are referred to him by other therapists (mainly those who also specialize in alien abductions); or they come up to him after he presents a lecture. I wonder if any patient arrives at his portal wholly ignorant of popular abduction accounts and the therapist's own methods and beliefs. Before any words are exchanged, they know a great deal about one another.

Another prominent therapist gives his patients his own articles on alien abductions to help them "remember" their experiences. He is gratified when what they eventually recall under hypnosis resembles what he describes in his papers. The similarity of the cases is one of his chief reasons for believing that abductions really occur.

A leading UFO scholar comments that "When the hypnotist does not have an adequate knowledge of the subject [of alien abductions] the true nature of the abduction may never be revealed." Can we discern in this remark how the patient might be led without the therapist realizing that he's leading?

—

Sometimes when "falling" asleep we have the sense of toppling from a height, and our limbs suddenly flail on their own. The startle reflex, it's called. Perhaps it's left over from when our ancestors slept in trees.

Why should we imagine we recollect (a wonderful word) any better than we know when we're on firm ground? Why should we suppose that, of the vast treasure of memories stored in our heads, none of it could have been implanted after the event—by how a question is phrased when we're in a suggestible frame of mind, by the pleasure of telling or hearing a good story, by confusion with something we once read or overheard?

Chapter 10

THE
DRAGON
IN MY
GARAGE

[M]agic, it must be remembered, is an art
which demands collaboration between
the artist and his public.

E. M. BUTLER,
The Myth of the Magus
(1948)

A fire-breathing dragon lives in my garage."

Suppose (I'm following a group therapy approach by the psychologist Richard Franklin) I seriously make such an assertion to you. Surely you'd want to check it out, see for yourself. There have been innumerable stories of dragons over the centuries, but no real evidence. What an opportunity!

"Show me," you say. I lead you to my garage. You look inside and see a ladder, empty paint cans, an old tricycle—but no dragon.

"Where's the dragon?" you ask.

"Oh, she's right here," I reply, waving vaguely. "I neglected to mention that she's an invisible dragon."

You propose spreading flour on the floor of the garage to capture the dragon's footprints.

"Good idea," I say, "but this dragon floats in the air."

Then you'll use an infrared sensor to detect the invisible fire.

"Good idea, but the invisible fire is also heatless."

You'll spray-paint the dragon and make her visible.

"Good idea, except she's an incorporeal dragon and the paint won't stick."

And so on. I counter every physical test you propose with a special explanation of why it won't work.

Now, what's the difference between an invisible, incorporeal, floating dragon who spits heatless fire and no dragon at all? If there's no way to disprove my contention, no conceivable experiment that would count against it, what does it mean to say that my dragon exists? Your inability to invalidate my hypothesis is not at all the same thing as proving it true. Claims that cannot be tested, assertions immune to disproof are veridically worthless, whatever value they may have in inspiring us or in exciting our sense of wonder. What I'm asking you to do comes down to believing, in the absence of evidence, on my say-so. The only thing you've really learned from my insistence that there's a dragon in my garage is that something funny is going on inside my head. You'd wonder, if no physical tests apply, what convinced me. The possibility that it was a dream or a hallucination would certainly enter your mind. But then why am I taking it so seriously? Maybe I need help. At the least, maybe I've seriously underestimated human fallibility.

Imagine that, despite none of the tests being successful, you wish to be scrupulously open-minded. So you don't outright reject the notion that there's a fire-breathing dragon in my garage. You merely put it on hold. Present evidence is strongly against it, but if a new body of data emerge you're prepared to examine it and see if it convinces you. Surely it's unfair of me to be offended at not being believed; or to criticize you for being stodgy and unimaginative—merely because you rendered the Scottish verdict of "not proved."

Imagine that things had gone otherwise. The dragon is invisible, all right, but footprints are being made in the flour as you watch. Your infrared detector reads off-scale. The spray paint reveals a jagged crest bobbing in the air before you. No matter how skeptical you might have been about the existence of dragons—to say nothing about invisible ones—you must now acknowledge that there's something here, and that in a preliminary way it's consistent with an invisible, fire-breathing dragon.

Now another scenario: Suppose it's not just me. Suppose that several people of your acquaintance, including people who you're pretty sure don't know each other, all tell you they have dragons in their garages—but in every case the evidence is maddeningly elusive. All of us admit we're disturbed at being gripped by so odd a conviction so ill-supported by the physical evidence. None of us is a lunatic. We speculate about what it would mean if invisible dragons were really hiding out in garages all over the world, with us humans just catching on. I'd rather it not be true, I tell you. But maybe all those ancient European and Chinese myths about dragons weren't myths at all. . .

Gratifyingly, some dragon-size footprints in the flour are now reported. But they're never made when a skeptic is looking. An alternative explanation presents itself: On close examination it seems clear that the footprints could have been faked. Another dragon enthusiast shows up with a burnt finger and attributes it to a rare physical manifestation of the dragon's fiery breath. But again, other possibilities exist. We understand that there are other ways to burn fingers besides the breath of invisible dragons. Such "evidence"—no matter how important the dragon advocates consider it—is far from compelling. Once again, the only sensible approach is tentatively to reject the dragon hypothesis, to be open to future physical data, and to wonder what the cause might be that so many apparently sane and sober people share the same strange delusion.

—

Magic requires tacit cooperation of the audience with the magician—an abandonment of skepticism, or what is sometimes described as the willing suspension of disbelief. It immediately follows that to penetrate the magic, to expose the trick, we must cease collaborating.

How can further progress be made in this emotionally laden, controversial, and vexing subject? Patients might exercise caution about therapists quick to deduce or confirm alien abductions. Those treating abductees might explain to their patients that hallucinations are normal, and that childhood sexual abuse is disconcertingly common. They might bear in mind that no client can be wholly uncontaminated by the aliens in popular culture. They might take scrupulous care not subtly to lead the witness. They might teach their clients skepticism. They might recharge their own dwindling reserves of the same commodity.

Purported alien abductions trouble many people and in more ways than one. The subject is a window into the internal lives of our fellows. If many falsely report being abducted, this is cause for worry. But much more worrisome is that so many therapists accept these reports at face value—with inadequate attention given to the suggestibility of clients and to unconscious cuing by their interlocutors.

I'm surprised that there are psychiatrists and others with at least some scientific training, who know the imperfections of the human mind, but who dismiss the idea that these accounts might be some species of hallucination, or some kind of screen memory. I'm even more surprised by claims that the alien abduction story represents true magic, that it is a challenge to our grip on reality, or that it constitutes support for a mystical view of the world. Or, as the matter is put by John Mack, "There are phenomena important enough to warrant serious research, and the metaphysics of the dominant Western scientific paradigm may be inadequate fully to support this research." In an interview with *Time* magazine, he goes on to say:

> I don't know why there's such a zeal to find a conventional physical explanation. I don't know why people have such trouble simply accepting the fact that something unusual is going on here. . . We've lost all that ability to know a world beyond the physical.*

But we know that hallucinations arise from sensory deprivation, drugs, illness and high fever, a lack of REM sleep, changes in brain chemistry, and so on. And even if, with Mack, we took the cases at face value, their remarkable aspects (slithering through walls and so on) are more readily attributable to something well within the realm of "the physical"—advanced alien technology—than to witchcraft.

A friend of mine claims that the only interesting question in the alien abduction paradigm is "Who's conning who?" Is the client deceiving the therapist, or vice versa? I disagree. For one thing, there are many other interesting questions about claims of alien abduction. For another, those two alternatives aren't mutually exclusive:

Something about the alien abduction cases tugged at *my* memory for years. Finally, I remembered. It was a 1954 book I had read in college, *The Fifty-Minute Hour*. The author, a psychoanalyst named Robert Lindner, had been called by the Los Alamos National Laboratory to treat a brilliant young nuclear physicist whose delusional system was beginning to interfere with his secret government research. The physicist (given the pseudonym Kirk Allen) had, it turned out, another life besides making nuclear weapons: In the far future, he confided, he piloted (or will pilot—the tenses get a little addled) interstellar spacecraft. He enjoyed rousing, swashbuckling adventures on planets of other stars. He was "lord" of many worlds. Perhaps they called him Captain Kirk. Not only could he "remember" this other life; he could also enter into it whenever he chose. By thinking in the right way, by wishing, he could transport himself across the light-years and the centuries.

---

* And then, in a sentence that reminds us how close the alien abduction paradigm is to messianic and chiliastic religion, Mack concludes, "I am a bridge between those two worlds."

In some way I could not comprehend, by merely desiring it to be so, I had crossed the immensities of space, broken out of time, and merged with—literally became—that distant and future self. . . . Don't ask me to explain. I can't, although God knows I've tried.

Lindner found him intelligent, sensitive, pleasant, polite, and perfectly able to deal with everyday human affairs. But—in reflecting on the excitement of his life among the stars—Allen had found himself a little bored with his life on Earth, even if it did involve building weapons of mass destruction. When admonished by his laboratory supervisors for distraction and dreaminess, he apologized; he would try, he assured them, to spend more time on this planet. That's when they contacted Lindner.

Allen had written 12,000 pages on his experiences in the future, and dozens of technical treatises on the geography, politics, architecture, astronomy, geology, life-forms, genealogy, and ecology of the planets of other stars. A flavor of the material is given by these monograph titles: "The Unique Brain Development of the Chrystopeds of Srom Norba X," "Fire Worship and Sacrifice on Srom Sodrat II," "The History of the Intergalactic Scientific Institute," and "The Application of Unified Field Theory and the Mechanics of the Stardrive to Space Travel." (That last is the one I'd like to see; after all, Allen was said to have been a first-rate physicist.) Fascinated, Lindner pored over the material.

Allen was not in the least shy about presenting his writings to Lindner or discussing them in detail. Unflappable and intellectually formidable, he seemed not to be yielding an inch to Lindner's psychiatric ministrations. When everything else failed, the psychiatrist attempted something different:

I tried . . . to avoid giving in any way the impression that I was entering the lists with him to prove that he was psychotic, that this was to be a tug of war over the question of his sanity. Instead, because it was obvious that both his temperament and training were scientific, I set myself to capitalize on the one quality he had demonstrated throughout his life . . . the quality that urged him toward a scientific career: his curiosity. . . . This meant . . . that at least for the time being I "accepted" the validity of his experiences. . . In a sudden flash of inspiration it came to me that in order to separate Kirk from his madness it was necessary for me to enter his fantasy and, from that position, to pry him loose from the psychosis.

Lindner highlighted certain apparent contradictions in the documents and asked Allen to resolve them. This required the physicist to re-enter the future to find the answers. Dutifully, Allen would arrive at the next session with a clarifying document written in his neat hand. Lindner found himself eagerly awaiting each interview, so he could be once more captivated by the vision of abundant life and intelligence in the Galaxy. Between them, they were able to resolve many problems of consistency.

Then a strange thing happened: "The materials of Kirk's psychosis and the Achilles heel of my personality met and meshed like the gears of a clock." The psychoanalyst became a co-conspirator in his patient's delusion. He began to reject psychological explanations of Allen's story. How sure are we that it couldn't really be true? He found himself defending the notion that another life, that of a spacefarer in the far future, could be entered into by a simple effort of the will.

At a startlingly rapid rate . . . larger and larger areas of my mind were being taken over by the fantasy. . . With Kirk's puzzled assistance I was taking part in cosmic adventures, sharing the exhilaration of the sweeping extravaganza he had plotted.

But eventually, an even stranger thing happened: Concerned for the well-being of his therapist, and mustering admirable reserves of integrity and courage, Kirk Allen confessed: He had made the whole thing up. It had roots in his lonely childhood and his unsuccessful relationships with women. He had shaded, and then forgotten, the boundary between reality and imagination. Filling in plausible details and weaving a rich tapestry about other worlds was challenging and exhilarating. He was sorry he had led Lindner down this primrose path.

"Why," the psychiatrist asked, "why did you pretend? Why did you keep on telling me. . . ?"

"Because I felt I had to," the physicist replied. "Because I felt you *wanted me to.*"

"Kirk and I reversed roles," Lindner explained,

and, in one of those startling denouements that make my work the unpredictable, wonderful and rewarding pursuit it is, the folly we shared collapsed. . . I employed the rationalization of clinical altruism for personal ends and thus fell into a trap that awaits all unwary therapists of the mind. . . Until Kirk Allen came into my life, I had never doubted my own stability. The aberrations of mind, so I had always thought, were for others. . . I am shamed by this smugness. But now, as I listen from my chair behind the couch, I know better. I know that my chair and the couch are separated only by a thin line. I know that it is, after all, but a happier combination of accidents that determines, finally, who shall lie on the couch, and who shall sit behind it.

I'm not sure from this account that Kirk Allen was truly delusional. Maybe he was just suffering from some character disorder which delighted in inventing charades at the expense of others. I don't know to what extent Lindner may have embellished or invented part of the story. While he wrote of "sharing" and of "entering" Allen's fantasy, there is nothing to suggest that the psychiatrist imagined he himself voyaged to the far future and partook of interstellar high adventure. Likewise, John Mack and the other alien abduction therapists do not suggest that they have been abducted; only their patients.

What if the physicist hadn't confessed? Might Lindner have convinced himself, beyond a reasonable doubt, that it really was possible to slip into a more romantic era? Would he have said he started out as a skeptic, but was convinced by the sheer weight of the evidence? Might he have advertised himself as an expert who assists space travelers from the future who are stranded in the twentieth century? Would the existence of such a psychiatric specialty encourage others to take fantasies or delusions of this sort seriously? After a few similar cases, would Lindner have impatiently resisted all arguments of the "Be reasonable, Bob" variety, and deduced he was penetrating some new level of reality?

His scientific training helped to save Kirk Allen from his madness. There was a moment when therapist and patient had exchanged roles. I like to think of it as the patient saving the therapist. Perhaps John Mack was not so lucky.

—

Consider a very different approach to finding aliens—the radio search for extraterrestrial intelligence. How is this different from fantasy and pseudoscience? In Moscow in the early 1960s, Soviet astronomers held a press conference in which they announced that the intense radio emission from a mysterious distant object called CTA-102 was varying regularly, like a sine wave, with a period of about 100 days. No periodic distant source had ever before been found. Why did they convene a press conference to announce so arcane a discovery? Because they thought they had detected an extraterrestrial civilization of immense powers. Surely, that's worth calling a press conference for. The report was briefly a media sensation, and the rock group the Byrds even composed and recorded a song about it. ["CTA-102, we're over here receiving you. / Signals tells us that you're there. / We can hear them loud and clear. . ."]

Radio emission from CTA-102? Certainly. But what *is* CTA-102? Today we know that CTA-102 is a distant quasar. At the time, the word "quasar" had not even been coined. We still don't know very well what quasars are; and there is more than one mutually exclusive explanation for them in the scientific literature. Nevertheless, no astronomers today—including those involved in that Moscow press conference—seriously contend that a quasar like CTA-102 is some extraterrestrial civilization billions of light-years away with access to immense power levels. Why not? Because we have alternative explanations of the properties of quasars that are consistent with known physical laws and that do not invoke alien life. Extraterrestrials represent a hypothesis of last resort. You reach for it only if everything else fails.

In 1967, British scientists found a much nearer intense radio source turning on and off with astonishing precision, its period constant to ten or more significant figures. What was it? Their first thought was that it was a message intended for us, or maybe an interstellar navigation and timing beacon for spacecraft that ply the space between the stars. They even gave it, among themselves at Cambridge University, the wry designation LGM-1—LGM standing for Little Green Men.

However, they were wiser than their Soviet counterparts. They did not call a press conference. It soon became clear that what they were observing was what is now called a "pulsar," the first pulsar. So, what's a pulsar? A pulsar is the end state of a massive star, a sun shrunk to the size of a city, held up as no other stars are, not by gas pressure, not by electron degeneracy, but by nuclear forces. It is in a certain sense an atomic nucleus ten miles or so across. Now *that*, I maintain, is a notion at least as bizarre as an interstellar navigation beacon. The answer to what a pulsar is has to be something mighty strange. It isn't an extraterrestrial civilization. It's something else: but a something else that opens our eyes and our minds and indicates unguessed possibilities in Nature. Anthony Hewish won the Nobel Prize in physics for the discovery of pulsars.

The original Ozma experiment (the first intentional radio search for extraterrestrial intelligence), the Harvard University/Planetary Society META (Megachannel Extraterrestrial Assay) program, the Ohio State University search, the SERENDIP Project of the University of California, Berkeley, and many other groups have all detected anomalous signals from space that make the observer's heart palpitate a little. We think for a moment that we've picked up a genuine signal of intelligent origin from far beyond our solar system. In reality, we have not the foggiest idea what it is, because the signal does not repeat. A few minutes later, or the next day, or years later you turn the same telescope to the same spot in the sky with the same frequency, bandpass, polarization, and everything else, and you don't hear a thing. You don't deduce, much less announce, aliens. It may have been a statistically inevitable electronic surge, or a malfunction in the detection system, or a spacecraft (from Earth), or a military aircraft flying by and broadcasting on channels that are supposed to be reserved for radio astronomy. Maybe it's even a garage door opener down the street or a radio station a hundred kilometers away. There are many possibilities. You must systematically check out all the alternatives, and see which ones can be eliminated. You don't declare that aliens have been found when your only evidence is an enigmatic nonrepeating signal.

And if the signal did repeat, would you then announce it to the press and the public? You would not. Maybe someone's hoaxing you. Maybe it's something you haven't been smart enough to figure out that's happening to your detection system. Maybe it's some previously unrecognized astrophysical source. Instead, you would call scientists at other radio observatories and inform them that at this particular spot in the sky, at this frequency and bandpass and all the rest, you seem to be getting something funny. Could they please see if they can confirm? Only if several independent observers—all of them fully aware of the complexity of Nature and the fallibility of observers—get the same kind of information from the same spot in the sky do you seriously consider that you have detected a genuine signal from alien beings.

There's a certain discipline involved. We can't just go off shouting "little green men" every time we detect something we don't at first understand, because we're going to look mighty silly—as the Soviet radio astronomers did with CTA-102—when it turns out to be something else. Special cautions are necessary when the stakes are high. We are not obliged to make up our minds before the evidence is in. It's permitted not to be sure.

I'm frequently asked, "Do you believe there's extraterrestrial intelligence?" I give the standard arguments—there are a lot of places out there, the molecules of life are everywhere, I use the word *billions*, and so on. Then I say it would be astonishing to me if there weren't extraterrestrial intelligence, but of course there is as yet no compelling evidence for it.

Often, I'm asked next, "What do you really think?"

I say, "I just told you what I really think."

"Yes, but what's your gut feeling?"

But I try not to think with my gut. If I'm serious about understanding the world, thinking with anything besides my brain, as tempting as that might be, is likely to get me into trouble. Really, it's okay to reserve judgment until the evidence is in.

———

I would be very happy if flying saucer advocates and alien abduction proponents were right and real evidence of extraterrestrial life were here for us to examine. They do not ask us, though, to believe on faith. They ask us to believe on the strength of their evidence. Surely it is our duty to scrutinize the purported evidence at least as closely and skeptically as radio astronomers do who are searching for alien radio signals.

No anecdotal claim—no matter how sincere, no matter how deeply felt, no matter how exemplary the lives of the attesting citizens—carries much weight on so important a question. As in the older UFO cases, anecdotal accounts are subject to irreducible error. This is not a personal criticism of those who say they've been abducted or of those who interrogate them. It is not tantamount to contempt for purported witnesses.* It is not—or should not be—arrogant dismissal of sincere and affecting testimony. It is merely a reluctant response to human fallibility.

If any powers whatever may be ascribed to the aliens—because their technology is so advanced—then we can account for any discrepancy, inconsistency, or implausibility. For instance, one academic UFOlogist suggests that both the aliens and the abductees are rendered invisible during the abduction (although not to each other); that's why more of the neighbors haven't noticed. Such "explanations" can explain anything, and therefore in fact nothing.

American police procedure concentrates on evidence and not anecdotes. As the European witch trials remind us, suspects can be intimidated during interrogation; people confess to crimes they never committed; eyewitnesses can be mistaken. This is also the linchpin of much detective fiction. But real, unfabricated evidence—powder burns, fingerprints, DNA samples, footprints, hair under the fingernails of the struggling victim—carry great weight. Criminalists employ something very close to the scientific method, and for the same reasons. So in the world of UFOs and alien abductions, it is fair to ask: Where is the evidence—the real, unambiguous physical evidence, the data that would convince a jury that hasn't already made up its mind?

Some enthusiasts argue that there are "thousands" of cases of "disturbed" soil where UFOs supposedly landed, and why isn't that good enough? It isn't good enough because there are ways of disturbing the soil other than by aliens in UFOs—humans with shovels is a possibility that springs readily to mind. One UFOlogist rebukes me for ignoring "4400 physical trace cases from 65 countries." But not one of these cases, so far as I know, has been analyzed, with results published in a peer-reviewed journal in physics or chemistry, metallurgy or soil science, showing that the "traces" could not have been generated by people. It's a modest enough scam—compared, say, with the crop circles of Wiltshire.

Likewise, not only can photographs easily be faked, but huge num-

---

* They cannot be called, simply, witnesses—because whether they witnessed anything (or, at least, anything in the outside world) is often the very point at issue.

bers of alleged photographs of UFOs have without a doubt been faked. Some enthusiasts go out night after night into a field looking for bright lights in the sky. When they see one, they flash their flashlights. Sometimes, they say, there's an answering flash. Well, maybe. But low-altitude aircraft make lights in the sky, and pilots are able, if so inclined, to blink their lights back. None of this constitutes anything approaching serious evidence.

Where is the physical evidence? As in satanic ritual abuse claims (and echoing "Devil's marks" in the witch trials), the most common physical evidence pointed to are scars and "scoop marks" on the bodies of abductees—who say they have no knowledge of where their scars came from. But this point is key: If the scars are within human capacity to generate, then they cannot be compelling physical evidence of abuse by aliens. Indeed, there are well-known psychiatric disorders in which people scoop, scar, tear, cut, and mutilate themselves (or others). And some of us with high pain thresholds and bad memories can injure ourselves accidentally with no recollection of the event.

One of John Mack's patients claims to have scars all over her body that are wholly baffling to her physicians. What do they look like? Oh, she can't show them; as in the witch mania, they're in private places. Mack considers this compelling evidence. Has he seen the scars? Can we have photographs of the scars taken by a skeptical physician? Mack knows, he says, a quadriplegic with scoop marks and considers this a *reductio ad absurdum* of the skeptical position; how can a quadriplegic scar himself? The argument is a good one only if the quadriplegic is hermetically sealed in a room to which no other human has access. Can we see his scars? Can an independent physician examine him? Another of Mack's patients says that the aliens have been taking eggs from her since she was sexually mature, and that her reproductive system baffles her gynecologist. Is it baffling enough to write the case up and submit a research paper to *The New England Journal of Medicine*? Apparently it's not that baffling.

Then we have the fact that one of his subjects made the whole thing up, as reported by *Time* magazine, and Mack didn't have a clue. He bought it hook, line, and sinker. What are his standards of critical scrutiny? If he allowed himself to be deceived by one subject, how do we know the same wasn't true of all?

Mack talks about these cases, the "phenomena," as posing a fundamental challenge to Western thinking, to science, to logic itself. Probably, he says, the abducting entities are not alien beings from our own universe, but visitors from "another dimension." Here's a typical, and revealing, passage from his book:

> When abductees call their experience "dreams," which they often do, close questioning can elicit that this may be a euphemism to cover what they are sure cannot be that, namely an event from which there was no awakening that occurred in another dimension.

Now the idea of higher dimensions did not arise from the brow of UFOlogy or the New Age. Instead, it is part and parcel of the physics of the twentieth century. Since Einstein's general relativity, a truism of cosmology is that space-time is bent or curved through a higher physical dimension. Kaluza-Klein theory posits an eleven-dimensional universe. Mack presents a thoroughly scientific idea as the key to "phenomena" beyond the reach of science.

We know something about how a higher-dimensional object would look in encountering our three-dimensional universe. For clarity, let's go down one dimension: An apple passing through a plane must change its shape as perceived by two-dimensional beings confined to the plane. First it seems to be a point, then larger apple cross-sections, then smaller ones, a point again—and finally, poof!, gone. Similarly, a fourth- or higher-dimensional object—provided it's not a very simple figure such as a hypercylinder passing through three dimensions along its axis—will wildly alter its geometry as we witness it passing through our universe. If aliens were systematically reported as shape-changers, I could at least see how Mack might pursue the notion of a higher-dimensional origin. (Another problem is trying to understand what a genetic cross between a three-dimensional and a four-dimensional being means. Are the offspring from the $3\frac{1}{2}$th dimension?)

What Mack really means when he talks about beings from other dimensions is that—despite his patients' occasional descriptions of their experiences as dreams and hallucinations—he hasn't the foggiest notion of what they are. But, tellingly, when he tries to describe them, he reaches for physics and mathematics. He wants it both ways—the language and credibility of science, but without being bound by its method and rules. He seems not to realize that the credibility is a consequence of the method.

The main challenge posed by Mack's cases is the old one of how to teach critical thinking more broadly and more deeply in a society—conceivably even including Harvard professors of psychiatry—awash in gullibility. The idea that critical thinking is the latest Western fad is silly. If you're buying a used car in Singapore or Bangkok—or a used chariot in ancient Susa or Rome—the same precautions will be useful as in Cambridge, Massachusetts.

When you buy a used car, you might very much want to believe what the salesman is saying: "So much car for so little money!" And anyway, it takes work to be skeptical; you have to know something about cars, and it's unpleasant to make the salesman angry at you. Despite all that, though, you recognize that the salesman might have a motive to shade the truth, and you've heard of other people in similar situations being taken. So you kick the tires, look under the hood, go for a test drive, ask searching questions. You might even bring along a mechanically inclined friend. You know that some skepticism is required, and you understand why. There is usually at least a small degree of hostile confrontation involved in the purchase of a used car and nobody claims it's an especially cheering experience. But if you don't exercise some minimal skepticism, if you have an absolutely untrammeled gullibility, there's a price you'll have to pay later. Then you'll wish you had made a small investment of skepticism early on.

Many homes in America now have moderately sophisticated burglar alarm systems, including infrared sensors and cameras triggered by motion. An authentic videotape, with time and date denoted, showing an alien incursion—especially as they slip through the walls—might be very good evidence. If millions of Americans have been abducted, isn't it strange that not one lives in such a home?

Some women, so the story goes, are impregnated by aliens or alien sperm; the fetuses are then removed by the aliens. Vast numbers of such cases are alleged. Isn't it odd that nothing anomalous has ever been seen in routine sonograms of such fetuses, or in amniocentesis, and that there has never been a miscarriage producing an alien hybrid? Or are medical personnel so doltish that they idly glance at the half-human, half-alien fetus and move on to the next patient? An epidemic of missing fetuses is something that would surely cause a stir among gynecologists, midwives, obstetrical nurses—especially in an age of heightened feminist awareness. But not a single medical record has been produced substantiating such claims.

Some UFOlogists consider it a telling point that women who claim to have been sexually inactive wind up pregnant, and attribute their state to alien impregnation. A goodly number appear to be teenagers. Taking their stories at face value is not the only option available to the serious investigator. Surely we can understand why, in the anguish of an unwanted pregnancy, a teenager living in a society flooded with accounts of alien visitation might invent such a story. Here, too, there are possible religious antecedents.

Some abductees say that tiny implants, perhaps metallic, were inserted into their bodies—high up their nostrils, for example. These implants, alien abduction therapists tell us, sometimes accidentally fall out, but "in all but a few of the cases the artifact has been lost or discarded." These abductees seem stupefyingly incurious. A strange object—possibly a transmitter sending telemetered data about the state of your body to an alien spaceship somewhere above the Earth—drops out of your nose; you idly examine it and then throw it in the garbage. Something like this is true, we are told, of the majority of abduction cases.

A few such "implants" have been produced and examined by experts. None has been confirmed as of unearthly manufacture. No components are made of unusual isotopes, despite the fact that other stars and other worlds are known to be constituted of different isotopic proportions than the Earth. There are no metals from the transuranic "island of stability," where physicists think there should be a new family of nonradioactive chemical elements unknown on Earth.

What abduction enthusiasts considered the best case was that of Richard Price, who claims that aliens abducted him when he was eight years old and implanted a small artifact in his penis. A quarter century later a physician confirmed a "foreign body" embedded there. After eight more years, it fell out. Roughly a millimeter in diameter and 4 millimeters long, it was carefully examined by scientists from MIT and Massachusetts General Hospital. Their conclusion? Collagen formed by the body at sites of inflammation plus cotton fibers from Price's underpants.

On August 28, 1995, television stations owned by Rupert Murdoch ran what was purported to be an autopsy of a dead alien, shot on 16-millimeter film. Masked pathologists in vintage radiation-protection suits (with rectangular glass windows to see out of) cut up a large-eyed 12-fingered figure and examined the internal organs. While the film was sometimes out of focus, and the view of the cadaver often blocked by the humans crowding around it, some viewers found the effect chilling. The *Times* of London, also owned by Murdoch, didn't know what to make of it, although it did quote one pathologist who thought the autopsy performed with unseemly and unrealistic haste (ideal, though, for television viewing). It was said to have been shot in New Mexico in 1947 by a participant, now in his eighties, who wished to remain anonymous. What appeared to be the clincher was the announcement that the leader of the film (its first few feet) contained coded information that Kodak, the manufacturer, dated to 1947. However, it turns out that the full film magazine was not presented to Kodak, but at most the cut leader. For all we know, the leader could have been cut from a 1947 newsreel, abundantly archived in America, and the "autopsy" staged and filmed separately and recently. There's a dragon footprint all right—but a fakable one. If this is a hoax, it requires not much more cleverness than crop circles and the MJ-12 document.

In none of these stories is there anything strongly suggestive of extraterrestrial origin. There is certainly no retrieval of cunning machinery far beyond current technology. No abductee has filched a page from the captain's logbook, or an examining instrument, or taken an authentic photograph of the interior of the ship, or come back with detailed and verifiable scientific information not hitherto available on Earth. Why not? These failures must tell us something.

Since the middle of the twentieth century, we've been assured by proponents of the extraterrestrial hypothesis that physical evidence—not star maps remembered from years ago, not scars, not disturbed soil, but real alien technology—was in hand. The analysis would be released momentarily. These claims go back to the earliest crashed-saucer scam of Newton and GeBauer. Now it's decades later and we're still waiting. Where are the articles published in the refereed scientific literature, in the metallurgical and ceramics journals, in publications of the Institute of Electrical and Electronic Engineers, in *Science* or *Nature*?

Such a discovery would be momentous. If there were real artifacts, physicists and chemists would be fighting for the privilege of discovering that there are aliens among us—who use, say, unknown alloys, or materials of extraordinary tensile strength or ductility or conductivity. The practical implications of such a finding—never mind the confirmation of an alien invasion—would be immense. Discoveries like this are what scientists live for. Their absence must tell us something.

—

Keeping an open mind is a virtue—but, as the space engineer James Oberg once said, not so open that your brains fall out. Of course we must be willing to change our minds when warranted by new evidence. But the evidence must be strong. Not all claims to knowledge have equal merit. The standard of evidence in most of the alien abduction cases is roughly what is found in cases of the apparition of the Virgin Mary in medieval Spain.

The pioneering psychoanalyst Carl Gustav Jung had much that was sensible to say on issues of this sort. He explicitly argued that UFOs were a kind of projection of the unconscious mind. In a related discussion of regression and what today is called "channeling," he wrote

> One can very well . . .take it simply as a report of psychological facts or a continuous series of communications from the unconscious. . . They have this in common with dreams; for dreams, too, are statements about the unconscious. . . The present state of affairs gives us reason enough to wait quietly until more impressive physical phenomena put in an appearance. If, after making allowance for conscious and unconscious falsification, self-deception, prejudice, etc., we should still find something positive behind them, then the exact sciences will surely conquer this field by experiment and verification, as has happened in every other realm of human experience.

Of those who accept such testimony at face value, he remarked

> These people are lacking not only in criticism but in the most elementary knowledge of psychology. At bottom they do not want to

be taught any better, but merely to go on believing—surely the naivest of presumptions in view of our human failings.

Perhaps some day there will be a UFO or alien abduction case that is well-attested, accompanied by compelling physical evidence, and explicable only in terms of extraterrestrial visitation. It's hard to think of a more important discovery. So far, though, there have been no such cases, nothing that comes close. So far, the invisible dragon has left no unfakable footprints.

Which, then, is more likely: that we're undergoing a massive but generally overlooked invasion by alien sexual abusers, or that people are experiencing some unfamiliar internal mental state they do not understand? Admittedly, we're very ignorant both about extraterrestrial beings, if any, and about human psychology. But if these really were the only two alternatives, which one would you pick?

And if the alien abduction accounts are mainly about brain physiology, hallucinations, distorted memories of childhood, and hoaxing, don't we have before us a matter of supreme importance—touching on our limitations, the ease with which we can be misled and manipulated, the fashioning of our beliefs, and perhaps even the origins of our religions? There is genuine scientific paydirt in UFOs and alien abductions—but it is, I think, of a distinctly homegrown and terrestrial character.

Chapter 11

———

THE CITY
OF GRIEF

. . . how alien, alas, are the streets
of the city of grief.

RAINER MARIA RILKE,
"The Tenth Elegy"
(1923)

A short summary of the argument in the preceding seven chapters appeared in *Parade* magazine on March 7, 1993. I was struck by how many letters it evoked, how passionate were the responses, and how much agony is associated with this strange experience—whatever its true explanation might be. Alien abduction accounts provide an unexpected window into the lives of some of our fellow citizens. Some letter writers reasoned, some asserted, some harangued, some were frankly perplexed, some were deeply troubled.

The article was also widely misunderstood. A television talk-show host, Geraldo Rivera, held up a copy of *Parade* and announced I thought we were being visited. A *Washington Post* videocassette reviewer quoted me as saying there's an abduction every few seconds, missing the ironical tone and the following sentence ("It's surprising more of the neighbors haven't noticed"). My description (Chapter 6) of on rare occasions seeming to hear the voices of my dead parents—what I described as "a lucid recollection"—were keynoted by Raymond Moody, in the *New Age Journal* and in the introduction of his book *Reunions*, as evidence that we "survive" death. Dr. Moody has spent his life trying to find evidence for life after death. If my testimony is worth quoting, it seems clear he hasn't found much. Many letter writers concluded that since I had worked on the possibility of extraterrestrial life, I must "believe" in UFOs; or conversely that, if I was skeptical about UFOs, I must embrace the absurd belief that humans are the only intelligent beings in the Universe. There's something about this subject unconducive to clear thinking.

Here, without further comment, is a representative sampling of my mail on the subject:

• I wonder how some of our fellow animals may describe their encounters with us. They see a large hovering object making a terrible noise above them. They begin to run and feel a sharp pain in their side. Suddenly they fall to the ground. . . Several man-creatures approach them carrying strange looking instruments. They examine your sexual organs and teeth. They place a net under you and then let it take you in the air with a strange device. After all the examinations, they then clamp a strange metal object to your ear. Then, just as suddenly as they had appeared, they are gone. Eventually, muscle control returns, and the poor disoriented creature staggers off into the forest, not knowing [whether] what just transpired was a nightmare or a reality.

• I was sexually abused as a child. In my recovery I have drawn many "space beings" and have felt many times I was being overpowered, held down, and the sensation of having left my body to float around the room. None of the abductee accounts really come as a surprise to someone who has dealt with childhood sexual abuse issues. . . Believe me, I would much rather have blamed my abuse on a space alien than have to face the truth about what happened to me with the adults I was supposed to be able to trust. It's been driving me crazy to hear some of my friends speak of their memories that imply they have been abducted by aliens. . . I keep saying to them that this is the ultimate victim role in which we as adults have no power when these little gray men come to us in our sleep! This is not real. The ultimate victim role is the one between an abusive parent and the victimized child.

• I don't know if these people are some sort of demons, or if they really don't exist. My daughter said she had sensors put in her body when she was small. I don't know. . . We keep our doors locked and bolted and it really scares me. I don't have the money to send her to a good doctor, and she can't work on account of all this. . . My daughter is hearing a voice on a tape. These go out at night and take kids and sexually abuse them. If you don't do as they say, someone in your family will be hurt. Who in their right mind would harm little children? They know everything that is said in the house. . . Somebody said long, long ago somebody put a curse on our family. If somebody did, how do you get the curse off? I know all this sounds strange and bizarre, but believe me it's scary.

• How many human females who had the misfortune of being raped had the foresight to take from their attacker an ID card, a picture of the rapist, or anything else which could be used as evidence as to an alleged rape?

• I for one will be sleeping with my Polaroid from now on, in hopes that the next time I'm abducted I can provide the proof needed. . . Why should it be up to the abductees to prove what's happening?

• I am living proof of Carl Sagan's claim of the possibility that alien abductions occur in the minds of people suffering from sleep paralysis. They truly believe it's real.

• In 2001 A.D. Starships from the 33 planets of the Interplanetary Confederation will land on earth carrying 33,000 Brothers! They are extraterrestrial teachers and scientists who will help to expand our understanding of interplanetary life, as our own earth planet will become the 33rd member of the Confederation!

• This is a grotesquely challenging arena. . . I studied UFOs for over 20 years. Finally I became quite disenchanted by the cult and the cult fringe groups.

• I am a 47-year-old grandmother who has been the victim of this phenomena since early childhood. I do not—nor have I ever—accepted it at face value. I do not—nor have I ever—claimed to understand what it is . . . I would gladly accept a diagnosis of schizophrenia, or some other understood pathology, in exchange for this unknown . . . The lack of physical evidence is, I fully agree, most frustrating for both victims and researchers. Unfortunately, the retrieval of such evidence is made extremely difficult by the manner in which the victims are abducted. Often I am removed either in my nightgown (which is later removed) or already naked. This condition makes it quite impossible to hide a camera . . . I have awakened with deep gashes, puncture wounds, scooped out tissue, eye damage, bleeding from the nose and ears, burns, and finger marks and bruises which persist for days after the event. I have had all of these examined by qualified physicians but none have been satisfactorily explained. I am not into self-mutilation; these are not stigmata . . . Please be aware that the majority of abductees claim to have had no interest in UFOs previously (I am one), have no history of childhood abuses (I am one), have no desire for publicity or notoriety (I am one), and, in fact, have gone to great lengths to avoid acknowledging any involvement whatsoever, assuming he or she is experiencing a nervous breakdown or other psychological disorder (I am one). Agreed, there are many self-proclaimed abductees (and contactees) who seek out publicity for monetary gain or to satisfy a need for attention. I would be the last to deny these people exist. What I do deny is that ALL abductees are imagining or falsifying these events to satisfy their own personal agendas.

• UFOs don't exist. I think that requires an eternal energy source, and this doesn't exist . . . I have spoken with Jesus. The commentary on the *Parade* magazine is very destructive, and it enjoys scaring society, I beg you to think more openly because our intelligent beings from outer spaces do exist and they are our creators. . . I too was an abductee. To be honest, these dear beings have done me more good than bad. They have saved my life. . . The trouble with Earth beings is that they want proof, proof, and proof!

• In the Bible it talks about terrestrial and celestial bodies. This is not to say that God is out for sexual abuse on people or that we're crazy.

• I have been strongly telepathic for twenty-seven years now. I do not receive—I transmit. . . Waves are coming from outer space somewhere—beaming through my head and transmitting thoughts, words, and images into the heads of anybody within range. . . Images will pop into my head that *I did not put there*, and vanish just as suddenly. Dreams are not dreams anymore—they are more like Hollywood productions. . . They are smart critters and they won't give up. . . Maybe all these little guys want to do is communicate. . . If I finally go psychotic from all this pressure—or have another heart attack—there goes your last sure evidence that there is life in space.

• I think I have found a plausible terrestrial scientific explanation for numerous UFO reports. [The writer then discusses ball lightning]. If you like my stuff, could you help me get it published?

• Sagan refuses to take seriously the witnesses' reports of anything that twentieth-century science can't explain.

• Now readers will feel free to treat abductees. . . as if they are victims of nothing more than an illusion. Abductees suffer the same sort of trauma a rape victim endures, and to have their experiences rejected by those closest to them is a second victimization that leaves them without any support system. Encounters with aliens is hard enough to cope with; victims need support, not rationalizations.

• My friend Frankie wants me to bring back an ashtray or a matchbook, but I think these visitors are probably much too intelligent to smoke.

• My own feeling is that the alien abduction phenomena is little more than a dreamlike sequence vicariously retrieved from memory storage. There are no more little green men or flying saucers than there are images of those things already stored in our brains.

• When alleged scientists conspire to censor and intimidate those who endeavor to offer new insightful hypotheses on conventional theories . . . they no longer should be considered scientists, but merely the insecure, self-serving impostors that they apparently are . . . In the same token, must we also still suppose that J. Edgar Hoover was a fine FBI director, rather than the homosexual tool of organized crime he was?

• Your conclusion that large numbers of people in this country, perhaps as many as five million, are all victims of an identical mass hallucination is asinine.

• Thanks to the Supreme Court. . . America is now wide open for the Eastern pagan religions, under the aegis of Satan and his demons, so now we have four-foot gray beings kidnapping Earthlings and performing all sorts of experiments on them, and are being propagated by those who are educated beyond their intelligence and should know better. . . Your question ["Are We Being Visited?"] is no problem for those who *know* the word of God, and are born-again Christians, and are looking for our Redeemer from Heaven, to rapture us out of this world of sin, sickness, war, AIDS, crime, abortion, homosexuality, New-Age-New-World-Order indoctrination, media brainwashing, perversion and subversion in government, education, business, finance, society, religion, etc. Those who reject the Creator God of the Bible are bound to fall for the kind of fairy tales which your article tries to propagate as being truth.

• If there is no reason to take the matter of alien visitation seriously, why is it the most highly classified subject in the U.S. government?

• Perhaps some vastly older alien race, from a relatively metal-deficient star system, is seeking to prolong its existence by taking over a younger, better world and blending with its inhabitants.

• If I were a betting man, I would give you odds that your mailbox will overflow with stories such as I just related. I suspect that the psychic [psyche] brings forth these demons and angels, lights and circles as a part of our development. They are part of our nature.

• Science has become the "magic that works." The UFOlogists are heretics to be excommunicated or burned at the stake.

• [Several readers wrote to say that aliens were demons sent by Satan, who is able to cloud our minds. One proposes that the insidious Satanic purpose is to make us worried about an alien invasion, so that when Jesus and his angels appear over Jerusalem we will be frightened rather than glad.] I do hope you will not dismiss me, [she writes,] as another religious crackpot. I am quite normal and well-known in my own little community.

• You, sir, are in a position to do one of two things: Know about the abductions and be covering them up, or feel that because you have not been abducted (perhaps they are not interested in you) they do not occur.

• A treason suit [was filed] against the President and Congress of the United States over a treaty made with aliens in the early '40s, who had later shown themselves to be hostile. . . The treaty agreed to protect the secrecy of the aliens in return for some of their technology [stealth aircraft and fiber optics, another correspondent reveals].

• Some of these beings are capable of intercepting the spiritual body when it is traveling.

• I am having communication with an alien being. This communication started early in 1992. What else can I say?

• The aliens can stay a step or two ahead of the thinking of scientists, and know how to leave insufficient clues behind that would satisfy the Sagan types, until society is better prepared mentally to face up to it all. . . Perhaps you share the view that what's going on with respect to UFOs and aliens, if deemed real, would be too traumatic to think about. However . . . they've shown themselves until back some 5,000–15,000 years or more ago when they were here for extended periods, spawning the god/goddess mythology of all cultures. The bottom line is that in all that time they haven't taken over Earth; they haven't subjected us or wiped us out.

• *Homo sapiens* was genetically fashioned, created initially to be substitute laborers and domestics for the SKY-LORDS (DINGIRS/ELOHIM/ANUNNAKI).

• The explosion that people saw was hydrogen fuel from a star cruiser, the landing sight was to be Northern California. . . The people on that star cruiser looked like Mr. Spock from the *Star Trek* TV series.

• Be the reports from the 15th century or the 20th, a common thread ties the reports. Individuals who have experienced sexual trauma have a great deal of difficulty understanding and coping with the trauma. The terms used to describe the [resulting] hallucinations can be incoherent and incomprehensible.

• We find we are not as intelligent as we thought although we are still stiff-necked and our greatest sin is our pride. And we do not even know we are being led to Armageddon. The star pin-pointed a single shed, moved across the sky leading wisemen to that shed, frightened shepherds with the words Fear not. Its spotlight was Ezekiel's glory of God, Paul's light that temporarily blinded him. . . It was the ship in which little men took off old Rip, the little men called brownies, fairies, elves, these "creations" of creators given specific duties. . . The God People are not yet ready to make themselves known to us. First, Armageddon, then, after we KNOW, we can go it alone. When we are humbled, when we do not shoot them down, God will return.

• The answer to these aliens from outer-space is simple. It comes from man. Man using drugs on people. In mental institutions all over the country, there are people who have no control over their emotions and behavior. To control these people, they are given a variety of antipsychotic drugs. . . If you have been drugged often . . . you will begin to have what is called "bleedthroughs." This will be flash images popping into your mind of strange-looking people coming up to your face. This will begin your search for the answer of what the aliens were doing to you. You will be one of the thousands of UFO abductees. People will call you crazy. The reason for the strange creatures you are seeing is because Thorazine distorts the vision of your subconscious mind. . . The writer was laughed at, ridiculed, had his life threatened [because of presenting these ideas].

• Hypnosis prepares the mind for the invasion of demons, devils, and little gray men. God wants us to be clothed and in our right minds. . . Anything your "little gray men" can do, Christ can do better!

• I hope that I never feel so superior that I cannot acknowledge that Creation is not limited to myself, but encompasses the Universe and all its entities.

• In 1977 an heavenly being spoke to me about an injury to my head that happened in 1968.

• [A letter from a man who had 24 separate encounters with] a silent hovering saucer-shaped vehicle [and who has in consequence] experienced an ongoing development and amplification of such mental functions as clairvoyance, telepathy, and the challenging [channeling] of universal life energy for the purpose of healing.

• Over the years I have seen and talked to "ghosts," been visited (though not yet abducted) by aliens, seen 3-dimensional heads floating by my bed, heard knocks on my door. . . These experiences seemed as real as life. I have never thought of these experiences as anything more than what they certainly are: my mind playing tricks on itself.*

• A hallucination might account for 99%, but can it ever account for 100%?

• UFOs are . . . a subject of deep fantasy which has no FACTUAL BASIS WHATSOEVER. I pray you won't lend your credence to a hoax.

• Dr. Sagan served on the Air Force committee that evaluated government investigations of UFOs, and yet he wants us to believe that there's no substantial proof that UFOs exist. Please explain why the government needed to be evaluated.

• I'm going to lobby my Representative to try to cancel funds for this program of listening for alien signals from space, because it would be a waste of money. They're already among us.

• The government spends millions of tax dollars for researching UFOs. The SETI project (search for extraterrestrial intelligence) would be a waste of money if the government truly believed UFOs were nonexistent. I am personally excited about the SETI project because it shows that we're moving in the right direction; toward communication with aliens, rather than being an unwilling observer.

• The succubi, which I identified more as astral rape, occurred from '78-'92. It was hard on a moral and seriously practicing Catholic, demoralizing, dehumanizing, and quite literally had me worried by the physical aftermath of disease effects.

• The space people are coming! They hope to remove whom they can, especially children who are the "seedlings" of the next humanity generation along with their cooperating parents, grandparents and other adults, to safety before the upcoming *major* sunspot/planetary peak, which is just over the horizon. The Space Ship is in sight every night and close in to assist us when the Major Solar Flares do, before turbulence starts in the atmosphere. The Polar Shift is due now as it

---

* From a letter received by *The Skeptical Inquirer*; courtesy, Kendrick Frazier.

moves to its new position for the Aquarian Age. . . [The authors also inform me that they are] working with the Ashtar Command, where Jesus Christ meets with those aboard for instructions. Many dignitaries are present, including archangels Michael and Gabriel.

• I have extensive experience in therapeutic energy work, which involves removing grid patterns, negative memory cords, and alien implants from human bodies and their surrounding energy fields. My work is primarily utilized as an adjunctive aide to psychotherapy. My clients range from businessmen, homemakers, professional artists, therapists, and children. . . The alien energy is very fluid, both within the body and after it is removed, and must be contained as soon as possible. The energy grids are most often locked around the heart or in a triangular formation across the shoulders.

• I don't know how, after such an experience, I could have just turned over and gone back to sleep.

• I believe in happy endings. I always have. Once you have seen a figure as tall as the room—with golden hair, and shining like a lighted Christmas tree, lifting up the little child beside us, how could you not? I understood the message the figure was relaying—to the little child— and it was me. We had always talked together. How else could life be bearable—in a place like this?. . . Unfamiliar mental states? You put your finger right on it.

• Who is *really* in charge of this planet?

Chapter 12
—

THE FINE ART
OF BALONEY
DETECTION

The human understanding is no dry light, but
receives infusion from the will and affections;
whence proceed sciences which may be called
"sciences as one would." For what a man had
rather were true he more readily believes.
Therefore he rejects difficult things from
impatience of research; sober things, because
they narrow hope; the deeper things of nature,
from superstition; the light of experience, from
arrogance and pride; things not commonly
believed, out of deference to the opinion of the
vulgar. Numberless in short are the ways, and
sometimes imperceptible, in which the
affections color and infect the understanding.

FRANCIS BACON,
Novum Organon
(1620)

My parents died years ago. I was very close to them. I still miss them terribly. I know I always will. I long to believe that their essence, their personalities, what I loved so much about them, are—really and truly—still in existence somewhere. I wouldn't ask very much, just five or ten minutes a year, say, to tell them about their grandchildren, to catch them up on the latest news, to remind them that I love them. There's a part of me—no matter how childish it sounds—that wonders how they are. "Is everything all right?" I want to ask. The last words I found myself saying to my father, at the moment of his death, were "Take care."

Sometimes I dream that I'm talking to my parents, and suddenly—still immersed in the dreamwork—I'm seized by the overpowering realization that they didn't really die, that it's all been some kind of horrible mistake. Why, here they are, alive and well, my father making wry jokes, my mother earnestly advising me to wear a muffler because the weather is chilly. When I wake up I go through an abbreviated process of mourning all over again. Plainly, there's something within me that's ready to believe in life after death. And it's not the least bit interested in whether there's any sober evidence for it.

So I don't guffaw at the woman who visits her husband's grave and chats him up every now and then, maybe on the anniversary of his death. It's not hard to understand. And if I have difficulties with the ontological status of who she's talking to, that's all right. That's not what this is about. This is about humans being human. More than a third of American adults believe that on some level they've made contact with the dead. The number seems to have jumped by 15 percent between 1977 and 1988. A quarter of Americans believe in reincarnation.

But that doesn't mean I'd be willing to accept the pretensions of a "medium," who claims to channel the spirits of the dear departed, when I'm aware the practice is rife with fraud. I know how much I want to believe that my parents have just abandoned the husks of their bodies, like insects or snakes molting, and gone somewhere else. I understand that those very feelings might make me easy prey even for an unclever con, or for normal people unfamiliar with their unconscious minds, or for those suffering from a dissociative psychiatric disorder. Reluctantly, I rouse some reserves of skepticism.

How is it, I ask myself, that channelers never give us verifiable information otherwise unavailable? Why does Alexander the Great never tell us about the exact location of his tomb, Fermat about his Last Theorem, John Wilkes Booth about the Lincoln assassination conspiracy, Hermann Göring about the Reichstag fire? Why don't Sophocles, Democritus, and Aristarchus dictate their lost books? Don't they wish future generations to have access to their masterpieces?

If some good evidence for life after death were announced, I'd be eager to examine it; but it would have to be real scientific data, not mere anecdote. As with the face on Mars and alien abductions, better the hard truth, I say, than the comforting fantasy. And in the final tolling it often turns out that the facts are more comforting than the fantasy.

The fundamental premise of "channeling," spiritualism, and other forms of necromancy is that when we die we don't. Not exactly. Some thinking, feeling, and remembering part of us continues. That whatever-it-is—a soul or spirit, neither matter nor energy, but something else—can, we are told, re-enter the bodies of human and other beings in the future, and so death loses much of its sting. What's more, we have an opportunity, if the spiritualist or channeling contentions are true, to make contact with loved ones who have died.

J. Z. Knight of the State of Washington claims to be in touch with a 35,000-year-old somebody called "Ramtha." He speaks English very well, using Knight's tongue, lips and vocal chords, producing what sounds to me to be an accent from the Indian Raj. Since most people know how to talk, and many—from children to professional actors—have a repertoire of voices at their command, the simplest hypothesis is that Ms. Knight makes "Ramtha" speak all by herself, and that she has no contact with disembodied entities from the Pleistocene Ice Age. If there's evidence to the contrary, I'd love to hear it. It would be considerably more impressive if Ramtha could speak by himself, without the assistance of Ms. Knight's mouth. Failing that, how might we test the claim? (The actress Shirley MacLaine attests that Ramtha was her brother in Atlantis, but that's another story.)

Suppose Ramtha were available for questioning. Could we verify whether he is who he says he is? How does he know that he lived 35,000 years ago, even approximately? What calendar does he employ? Who is keeping track of the intervening millennia? Thirty-five thousand plus or minus what? What were things like 35,000 years ago? Either Ramtha really is 35,000 years old, in which case we discover something about that period, or he's a phony and he'll (or rather she'll) slip up.

Where did Ramtha live? (I know he speaks English with an Indian accent, but where 35,000 years ago did they do that?) What was the climate? What did Ramtha eat? (Archaeologists know something about what people ate back then.) What were the indigenous languages, and social structure? Who else did Ramtha live with—wife, wives, children, grandchildren? What was the life cycle, the infant mortality rate, the life expectancy? Did they have birth control? What clothes did they wear? How were the clothes manufactured? What were the most dangerous predators? Hunting and fishing implements and strategies? Weapons? Endemic sexism? Xenophobia and ethnocentrism? And if Ramtha came from the "high civilization" of Atlantis, where are the linguistic, technological, historical and other details? What was their writing like? Tell us. Instead, all we are offered are banal homilies.

Here, to take another example, is a set of information channeled not from an ancient dead person, but from unknown non-human entities who make crop circles, as recorded by the journalist Jim Schnabel:

> We are so anxious at this sinful nation spreading lies about us. We do not come in machines, we do not land on your earth in machines. . . We come like the wind. We are Life Force. Life Force from the ground. . . Come here. . . We are but a breath away . . . a breath away . . . we are not a million miles away . . . a Life Force that is larger than the energies in your body. But we meet at a higher level of life. . . We need no name. We are parallel to your world, alongside your world. . . The walls are broken. Two men will rise from the past . . . the great bear . . . the world will be at peace.

People pay attention to these puerile marvels mainly because they promise something like old-time religion, but especially life after death, even life eternal.

A very different prospect for something like eternal life was once proposed by the versatile British scientist J.B.S. Haldane, who was, among many other things, one of the founders of population genetics. Haldane imagined a far future when the stars have darkened and space is mainly filled with a cold, thin gas. Nevertheless, if we wait long enough statistical fluctuations in the density of this gas will occur. Over immense periods of time the fluctuations will be sufficient to reconstitute a Universe something like our own. If the Universe is infinitely old, there will be an infinite number of such reconstitutions, Haldane pointed out.

So in an infinitely old universe with an infinite number of appearances of galaxies, stars, planets, and life, an identical Earth must reappear on which you and all your loved ones will be reunited. I'll be able to see my parents again and introduce them to the grandchildren they never knew. And all this will happen not once, but an infinite number of times.

Somehow, though, this does not quite offer the consolations of religion. If none of us is to have any recollection of what happened *this* time around, the time the reader and I are sharing, the satisfactions of bodily resurrection, in my ears at least, ring hollow.

But in this reflection I have underestimated what infinity means. In Haldane's picture, there will be universes, indeed an infinite number of them, in which our brains will have full recollection of many previous rounds. Satisfaction is at hand—tempered, though, by the thought of all those other universes which will also come into existence (again, not once but an infinite number of times) with tragedies and horrors vastly outstripping anything I've experienced this turn.

The Consolation of Haldane depends, though, on what kind of universe we live in, and maybe on such arcana as whether there's enough matter to eventually reverse the expansion of the universe, and the character of vacuum fluctuations. Those with a deep longing for life after death might, it seems, devote themselves to cosmology, quantum gravity, elementary particle physics, and transfinite arithmetic.

—

Clement of Alexandria, a Father of the early Church, in his *Exhortations to the Greeks* (written around the year 190) dismissed pagan beliefs in words that might today seem a little ironic:

> Far indeed are we from allowing grown men to listen to such tales. Even to our own children, when they are crying their heart out, as the saying goes, we are not in the habit of telling fabulous stories to soothe them.

In our time we have less severe standards. We tell children about Santa Claus, the Easter Bunny, and the Tooth Fairy for reasons we think emotionally sound, but then disabuse them of these myths before they're grown. Why retract? Because their well-being as adults depends on them knowing the world as it really is. We worry, and for good reason, about adults who still believe in Santa Claus.

On doctrinaire religions, "Men dare not avow, even to their own hearts," wrote the philosopher David Hume,

> the doubts which they entertain on such subjects. They make a merit of implicit faith; and disguise to themselves their real infidelity, by the strongest asseverations and the most positive bigotry.

This infidelity has profound moral consequences, as the American revolutionary Tom Paine wrote in *The Age of Reason*:

> Infidelity does not consist in believing, or in disbelieving; it consists in professing to believe what one does not believe. It is impossible to calculate the moral mischief, if I may so express it, that mental lying has produced in society. When man has so far corrupted and prostituted the chastity of his mind, as to subscribe his professional belief to things he does not believe, he has prepared himself for the commission of every other crime.

T. H. Huxley's formulation was

> The foundation of morality is to . . . give up pretending to believe that for which there is no evidence, and repeating unintelligible propositions about things beyond the possibilities of knowledge.

Clement, Hume, Paine, and Huxley were all talking about religion. But much of what they wrote has more general applications—for example to the pervasive background importunings of our commercial civilization: There is a class of aspirin commercials in which actors pretending to be doctors reveal the competing product to have only so much of the painkilling ingredient that doctors recommend most—they don't tell you what the mysterious ingredient is. Whereas *their* product has a dramatically larger amount (1.2 to 2 times more per tablet). So buy their product. But why not just take two of the competing tablets? Or consider the analgesic that works better than the "regular-strength" product of the competition. Why not then take the "extra-strength" competitive product? And of course they do not tell us of the more than a thousand deaths each year in the United States from the use of aspirin, or the apparent 5,000 annual cases of kidney failure from the use of acetaminophen, of which the best-selling brand is Tylenol. (This, however, may represent a case of correlation without causation.) Or who cares which breakfast cereal has more vitamins when we can take a vitamin pill with breakfast? Likewise, why should it matter whether an antacid contains calcium if the calcium is for nutrition and irrelevant for gastritis? Commercial culture is full of similar misdirections and evasions at the expense of the consumer. You're not supposed to ask. Don't think. Buy.

Paid product endorsements, especially by real or purported experts, constitute a steady rainfall of deception. They betray contempt for the intelligence of their customers. They introduce an insidious corruption of popular attitudes about scientific objectivity. Today there are even commercials in which real scientists, some of considerable distinction, shill for corporations. They teach that scientists too will lie for money. As Tom Paine warned, inuring us to lies lays the groundwork for many other evils.

I have in front of me as I write the program of one of the annual Whole Life Expos, New Age expositions held in San Francisco. Typically, tens of thousands of people attend. Highly questionable experts tout highly questionable products. Here are some of the presentations: "How Trapped Blood Proteins Produce Pain and Suffering." "Crystals, Are They Talismans or Stones?" (I have an opinion myself.) It continues: "As a crystal focuses sound and light waves for radio and television"—this is a vapid misunderstanding of how radio and television work—"so may it amplify spiritual vibrations for the attuned human." Or here's one: "Return of the Goddess, a Presentational Ritual." Another: "Synchronicity, the Recognition Experience." That one is given by "Brother Charles." Or, on the next page, "You, Saint-Germain, and Healing Through the Violet Flame." It goes on and on, with plenty of ads about "opportunities"—running the short gamut from the dubious to the spurious—that are available at the Whole Life Expo.

Distraught cancer victims make pilgrimages to the Philippines, where "psychic surgeons," having palmed bits of chicken liver or goat heart, pretend to reach into the patient's innards and withdraw the diseased tissue, which is then triumphantly displayed. Leaders of Western democracies regularly consult astrologers and mystics before making decisions of state. Under public pressure for results, police with an unsolved murder or a missing body on their hands consult ESP "experts" (who never guess better than expected by common sense, but the police, the ESPers say, keep calling). A clairvoyance gap with adversary nations is announced, and the Central Intelligence Agency, under Congressional prodding, spends tax money to find out whether submarines in the ocean depths can be located by thinking hard at them. A "psychic"—using pendulums over maps and dowsing rods in airplanes—purports to find new mineral deposits; an Australian mining company pays him top dollar up front, none of it returnable in the event of failure, and a share in the exploitation of ores in the event of success. Nothing is discovered. Statues of Jesus or murals of Mary are spotted with moisture, and thousands of kind-hearted people convince themselves that they have witnessed a miracle.

These are all cases of proved or presumptive baloney. A deception arises, sometimes innocently but collaboratively, sometimes with cynical premeditation. Usually the victim is caught up in a powerful emotion—wonder, fear, greed, grief. Credulous acceptance of baloney can cost you money; that's what P. T. Barnum meant when he said, "There's a sucker born every minute." But it can be much more dangerous than that, and when governments and societies lose the capacity for critical thinking, the results can be catastrophic—however sympathetic we may be to those who have bought the baloney.

In science we may start with experimental results, data, observations, measurements, "facts." We invent, if we can, a rich array of possible explanations and systematically confront each explanation with the facts. In the course of their training, scientists are equipped with a baloney detection kit. The kit is brought out as a matter of course whenever new ideas are offered for consideration. If the new idea survives examination by the tools in our kit, we grant it warm, although tentative, acceptance. If you're so inclined, if you don't want to buy baloney even when it's reassuring to do so, there are precautions that can be taken; there's a tried-and-true, consumer-tested method.

What's in the kit? Tools for skeptical thinking.

What skeptical thinking boils down to is the means to construct, and to understand, a reasoned argument and—especially important—to recognize a fallacious or fraudulent argument. The question is not whether we *like* the conclusion that emerges out of a train of reasoning, but whether the conclusion *follows* from the premise or starting point and whether that premise is true.

Among the tools:

• Wherever possible there must be independent confirmation of the "facts."

• Encourage substantive debate on the evidence by knowledgeable proponents of all points of view.

• Arguments from authority carry little weight—"authorities" have made mistakes in the past. They will do so again in the future. Perhaps a better way to say it is that in science there are no authorities; at most, there are experts.

• Spin more than one hypothesis. If there's something to be explained, think of all the different ways in which it *could* be explained. Then think of tests by which you might systematically disprove each of the alternatives. What survives, the hypothesis that resists disproof in this Darwinian selection among "multiple working hypotheses," has a much better chance of being the right answer than if you had simply run with the first idea that caught your fancy.*

• Try not to get overly attached to a hypothesis just because it's yours. It's only a way station in the pursuit of knowledge. Ask yourself why you like the idea. Compare it fairly with the alternatives. See if you can find reasons for rejecting it. If you don't, others will.

---

* This is a problem that affects jury trials. Retrospective studies show that some jurors make up their minds very early—perhaps during opening arguments—and then retain the evidence that seems to support their initial impressions and reject the contrary evidence. The method of alternative working hypotheses is not running in their heads.

• Quantify. If whatever it is you're explaining has some measure, some numerical quantity attached to it, you'll be much better able to discriminate among competing hypotheses. What is vague and qualitative is open to many explanations. Of course there are truths to be sought in the many qualitative issues we are obliged to confront, but finding *them* is more challenging.

• If there's a chain of argument, *every* link in the chain must work (including the premise)—not just most of them.

• Occam's Razor. This convenient rule-of-thumb urges us when faced with two hypotheses that explain the data *equally well* to choose the simpler.

• Always ask whether the hypothesis can be, at least in principle, falsified. Propositions that are untestable, unfalsifiable are not worth much. Consider the grand idea that our Universe and everything in it is just an elementary particle—an electron, say—in a much bigger Cosmos. But if we can never acquire information from outside our Universe, is not the idea incapable of disproof? You must be able to check assertions out. Inveterate skeptics must be given the chance to follow your reasoning, to duplicate your experiments and see if they get the same result.

The reliance on carefully designed and controlled experiments is key, as I tried to stress earlier. We will not learn much from mere contemplation. It is tempting to rest content with the first candidate explanation we can think of. One is much better than none. But what happens if we can invent several? How do we decide among them? We don't. We let experiment do it. Francis Bacon provided the classic reason:

> Argumentation cannot suffice for the discovery of new work, since the subtlety of Nature is greater many times than the subtlety of argument.

Control experiments are essential. If, for example, a new medicine is alleged to cure a disease 20 percent of the time, we must make sure that a control population, taking a dummy sugar pill which as far as the subjects know might be the new drug, does not also experience spontaneous remission of the disease 20 percent of the time.

Variables must be separated. Suppose you're seasick, and given both an acupressure bracelet and 50 milligrams of meclizine. You find the unpleasantness vanishes. What did it—the bracelet or the pill? You can tell only if you take the one without the other, next time you're seasick. Now imagine that you're not so dedicated to science as to be willing to be seasick. Then you won't separate the variables. You'll take both remedies again. You've achieved the desired practical result; further knowledge, you might say, is not worth the discomfort of attaining it.

Often the experiment must be done "double-blind," so that those hoping for a certain finding are not in the potentially compromising position of evaluating the results. In testing a new medicine, for example, you might want the physicians who determine which patients' symptoms are relieved not to know which patients have been given the new drug. The knowledge might influence their decision, even if only unconsciously. Instead the list of those who experienced remission of symptoms can be compared with the list of those who got the new drug, each independently ascertained. Then you can determine what correlation exists. Or in conducting a police lineup or photo identification, the officer in charge should not know who the prime suspect is, so as not consciously or unconsciously to influence the witness.

—

In addition to teaching us what to do when evaluating a claim to knowledge, any good baloney detection kit must also teach us what *not* to do. It helps us recognize the most common and perilous fallacies of logic and rhetoric. Many good examples can be found in religion and politics, because their practitioners are so often obliged to justify two contradictory propositions. Among these fallacies are:

- *ad hominem*—Latin for "to the man," attacking the arguer and not the argument (e.g., *The Reverend Dr. Smith is a known Biblical fundamentalist, so her objections to evolution need not be taken seriously*);
- argument from authority (e.g., *President Richard Nixon should be re-elected because he has a secret plan to end the war in Southeast Asia*—but because it was secret, there was no way for the electorate to evaluate it on its merits; the argument amounted to trusting him because he was President: a mistake, as it turned out);
- argument from adverse consequences (e.g., *A God meting out punishment and reward must exist, because if He didn't, society would be*

*much more lawless and dangerous—perhaps even ungovernable.** Or: *The defendant in a widely publicized murder trial must be found guilty; otherwise, it will be an encouragement for other men to murder their wives*);

• appeal to ignorance—the claim that whatever has not been proved false must be true, and vice versa (e.g., *There is no compelling evidence that UFOs are not visiting the Earth; therefore UFOs exist—and there is intelligent life elsewhere in the Universe. Or: There may be seventy kazillion other worlds, but not one is known to have the moral advancement of the Earth, so we're still central to the Universe.*) This impatience with ambiguity can be criticized in the phrase: absence of evidence is not evidence of absence.

• special pleading, often to rescue a proposition in deep rhetorical trouble (e.g., *How can a merciful God condemn future generations to torment because, against orders, one woman induced one man to eat an apple? Special plead: you don't understand the subtle Doctrine of Free Will. Or: How can there be an equally godlike Father, Son, and Holy Ghost in the same Person? Special plead: You don't understand the Divine Mystery of the Trinity. Or: How could God permit the followers of Judaism, Christianity, and Islam—each in their own way enjoined to heroic measures of loving kindness and compassion—to have perpetrated so much cruelty for so long? Special plead: You don't understand Free Will again. And anyway, God moves in mysterious ways.*)

• begging the question, also called assuming the answer (e.g., *We must institute the death penalty to discourage violent crime. But does the violent crime rate in fact fall when the death penalty is imposed? Or: The stock market fell yesterday because of a technical adjustment and profit-taking by investors*—but is there any *independent* evidence for the causal role of "adjustment" and profit-taking; have we learned anything at all from this purported explanation?);

• observational selection, also called the enumeration of favorable circumstances, or as the philosopher Francis Bacon described it,

---

* A more cynical formulation by the Roman historian Polybius:

Since the masses of the people are inconstant, full of unruly desires, passionate, and reckless of consequences, they must be filled with fears to keep them in order. The ancients did well, therefore, to invent gods, and the belief in punishment after death.

counting the hits and forgetting the misses* (e.g., *A state boasts of the Presidents it has produced, but is silent on its serial killers*);

• statistics of small numbers—a close relative of observational selection (e.g., *"They say 1 out of every 5 people is Chinese. How is this possible? I know hundreds of people, and none of them is Chinese. Yours truly."* Or: *"I've thrown three sevens in a row. Tonight I can't lose."*);

• misunderstanding of the nature of statistics (e.g., *President Dwight Eisenhower expressing astonishment and alarm on discovering that fully half of all Americans have below average intelligence*);

• inconsistency (e.g., *Prudently plan for the worst of which a potential military adversary is capable, but thriftily ignore scientific projections on environmental dangers because they're not "proved." Or: Attribute the declining life expectancy in the former Soviet Union to the failures of communism many years ago, but never attribute the high infant mortality rate in the United States (now highest of the major industrial nations) to the failures of capitalism. Or: Consider it reasonable for the Universe to continue to exist forever into the future, but judge absurd the possibility that it has infinite duration into the past*);

• *non sequitur*—Latin for "It doesn't follow" (e.g., *Our nation will prevail because God is great.* But nearly every nation pretends this to be true; the German formulation was *"Gott mit uns"*). Often those falling into the *non sequitur* fallacy have simply failed to recognize alternative possibilities;

---

* My favorite example is this story, told about the Italian physicist Enrico Fermi, newly arrived on American shores, enlisted in the Manhattan nuclear weapons Project, and brought face-to-face in the midst of World War II with U.S. flag officers:

> So-and-so is a great general, he was told.
> What is the definition of a great general? Fermi characteristically asked.
> I guess it's a general who's won many consecutive battles.
> How many?
> After some back and forth, they settled on five.
> What fraction of American generals are great?
> After some more back and forth, they settled on a few percent.

But imagine, Fermi rejoined, that there is no such thing as a great general, that all armies are equally matched, and that winning a battle is purely a matter of chance. Then the chance of winning one battle is one out of two, or 1/2; two battles 1/4, three 1/8, four 1/16, and five consecutive battles 1/32—which is about 3 percent. You would *expect* a few percent of American generals to win five consecutive battles—purely by chance. Now, has any of them won *ten* consecutive battles . . . ?

• *post hoc, ergo propter hoc*—Latin for "It happened after, so it was caused by" (e.g., Jaime Cardinal Sin, Archbishop of Manila: *"I know of . . . a 26-year-old who looks 60 because she takes [contraceptive] pills."* Or: *Before women got the vote, there were no nuclear weapons)*;

• meaningless question (e.g., *What happens when an irresistible force meets an immovable object?* But if there is such a thing as an irresistible force there can be no immovable objects, and vice versa);

• excluded middle, or false dichotomy—considering only the two extremes in a continuum of intermediate possibilities (e.g., *"Sure, take his side; my husband's perfect; I'm always wrong."* Or: *"Either you love your country or you hate it."* Or: *"If you're not part of the solution, you're part of the problem"*);

• short-term vs. long-term—a subset of the excluded middle, but so important I've pulled it out for special attention (e.g., *We can't afford programs to feed malnourished children and educate pre-school kids. We need to urgently deal with crime on the streets.* Or: *Why explore space or pursue fundamental science when we have so huge a budget deficit?*);

• slippery slope, related to excluded middle (e.g., *If we allow abortion in the first weeks of pregnancy, it will be impossible to prevent the killing of a full-term infant.* Or, conversely: *If the state prohibits abortion even in the ninth month, it will soon be telling us what to do with our bodies around the time of conception)*;

• confusion of correlation and causation (e.g., *A survey shows that more college graduates are homosexual than those with lesser education; therefore education makes people gay.* Or: *Andean earthquakes are correlated with closest approaches of the planet Uranus; therefore—despite the absence of any such correlation for the nearer, more massive planet Jupiter—the latter causes the former*)*;

• straw man—caricaturing a position to make it easier to attack (e.g.,

---

* Or: Children who watch violent TV programs tend to be more violent when they grow up. But did the TV cause the violence, or do violent children preferentially enjoy watching violent programs? Very likely both are true. Commercial defenders of TV violence argue that anyone can distinguish between television and reality. But Saturday morning children's programs now average 25 acts of violence per hour. At the very least this desensitizes young children to aggression and random cruelty. And if impressionable adults can have false memories implanted in their brains, what are we implanting in our children when we expose them to some 100,000 acts of violence before they graduate from elementary school?

Scientists suppose that living things simply fell together by chance—a formulation that willfully ignores the central Darwinian insight, that Nature ratchets up by saving what works and discarding what doesn't. Or—this is also a short-term/long-term fallacy—*environmentalists care more for snail darters and spotted owls than they do for people*);

• suppressed evidence, or half-truths (e.g., *An amazingly accurate and widely quoted "prophecy" of the assassination attempt on President Reagan is shown on television;* but—an important detail—was it recorded before or after the event? Or: *These government abuses demand revolution, even if you can't make an omelette without breaking some eggs.* Yes, but is this likely to be a revolution in which far more people are killed than under the previous regime? What does the experience of other revolutions suggest? Are all revolutions against oppressive regimes desirable and in the interests of the people?);

• weasel words (e.g., The separation of powers of the U.S. Constitution specifies that the United States may not conduct a war without a declaration by Congress. On the other hand, Presidents are given control of foreign policy and the conduct of wars, which are potentially powerful tools for getting themselves re-elected. Presidents of either political party may therefore be tempted to arrange wars while waving the flag and calling the wars something else—"police actions," "armed incursions," "protective reaction strikes," "pacification," "safeguarding American interests," and a wide variety of "operations," such as "Operation Just Cause." Euphemisms for war are one of a broad class of reinventions of language for political purposes. Talleyrand said, "An important art of politicians is to find new names for institutions which under old names have become odious to the public").

Knowing the existence of such logical and rhetorical fallacies rounds out our toolkit. Like all tools, the baloney detection kit can be misused, applied out of context, or even employed as a rote alternative to thinking. But applied judiciously, it can make all the difference in the world—not least in evaluating our own arguments before we present them to others.

———

The American tobacco industry grosses some $50 billion per year. There is a statistical correlation between smoking and cancer, the tobacco industry admits, but not, they say, a causal relation. A logical fallacy, they imply, is being committed. What might this mean? Maybe people with hereditary propensities for cancer also have hereditary propensities to take addictive drugs—so cancer and smoking might be correlated, but the cancer would not be caused by the smoking. Increasingly farfetched connections of this sort can be contrived. This is exactly one of the reasons science insists on control experiments.

Suppose you paint the backs of large numbers of mice with cigarette tar, and also follow the health of large numbers of nearly identical mice that have not been painted. If the former get cancer and the latter do not, you can be pretty sure that the correlation is causal. Inhale tobacco smoke, and the chance of getting cancer goes up; don't inhale, and the rate stays at the background level. Likewise for emphysema, bronchitis, and cardiovascular diseases.

When the first work was published in the scientific literature in 1953 showing that the substances in cigarette smoke when painted on the backs of rodents produce malignancies, the response of the six major tobacco companies was to initiate a public relations campaign to impugn the research, sponsored by the Sloan Kettering Foundation. This is similar to what the Du Pont Corporation did when the first research was published in 1974 showing that their Freon product attacks the protective ozone layer. There are many other examples.

You might think that before they denounce unwelcome research findings, major corporations would devote their considerable resources to checking out the safety of the products they propose to manufacture. And if they missed something, if independent scientists suggest a hazard, why would the companies protest? Would they rather kill people than lose profits? If, in an uncertain world, an error must be made, shouldn't it be biased toward protecting customers and the public? And, incidentally, what do these cases say about the ability of the free enterprise system to police itself? Aren't these instances where at least some government intrusion is in the public interest?

A 1971 internal report of the Brown and Williamson Tobacco Corporation lists as a corporate objective "to set aside in the minds of millions the false conviction that cigarette smoking causes lung cancer and other diseases; a conviction based on fanatical assumptions, fallacious rumors, unsupported claims and the unscientific statements and conjectures of publicity-seeking opportunists." They complain of the incredible, unprecedented and nefarious attack against the cigarette, constituting the greatest libel and slander ever perpetrated against any product in the history of free enterprise; a criminal libel of such major proportions and implications that one wonders how such a crusade of calumny can be reconciled under the Constitution can be so flouted and violated [sic].

This rhetoric is only slightly more inflamed than what the tobacco industry has from time to time uttered for public consumption.

There are many brands of cigarettes that advertise low "tar" (ten milligrams or less per cigarette). Why is this a virtue? Because it is the refractory tars in which polycyclic aromatic hydrocarbons and some other carcinogens are concentrated. Aren't the low-tar ads a tacit admission by the tobacco companies that cigarettes indeed cause cancer?

Healthy Buildings International is a for-profit organization, recipient of millions of dollars over the years from the tobacco industry. It performs research on second-hand smoke, and testifies for the tobacco companies. In 1994, three of its technicians complained that senior executives had faked data on inhalable cigarette particles in the air. In every case, the invented or "corrected" data made tobacco smoke seem safer than the technicians' measurements had indicated. Do corporate research departments or outside research contractors ever find a product to be more dangerous than the tobacco corporation has publicly declared? If they do, is their employment continued?

Tobacco is addictive; by many criteria more so than heroin and cocaine. There was a reason people would, as the 1940s ad put it, "walk a mile for a Camel." More people have died of tobacco than in all of World War II. According to the World Health Organization, smoking kills three million people every year worldwide. This will rise to ten million annual deaths by 2020—in part because of a massive advertising campaign to portray smoking as advanced and fashionable to young women in the developing world. Part of the success of the tobacco industry in purveying this brew of addictive poisons can be attributed to widespread unfamiliarity with baloney detection, critical thinking, and the scientific method. Gullibility kills.

Chapter 13

OBSESSED
WITH REALITY

A shipowner was about to send to sea an emigrant ship. He knew that she was old, and not overwell built at the first; that she had seen many seas and climes, and often had needed repairs. Doubts had been suggested to him that possibly she was not seaworthy. These doubts preyed upon his mind, and made him unhappy; he thought that perhaps he ought to have her thoroughly overhauled and refitted, even though this should put him to great expense. Before the ship sailed, however, he succeeded in overcoming these melancholy reflections. He said to himself that she had gone safely through so many voyages and weathered so many storms, that it was idle to suppose that she would not come safely home from this trip also. He would put his trust in Providence, which could hardly fail to protect all these unhappy families that were leaving their fatherland to seek for better times elsewhere. He would dismiss from his mind all ungenerous suspicions about the honesty of builders and contractors. In such ways he acquired a sincere and comfortable conviction that his vessel was thoroughly safe and seaworthy; he watched her departure with a light heart, and benevolent wishes for the success of the exiles in their strange new home that was to be; and he got his insurance money when she went down in mid-ocean and told no tales.

What shall we say of him? Surely this, that he was verily guilty of the death of those men. It is admitted that he did sincerely believe in the soundness of his ship; but the sincerity of his conviction can in nowise help him, because *he had no right to believe on such evidence as was before him.* He had acquired his belief not by honestly earning it in patient investigation, but by stifling his doubts. . .

WILLIAM K. CLIFFORD,
The Ethics of Belief
(1874)

At the borders of science—and sometimes as a carry-over from prescientific thinking—lurks a range of ideas that are appealing, or at least modestly mind-boggling, but that have not been conscientiously worked over with a baloney detection kit, at least by their advocates: the notion, say, that the Earth's surface is on the inside, not the outside, of a sphere; or claims that you can levitate yourself by meditating and that ballet dancers and basketball players routinely get up so high by levitating; or the proposition that I have something called a soul, made not of matter or energy, but of something else for which there is no other evidence, and which after my death might return to animate a cow or a worm.

Typical offerings of pseudoscience and superstition—this is merely a representative, not a comprehensive list—are astrology; the Bermuda Triangle; "Big Foot" and the Loch Ness monster; ghosts; the "evil eye"; multicolored halolike "auras" said to surround the heads of everyone (with colors personalized); extrasensory perception (ESP), such as telepathy, precognition, telekinesis, and "remote viewing" of distant places; the belief that 13 is an "unlucky" number (because of which many no-nonsense office buildings and hotels in America pass directly from the 12th to the 14th floors—why take chances?); bleeding statues; the conviction that carrying the severed foot of a rabbit around with you brings good luck; divining rods, dowsing, and water witching; "facilitated communication" in autism; the belief that razor blades stay sharper when kept inside small cardboard pyramids, and other tenets of "pyramidology"; phone calls (none of them collect) from the dead; the prophecies of Nostradamus; the alleged discovery that untrained flatworms can learn a task by eating the ground-up remains of other, better educated flatworms; the notion that more crimes are committed when the Moon is full; palmistry; numerology; polygraphy; comets, tea leaves, and "monstrous" births as harbingers of future events (plus the divinations fashionable in earlier epochs, accomplished by viewing entrails, smoke, the shapes of flames, shadows, and excrement; listening to gurgling stomachs; and even, for a brief period, examining tables of logarithms); "photography" of past events, such as the crucifixion of Jesus; a Russian elephant that speaks fluently; "sensitives" who, when carelessly blindfolded, read books with their fingertips; Edgar Cayce (who predicted that in the 1960s the "lost" continent of Atlantis would "rise") and other "prophets," sleeping and awake; diet quackery; out-of-body (e.g., near-death) experiences interpreted as real events in the external world; faith-healer fraud; Ouija boards; the emotional lives of geraniums, uncovered by intrepid use of a "lie detector"; water remembering what molecules used to be dissolved in it; telling character from facial features or bumps on the head; the "hundredth monkey" confusion and other claims that whatever a small fraction of us wants to be true really is true; human beings spontaneously bursting into flame and being burned to a crisp; 3-cycle biorhythms; perpetual motion machines, promising unlimited supplies of energy (but all of which, for one reason or another, are withheld from close examination by skeptics); the systematically inept predictions of Jeane Dixon (who "predicted" a 1953 Soviet invasion of Iran and in 1965 that the USSR would beat the U.S. to put the first human on the Moon*) and other professional "psychics"; the Jehovah's Witnesses' prediction that the world would end in 1917, and many similar prophecies; dianetics and Scientology; Carlos Castaneda and "sorcery"; claims of finding the remains of Noah's Ark; the "Amityville Horror" and other hauntings; and accounts of a small brontosaurus crashing through the rain forests of the Congo Republic in our time. [An in-depth discussion of many such claims can be found in *Encyclopedia of the Paranormal*, Gordon Stein, ed., Buffalo: Prometheus Books, 1996.]

Many of these doctrines are rejected out of hand by fundamentalist Christians and Jews because the Bible so enjoins. Deuteronomy (18:10,11) reads (in the King James translation):

> There shall not be found among you any one that maketh his son
> or his daughter to pass through the fire, or that useth divination, or

---

* Violating the rules for "Oraclers and Wizards" given by Thomas Ady in 1656: "In doubtful things, they gave doubtful answers . . . Where were more certain probabilities, there they gave more certain answers."

an observer of times, or an enchanter, or a witch. Or a charmer, or
a consulter with familiar spirits, or a wizard, or a necromancer.

Astrology, channeling, Ouija boards, predicting the future, and much
else is forbidden. The author of Deuteronomy does not argue that
such practices fail to deliver what they promise. But they are "abominations"—perhaps suitable for other nations, but not for the followers
of God. And even the Apostle Paul, so credulous on so many matters,
counsels us to "prove all things."

The twelfth century Jewish philosopher Moses Maimonides goes
further than Deuteronomy, in that he makes explicit that these pseudosciences don't work:

> It is forbidden to engage in astrology, to utilize charms, to whisper
> incantations. . . All of these practices are nothing more than lies
> and deceptions used by ancient pagan peoples to deceive the
> masses and lead them astray. . . Wise and intelligent people know
> better. [From the *Mishneh Torah, Avodah Zara,* Chapter 11.]

Some claims are hard to test—for example, if an expedition fails to
find the ghost or the brontosaurus, that doesn't mean it doesn't exist.
Absence of evidence is not evidence of absence. Others are easier—for
example, flatworm cannibalistic learning or the announcement that
colonies of bacteria subjected to an antibiotic on an agar dish thrive
when their prosperity is prayed for (compared to control bacteria unre-
deemed by prayer). A few—for example, perpetual motion machines—can be excluded on grounds of fundamental physics. Except
for them, it's not that we know *before* examining the evidence that the
notions are false; stranger things are routinely incorporated into the
corpus of science.

The question, as always, is how good is the evidence? The burden
of proof surely rests on the shoulders of those who advance such
claims. Revealingly, some proponents hold that skepticism is a liabil-
ity, that true science is inquiry *without* skepticism. They are perhaps
halfway there. But halfway doesn't do it.

Parapsychologist Susan Blackmore describes one of the steps in
her transformation to a more skeptical attitude on "psychic" phenomena:

A mother and daughter from Scotland asserted they could pick up images from each other's minds. They chose to use playing cards for the tests because that is what they used at home. I let them choose the room in which they would be tested and insured that there was no normal way for the "receiver" to see the cards. They failed. They could not get more right than chance predicted and they were terribly disappointed. They had honestly believed they could do it and I began to see how easy it is to be fooled by your own desire to believe.

I had similar experiences with several dowsers, children who claimed they could move objects psychokinetically, and several who said they had telepathic powers. They all failed. Even now I have a five-digit number, a word, and a small object in my kitchen at home. The place and items were chosen by a young man who intends to "see" them while traveling out of his body. They have been there (though regularly changed) for three years. So far, though, he has had no success.

"Telepathy" literally means to feel at a distance—just as "telephone" is to hear at a distance and "television" is to see at a distance. The word suggests the communication not of thoughts but of feelings, emotions. Around a quarter of all Americans believe they've experienced something like telepathy. People who know each other very well, who live together, who are practiced in one another's feeling tones, associations, and thinking styles can often anticipate what the partner will say. This is merely the usual five senses plus human empathy, sensitivity, and intelligence in operation. It may feel extrasensory, but it's not at all what's intended by the word "telepathy." If something like this *were* ever conclusively demonstrated, it would, I think, have discernible physical causes—perhaps electrical currents in the brain. Pseudoscience, rightly or wrongly labeled, is by no means the same thing as the supernatural, which is by definition something somehow outside of Nature.

It is barely possible that a few of these paranormal claims might one day be verified by solid scientific data. But it would be foolish to accept any of them without adequate evidence. In the spirit of garage dragons, it is much better, for those claims not already disproved or adequately explained, to contain our impatience, to nurture a tolerance for ambiguity, and to await—or, much better, to seek—supporting or disconfirming evidence.

—

In a far-off land in the South Seas, the word went out about a wise man, a healer, an embodied spirit. He could speak across time. He was an Ascended Master. He was coming, they said. He was coming. . . .

In 1988, Australian newspapers, magazines, and television stations began to receive the good news via press kits and videotape. One broadside read:

CARLOS
TO APPEAR IN AUSTRALIA

Those who have seen it will never forget. The brilliant young artist who has been talking to them suddenly seems to falter, his pulse slows dangerously and virtually stops at the point of death. The qualified medical attendant, who has been assigned to keep constant watch, is about to sound the alarm.

But then, with a heart-stirring burst, the pulse is felt again—faster and stronger than ever before. The life force clearly has returned to the body—but the entity inside that body is no longer Jose Luis Alvarez, the 19-year-old whose unique painted ceramics are featured in some of the wealthiest homes in America. Instead, the body has been taken over by Carlos, an ancient soul, whose teachings will come as both a shock and an inspiration. One being going through a form of death to make way for another: that is the phenomenon that has made Carlos, as channelled through Jose Luis Alvarez, the dominant new figure in New Age consciousness. As even one sceptical New York critic puts it: "The first and only case of a channeller offering tangible, physical proof of some mysterious change within his human physiology."

Now Jose, who has gone through more than 170 of these little deaths and transformations, has been told by Carlos to visit Australia—in the words of the Master, "the old new land" which is to be the source of a special revelation. Carlos already has foretold that in 1988 catastrophes will sweep the earth, two major world leaders will die and, later in the year, Australians will be among the first to see the rising of a great star which will deeply influence future life on earth.

SUNDAY 21ST

—3 PM—

OPERA HOUSE

DRAMA THEATRE

Following a 1986 motorcycle accident, the press kit explained, Jose Alvarez—then 17 years old—suffered a mild concussion. After he recovered, those who knew him could tell that he had changed. A very different voice sometimes emanated from him. Bewildered, Alvarez sought help from a psychotherapist, a specialist in multiple-personality disorders. The psychiatrist "discovered that Jose was channelling a distinct entity who was known as Carlos. This entity takes over the body of Alvarez when the body's life force is relaxed to the right degree." Carlos, it turns out, is a two-thousand-year-old spirit disincarnate, a ghost without bodily form, who last invaded a human body in Caracas, Venezuela in 1900. Unfortunately, that body died at age 12 in a fall from a horse. This may be why, the therapist explained, Carlos could enter Alvarez's body following the motorcycle accident. When Alvarez goes into his trance, the spirit of Carlos, focused by a large and rare crystal, enters him and utters the wisdom of the ages.

Included in the press kit was a list of major appearances in American cities, a videotape of the tumultuous reception that Alvarez/Carlos received at a Broadway theater, his interview on New York radio station WOOP, and other indications that here was a formidable American New Age phenomenon. Two small substantiating details: An article from a South Florida newspaper read, "THEATER NOTE: The three-day stay of channeler CARLOS has been extended at the War Memorial Auditorium ... in response to the requests for further appearances," and an excerpt from a television program guide listed a special on "THE ENTITY CARLOS: This in-depth study reveals the facts behind one of today's most popular and controversial personalities."

Alvarez and his manager arrived in Sydney first class on Qantas. They traveled everywhere in an enormous white stretch limousine. They occupied the Presidential Suite of one of the city's most prestigious hotels. Alvarez was attired in an elegant white gown with a golden medallion. In his first press conference, Carlos quickly emerged. The entity was forceful, literate, commanding. Australian television programs quickly lined up for appearances by Alvarez, his manager, and his nurse (to check his pulse and announce the presence of Carlos).

On Australia's *Today Show*, they were interviewed by the host, George Negus. When Negus posed a few reasonable and skeptical questions, the New Agers exhibited very thin skins. Carlos laid a curse on the anchorman. His manager doused Negus with a glass of water. Both stalked off the set. It was a sensation in the tabloid press, its significance rehashed on Australian television. "TV Outburst: Water Thrown at Negus" was the front-page headline in the February 16, 1988 *Daily Mirror*. Television stations were flooded with calls. One Sydney citizen advised taking the curse on Negus very seriously: The army of Satan had already assumed control of the United Nations, he said, and Australia might be next.

Carlos's next appearance was on the Australian version of *A Current Affair*. A skeptic was brought in who described a magician's trick by which the pulse in one hand is made briefly to stop: You put a rubber ball in your armpit and squeeze. When Carlos's authenticity was questioned, he was outraged: "This interview is terminated!" he thundered.

On the appointed day, the Drama Theatre of the Sydney Opera House was nearly filled. An excited crowd, young and old, milled about expectantly. Entrance was free—which reassured those who vaguely wondered if it might be some sort of scam. Alvarez seated himself on a low couch. His pulse was monitored. Suddenly it stopped. Seemingly, he was near death. Low, guttural noises emanated from deep within him. The audience gasped in wonder and awe. Suddenly, Alvarez's body took on power. His posture radiated confidence. A broad, humane, spiritual perspective flowed out of Alvarez's mouth. Carlos was here! Interviewed afterwards, many members of the audience described how they had been moved and delighted.

The following Sunday, Australia's most popular TV program—named *Sixty Minutes* after its American counterpart—revealed that the Carlos affair was a hoax, front to back. The producers thought it would be instructive to explore how easily a faith healer or guru could be created to bamboozle the public and the media. So naturally, they contacted one of the world's leading experts on deceiving the public (at least among those not holding or advising political office)—the magician James Randi.

—

"[T]here being so many disorders which cure themselves and such a disposition in mankind to deceive themselves and one another"— wrote Benjamin Franklin in 1784—

> ... and living long having given me frequent opportunities of seeing certain remedies cried up as curing everything, and yet soon after totally laid aside as useless, I cannot but fear that the expectation of great advantage from the new method of treating diseases will prove a delusion. That delusion may however in some cases be of use while it lasts.

He was referring to mesmerism. But "every age has its peculiar folly."

Unlike Franklin, most scientists feel it's not their job to expose pseudoscientific bamboozles—much less, passionately held self-deceptions. They tend not to be very good at it either. Scientists are used to struggling with Nature, who may surrender her secrets reluctantly but who fights fair. Often they are unprepared for those unscrupulous practitioners of the "paranormal" who play by different rules. Magicians, on the other hand, are in the deception business. They practice one of the many occupations—such as acting, advertising, bureaucratic religion, and politics—where what a naïve observer might misunderstand as lying is socially condoned as in the service of a higher good. Many magicians pretend they don't cheat, and hint at powers conferred by mystic sources or, lately, by alien largesse. Some use their knowledge to expose charlatans in and out of their ranks. A thief is set to catch a thief.

Few rise to this challenge as energetically as James "The Amazing" Randi, accurately self-described as an angry man. He is angry not so much about the survival into our day of antediluvian mysticism and superstition, but about how uncritical acceptance of mysticism and superstition works to defraud, to humiliate, and sometimes even to kill. Like all of us, he is imperfect: Sometimes Randi is intolerant and condescending, lacking in empathy for the human frailties that underlie credulity. He is routinely paid for his speeches and performances, but nothing compared to what he could receive if he declared that his tricks derived from psychic powers or divine or extraterrestrial influences. (Most professional conjurors, worldwide, seem to believe in the reality of psychic phenomena—according to polls of their views.) As a conjuror, he has done much to expose remote viewers, "telepaths," and faith healers who have bilked the public. He demonstrated the simple deceptions and misdirections by which psychic spoonbenders had conned prominent theoretical physicists into deducing new physical phenomena. He has received wide recognition among scientists and is a recipient of the MacArthur Foundation (so-called "genius") Prize Fellowship. One critic castigated him for being "obsessed with reality." I wish the same could be said of our nation and our species.

Randi has done more than anyone else in recent times to expose pretension and fraud in the lucrative business of faith healing. He sifts refuse. He reports gossip. He listens in on the stream of "miraculous" information coming to the itinerant healer—not by spiritual inspiration from God, but at the radio frequency of 39.17 megahertz, transmitted by his wife backstage.* He discovers that those who rise from their wheelchairs and are declared healed had never before been confined to wheelchairs—they were invited by an usher to sit in them. He challenges the faith healers to provide serious medical evidence for the validity of their claims. He invites local and federal government agencies to enforce the laws against fraud and medical malpractice. He chastises the news media for their studied avoidance of the issue. He exposes the profound contempt of these faith healers for their patients and parishioners. Many are conscious charlatans—using Christian evangelical or New Age language and symbols to prey on human frailty. Perhaps there are some with motives that are not venal.

Or am I being too harsh? How is the occasional charlatan in faith-healing different from the occasional fraud in science? Is it fair to be suspicious of an entire profession because of a few bad apples? There are at least two important differences, it seems to me. First, no one

---

* Whose minions had interviewed the gullible patients only an hour or two earlier. How, except through God, could the preacher know their symptoms and street addresses? This scam by the Christian fundamentalist faith-healer Peter Popoff, and exposed by Randi, was thinly fictionalized in the 1993 film *Leap of Faith*.

doubts that science actually works, whatever mistaken and fraudulent claim may from time to time be offered. But whether there are *any* "miraculous" cures from faith-healing, beyond the body's own ability to cure itself, is very much at issue. Secondly, the exposé of fraud and error in science is made almost exclusively by science. The discipline polices itself—meaning that scientists are aware of the potential for charlatanry and mistakes. But the exposure of fraud and error in faith-healing is almost never done by other faith-healers. Indeed, it is striking how reluctant the churches and synagogues are in condemning demonstrable deception in their midst.

When conventional medicine fails, when we must confront pain and death, of course we are open to other prospects for hope. And, after all, some illnesses are psychogenic. Many can be at least ameliorated by a positive cast of mind. Placebos are dummy drugs, often sugar pills. Drug companies routinely compare the effectiveness of their drugs against placebos given to patients with the same disease who had no way to tell the difference between the drug and the placebo. Placebos can be astonishingly effective, especially for colds, anxiety, depression, pain, and symptoms that are plausibly generated by the mind. Conceivably, endorphins—the small brain proteins with morphinelike effects—can be elicited by belief. A placebo works only if the patient believes it's an effective medicine. Within strict limits, hope, it seems, can be transformed into biochemistry.

As a typical example, consider the nausea and vomiting that frequently accompany the chemotherapy given to cancer and AIDS patients. Nausea and vomiting can also be caused psychogenically—for instance by fear. The drug ondansetron hydrochloride greatly reduces the incidence of these symptoms; but is it actually the drug or the expectation of relief? In a double-blind study 96 percent of the patients rated the drug effective. So did 10 percent of the patients taking an identical looking placebo.

In an application of the fallacy of observational selection, unanswered prayers may tend to be forgotten or dismissed. There is a real toll, though: Some patients who are not cured by faith reproach themselves—perhaps it's their own fault, perhaps they didn't believe hard enough. Skepticism, they are rightly told, is an impediment both to faith and to (placebo) healing.

Nearly half of all Americans believe there is such a thing as psychic or spiritual healing. Miraculous cures have been associated with a wide variety of healers, real and imagined, throughout human history. Scrofula, a kind of tuberculosis, was in England called the "King's evil," and was supposedly curable only by the King's touch. Victims patiently lined up to be touched; the monarch briefly submitted to another burdensome obligation of high office, and—despite no one, it seems, actually being cured—the practice continued for centuries.

A famous Irish faith healer of the seventeenth century was Valentine Greatraks. He found, somewhat to his surprise, that he had the power to cure disease, including colds, ulcers, "soreness," and epilepsy. The demand for his services became so great that he had no time for anything else. He was *forced* to become a healer, he complained. His method was to cast out the demons responsible for disease. All diseases, he asserted, were caused by evil spirits—many of whom he recognized and called by name. A contemporary chronicler, cited by Mackay, noted that

> he boasted of being much better acquainted with the intrigues of demons than he was with the affairs of men. . . So great was the confidence in him, that the blind fancied they saw the light which they did not see—the deaf imagined that they heard—the lame that they walked straight, and the paralytic that they had recovered the use of their limbs. An idea of health made the sick forget for awhile their maladies; and imagination, which was not less active in those merely drawn by curiosity than in the sick, gave a false view to the one class, from the desire of seeing, as it operated a false cure on the other from the strong desire of being healed.

There are countless reports in the world literature of exploration and anthropology not only of sicknesses being cured by faith in the healer, but also of people wasting away and dying when cursed by a sorcerer. A more or less typical example is told by Alvar Nuñez Cabeza de Vaca, who with a few companions and under conditions of terrible privation wandered on land and sea, from Florida to Texas to Mexico in 1528–1536. The many different communities of Native Americans he met longed to believe in the supernatural healing powers of the strange light-skinned, black-bearded foreigners and their black-skinned companion from Morocco, Estebanico. Eventually whole villages came out to meet them, depositing all their wealth at the feet of the Spaniards and humbly imploring cures. It began modestly enough:

> [T]hey tried to make us into medicine men, without examining us or asking for credentials, for they cure illnesses by blowing on the sick person . . . and they ordered us to do the same and be of some use. . . The way in which we cured was by making the sign of the cross over them and blowing on them and reciting a Pater Noster and an Ave Maria. . . [A]s soon as we made the sign of the cross over them, all those for whom we prayed told the others that they were well and healthy. . .

Soon they were curing cripples. Cabeza de Vaca reports he raised a man from the dead. After that,

> we were very much hampered by the large number of people who were following us . . . their eagerness to come and touch us was very great and their importunity so extreme that three hours would pass without our being able to persuade them to leave us alone.

When a tribe begged the Spaniards not to leave them, Cabeza de Vaca and his companions became angry. Then,

> a strange thing happened. . . [M]any of them fell ill, and eight men died the next day. All over the land, in the places where this became known, they were so afraid of us that it seemed that the very sight of us made them almost die of fear.
>
> They implored us not to be angry, nor to wish for any more of them to die; and they were altogether convinced that we killed them simply by wishing to.

In 1858, an apparition of the Virgin Mary was reported in Lourdes, France; the Mother of God confirmed the dogma of her immaculate conception which had been proclaimed by Pope Pius IX just four years earlier. Something like a hundred million people have come to Lourdes since then in the hope of being cured, many with illnesses that the medicine of the time was helpless to defeat. The Roman

Catholic Church rejected the authenticity of large numbers of claimed miraculous cures, accepting only 65 in nearly a century and a half (of tumors, tuberculosis, opthalmitis, impetigo, bronchitis, paralysis and other diseases, but not, say, the regeneration of a limb or a severed spinal cord). Of the 65, women outnumber men ten to one. The odds of a miraculous cure at Lourdes, then, are about one in a million; you are roughly as likely to recover after visiting Lourdes as you are to win the lottery, or to die in the crash of a regularly scheduled airplane flight—including the one taking you to Lourdes.

The spontaneous remission rate of all cancers, lumped together, is estimated to be something between one in ten thousand and one in a hundred thousand. If no more than 5 percent of those who come to Lourdes were there to treat their cancers, there should have been something between 50 and 500 "miraculous" cures of cancer alone. Since only three of the attested 65 cures are of cancer, the rate of spontaneous remission at Lourdes seems to be lower than if the victims had just stayed at home. Of course, if you're one of the 65, it's going to be very hard to convince you that your trip to Lourdes wasn't the cause of the remission of your disease. . . *Post hoc, ergo propter hoc.* Something similar seems true of individual faith healers.

After hearing much from his patients about alleged faith healing, a Minnesota physician named William Nolen spent a year and a half trying to track down the most striking cases. Was there clear medical evidence that the disease was really present before the "cure"? If so, had the disease *actually* disappeared after the cure, or did we just have the healer's or the patient's say-so? He uncovered many cases of fraud, including the first exposure in America of "psychic surgery." But he found not one instance of cure of any serious organic (non-psychogenic) disease. There were no cases where gallstones or rheumatoid arthritis, say, were cured, much less cancer or cardiovascular disease. When a child's spleen is ruptured, Nolen noted, perform a simple surgical operation and the child is completely better. But take that child to a faith healer and she's dead in a day. Dr. Nolen's conclusion:

> When [faith] healers treat serious organic disease, they are responsible for untold anguish and unhappiness. . . The healers become killers.

Even a recent book advocating the efficacy of prayer in treating disease (Larry Dossey, *Healing Words*) is troubled by the fact that some diseases are more easily cured or mitigated than others. If prayer works, why can't God cure cancer or grow back a severed limb? Why so much avoidable suffering that God could so readily prevent? Why does God have to be prayed to at all? Doesn't He already know what cures need to be performed? Dossey also begins with a quote from Stanley Krippner, M.D. (described as "one of the most authoritative investigators of the variety of unorthodox healing methods used around the world"):

> [T]he research data on distant, prayer-based healing are promising, but too sparse to allow any firm conclusion to be drawn.

This after many trillions of prayers over the millennia.

As Cabeza de Vaca's experience suggests, the mind can *cause* certain diseases, even fatal diseases. When blindfolded patients are deceived into believing they're being touched by a leaf such as poison ivy or poison oak, they produce an ugly red contact dermatitis. What faith healing characteristically may help are mind-mediated or placebo diseases: some back and knee pains, headaches, stuttering, ulcers, stress, hay fever, asthma, hysterical paralysis and blindness, and false pregnancy (with cessation of menstrual periods and abdominal swelling). These are all diseases in which the state of mind may play a key role. In the late medieval cures associated with apparitions of the Virgin Mary, most were of sudden, short-lived, whole-body or partial paralyses that are plausibly psychogenic. It was widely held, moreover, that only devout believers could be so cured. It's no surprise that appeals to a state of mind called faith can relieve symptoms caused, at least in part, by another, perhaps not very different, state of mind.

But there's something more: The Harvest Moon festival is an important holiday in traditional Chinese communities in America. In the week preceding the festival, the death rate in the community is found to fall by 35 percent. In the following week the death rate jumps by 35 percent. Control groups of non-Chinese show no such effect. You might think that suicides are responsible, but only deaths from natural causes are counted. You might think that stress or overeating might account for it, but this could hardly explain the fall in death rate before the harvest moon. The largest effect is for people with cardiovascular disease, which is known to be influenced by stress. Cancer showed a smaller effect. On more detailed study, it turned out that the fluctuations in death rate occurred exclusively among women 75 years old or older. The Harvest Moon Festival is presided over by the oldest women in the households. They were able to stave off death for a week or two to perform their ceremonial responsibilities. A similar effect is found among Jewish men in the weeks centered on Passover—a ceremony in which older men play a leading role—and likewise, worldwide for birthdays, graduation ceremonies and the like.

In a more controversial study, Stanford University psychiatrists divided 86 women with metastatic breast cancer into two groups—one in which they were encouraged to examine their fears of dying and to take charge of their lives, and the other given no special psychiatric support. To the surprise of the researchers, not only did the support group experience less pain, but they also lived longer—on average, 18 months longer.

The leader of the Stanford study, David Spiegel, speculates that the cause may by cortisol and other "stress hormones" which impair the body's protective immune system. Severely depressed people, students during exam periods, and the bereaved all have reduced white blood cell counts. Good emotional support may not have much effect on advanced forms of cancer, but it may work to reduce the chances of secondary infections in a person already much weakened by the disease or its treatment.

In his nearly forgotten 1903 book *Christian Science*, Mark Twain wrote

> The power which a man's imagination has over his body to heal it or make it sick is a force which none of us is born without. The first man had it, the last one will possess it.

Occasionally, some of the pain and anxiety, or other symptoms, of more serious diseases can be relieved by faith healers—however, without arresting the progress of the disease. But this is no small benefit. Faith and prayer may be able to relieve some symptoms of disease and their treatment, ease the suffering of the afflicted, and even prolong lives a little. In assessing the religion called Christian Science, Mark

Twain—its severest critic of the time—nevertheless allowed that the bodies and lives it had "made whole" by the power of suggestion more than compensated for those it had killed by withholding medical treatment in favor of prayer.

After his death, assorted Americans reported contact with the ghost of President John F. Kennedy. Before home shrines bearing his picture, miraculous cures began to be reported. "He gave his life for his people," one adherent of this stillborn religion explained. According to the *Encyclopedia of American Religions*, "To believers, Kennedy is thought of as a god." Something similar can be seen in the Elvis Presley phenomenon, and the heartfelt cry: "The King lives." If such belief systems could arise spontaneously, think how much more could be done by a well-organized, and especially an unscrupulous, campaign.

—

In response to their inquiry, Randi suggested to Australia's *Sixty Minutes* that they generate a hoax from scratch—using someone with no training in magic or public speaking, and no experience on the pulpit. As he was thinking the scam through, his eye fell upon Jose Luis Alvarez, a young performance sculptor who was Randi's tenant. Why not? answered Alvarez, who when I met him seemed bright, good-humored, and thoughtful. He went through intensive training, including mock TV appearances and press conferences. He didn't have to think up the answers, though because he had a nearly invisible radio receiver in his ear, through which Randi prompted. Emissaries from *Sixty Minutes* checked Alvarez's performance. The Carlos persona was Alvarez's invention.

When Alvarez and his "manager"—likewise recruited for the job with no previous experience—arrived in Sydney, there was James Randi, slouching and inconspicuous, whispering into his transmitter, at the periphery of the action. The substantiating documentation had all been faked. The curse, the water-throwing, and all the rest were rehearsed to attract media attention. They did. Many of the people who showed up at the Opera House had done so because of the television and press attention. One Australian newspaper chain even printed verbatim handouts from the "Carlos Foundation."

After *Sixty Minutes* aired, the rest of the Australian media was furious. They had been used, they complained, lied to. "Just as there are legal guidelines concerning the police use of provocateurs," thundered Peter Robinson in the *Australian Financial Review*,

> there must be limits to how far the media can go in setting up a misleading situation. . . I, for one, can simply not accept that telling a lie is an acceptable way of reporting the truth. . . Every poll of public opinion shows that there is a suspicion among the general public that the media do not tell the whole truth, or that they distort things, or that they exaggerate, or that they are biased.

Mr. Robinson feared that Carlos might have lent credence to this widespread misperception. Headlines ranged from "How Carlos Made Fools of Them All" to "Hoax Was Just Dumb." Newspapers that had not trumpeted Carlos patted themselves on the back for their restraint. Negus said of *Sixty Minutes*, "Even people of integrity can make mistakes," and denied being duped. Anyone calling himself a channeler, he said, is "a fraud by definition."

Sixty Minutes and Randi stressed that the Australian media had made no serious effort to check any of "Carlos's " bona fides. He had never appeared in any of the cities listed. The videotape of Carlos on the stage of a New York theater had been a favor granted by the magicians Penn and Teller, who were appearing there. They asked the audience just to give a big hand of applause; Alvarez, in smock and medallion, walked on; the audience dutifully applauded, Randi got his videotape, Alvarez waved goodbye, the show went on. And there is no New York City radio station with call letters WOOP.

Other reasons for suspicion could readily be mined in Carlos's writings. But because the intellectual currency has been so debased, because credulity—New Age and Old—is so rampant, because skeptical thinking is so rarely practiced, no parody is too implausible. The Carlos Foundation offered for sale (they were scrupulously careful not actually to sell anything) an "ATLANTIS CRYSTAL":

> Five of these unique crystals have so far been found by the ascended master during his travels. Unexplained by science, each crystal harnesses almost pure energy. . . [and has] enormous healing powers. The forms are actually fossilized spiritual energy and

are a great boon to the preparation of the Earth for the New Age
. . . Of the Five, the ascended master wears one Atlantis crystal at
all times close to his body for protection and to enhance all spiritual activities. Two have been acquired by kindly supplicants in
the United States of America in exchange for the substantial contribution the ascended master requests.

Or, under the heading "THE WATERS OF CARLOS":

The ascended master finds occasionally water of such purity that
he undertakes to energize a quantity of it for others to benefit, an
intensive process. To produce what is always too little, the ascended master purifies himself and a quantity of pure quartz crystal fashioned into flasks. He then places himself and the crystals
into a large copper bowl, polished and kept warm. For a twenty-
four hour period the ascended master pours energy into the spiritual repository of the water. . . . The water need not be removed
from the flask to be utilized spiritually. Simply holding the flask
and concentrating on healing a wound or illness will produce as-
tounding results. However, if serious mischance befalls you or a
close one, a tiny dab of the energized water will immediately assist
recovery.

Or, "TEARS OF CARLOS":

The red colour imparted to the holding flasks that the ascended
master has fashioned for the tears is proof enough of their power,
but their affect [sic] during meditation has been described by
those who have experienced it as "a glorious Oneness."

Then there is a little book, *The Teachings of Carlos*, which begins:

I AM CARLOS.

I HAVE COME TO YOU
FROM MANY PAST
INCARNATIONS.

I HAVE A GREAT LESSON TO
TEACH YOU.

LISTEN CAREFULLY.
READ CAREFULLY.
THINK CAREFULLY.

THE TRUTH IS HERE.

The first teaching asks, "Why are we here. . . ?" The answer: "**Who can say what is the one answer?** *There are many answers to any question, and all the answers are right answers. It is so. Do you see?*"

The book enjoins us not to turn to the next page until we have understood the page we are on. This is one of several factors that makes finishing it difficult.

"Of doubters," it reveals later, "I can say only this: let them take from the matter just what they wish. They end up with nothing—a handful of space, perhaps. And what does the believer have? EVERYTHING! All questions are answered, since all and any answers are correct answers. And the answers are right! Argue that, doubter."

Or: "Don't ask for explanations of everything. Westerners, in particular, are always demanding long-winded descriptions of why this, and why that. Most of what is asked is obvious. Why bother with probing into these matters? . . . By belief, all things become true."

The last page of the book displays a single word in large letters: We are exhorted to "THINK!"

The full text of *The Teachings of Carlos* was written by Randi. He and Alvarez dashed it off on a laptop computer in a few hours.

The Australian media felt betrayed by one of their own. The leading television program in the country had gone out of its way to expose shoddy standards of fact-checking and widespread gullibility in institutions devoted to news and public affairs. Some media analysts excused it on the grounds that it obviously wasn't important; if it *had* been important, they would have checked it out. There were few mea culpas. None who had been taken in were willing to appear on a retrospective of the "Carlos Affair" scheduled for the following Sunday on *Sixty Minutes*.

Of course, there's nothing special about Australia in all of this. Alvarez, Randi, and their co-conspirators could have chosen any nation on Earth and it would have worked. Even those who gave Carlos a national television audience knew enough to ask some skeptical questions—but they couldn't resist inviting him to appear in the first place. The internecine struggle within the media dominated the headlines after Carlos's departure. Puzzled commentaries were written about the exposé. What was the point? What was proved?

Alvarez and Randi proved how little it takes to tamper with our beliefs, how readily we are led, how easy it is to fool the public when people are lonely and starved for something to believe in. If Carlos had stayed longer in Australia and concentrated more on healing—by prayer, by believing in him, by wishing on his bottled tears, by stroking his crystals—there's no doubt that people would have reported being cured of many illnesses, especially psychogenic ones. Even with nothing more fraudulent than his appearance, sayings, and ancillary products, some people would have gotten better because of Carlos.

This, again, is the placebo effect found with almost every faith healer. We believe we're taking a potent medicine and the pain goes away—for a time at least. And when we believe we've received a potent spiritual cure, the disease sometimes also goes away—for a time at least. Some people spontaneously announce that they've been cured even when they haven't. Detailed follow-ups by Nolen, Randi, and many others of those who have been told they were cured, and agreed that they were—in, say, televised services by American faith healers—were unable to find even one person with serious organic disease who was in fact cured. Even significant improvement in their condition is dubious. As the Lourdes experience suggests, you may have to go through ten thousand to a million cases before you find one truly startling recovery.

A faith healer may or may not start out with fraud in mind. But to his amazement, his patients actually seem to be improving. Their emotions are genuine, their gratitude heart-felt. When the healer is criticized, such people rush to his defense. Several elderly attendees of the channeling at the Sydney Opera House were incensed after the *Sixty Minutes* exposé: "Never mind what they say," they told Alvarez, "we believe in you."

These successes may be enough to convince many charlatans—no matter how cynical they were at the beginning—that they actually *have* mystical powers. Maybe they're not successful every time. The powers come and go, they tell themselves. They have to cover the down time. If they must cheat a little now and then, it serves a higher purpose, they tell themselves. Their spiel is consumer-tested. It works.

Most of these figures are only after your money. That's the good news. But what worries me is that a Carlos will come along with bigger fish to fry—attractive, commanding, patriotic, exuding leadership. All of us long for a competent, uncorrupt, charismatic leader. We will leap at the opportunity to support, to believe, to feel good. Most reporters, editors, and producers—swept up with the rest of us—will shy away from real skeptical scrutiny. He won't be selling you prayers or crystals or tears. Perhaps he'll be selling you a war, or a scapegoat, or a much more all-encompassing bundle of beliefs than Carlos's. Whatever it is, it will be accompanied by warnings about the dangers of skepticism.

In the celebrated film *The Wizard of Oz*, Dorothy, the Scarecrow, the Tin Woodsman, and the Cowardly Lion are intimidated—indeed awed—by the outsized oracular figure called the Great Oz. But Dorothy's little dog Toto snaps at a concealing curtain and reveals that the Great Oz is in fact a machine run by a small, tubby, frightened man, as much an exile in this strange land as they.

I think we're lucky that James Randi is tugging at the curtain. But it would be as dangerous to rely on him to expose all the quacks, humbugs, and bunkum in the world as it would be to believe those same charlatans. If we don't want to get taken, we need to do this job for ourselves.

———

One of the saddest lessons of history is this: If we've been bamboozled long enough, we tend to reject any evidence of the bamboozle. We're no longer interested in finding out the truth. The bamboozle has captured us. It's simply too painful to acknowledge, even to ourselves, that we've been taken. Once you give a charlatan power over you, you almost never get it back. So the old bamboozles tend to persist as the new ones rise.

Séances occur only in darkened rooms, where the ghostly visitors can be seen dimly at best. If we turn up the lights a little, so we have a chance to see what's going on, the spirits vanish. They're shy, we're told, and some of us believe it. In twentieth-century parapsychology laboratories, there is the "observer effect": Those described as gifted psychics find that their powers diminish markedly whenever skeptics arrive, and disappear altogether in the presence of a conjurer as skilled as James Randi. What they need is darkness and gullibility.

A little girl who had been a co-conspirator in a famous nineteenth-century flimflam—spirit-rapping, in which ghosts answer questions by loud thumping—grew up and confessed it was an imposture. She was cracking the joint in her big toe. She demonstrated how it was done. But the public apology was largely ignored and, when acknowledged, denounced. Spirit-rapping was too reassuring to be abandoned merely on the say-so of a self-confessed rapper, even if she started the whole business in the first place. The story began to circulate that the confession was coerced out of her by fanatical rationalists.

As I described earlier, British hoaxers confessed to having made "crop circles," geometrical figures generated in grain fields. It wasn't alien artists working in wheat as their medium, but two blokes with a board, a rope, and a taste for whimsy. Even when they demonstrated how they did it, though, believers were unimpressed. Maybe *some* of the crop circles are hoaxes, they argued, but there are too many of them, and some of the pictograms are too complex. Only extraterrestrials could do it. Then others in Britain confessed. But crop circles abroad, it was objected, in Hungary for example, how can you explain *that*? Then copycat Hungarian teenagers confessed. But what about. . . ?

To test the credulity of an alien abduction psychiatrist, a woman poses as an abductee. The therapist is enthusiastic about the fantasies she spins. But when she announces it was all a fake, what is his response? To re-examine his protocols or his understanding of what these cases mean? No. On various days he suggests (1) even if she isn't herself aware of it, she was in fact abducted; or (2) she's crazy—after all, she went to a psychiatrist, didn't she?; or (3) he was on top of the hoax from the beginning and just gave her enough rope to hang herself.

If it's sometimes easier to reject strong evidence than to admit that we've been wrong, this is also information about ourselves worth having.

A scientist places an ad in a Paris newspaper offering a free horoscope. He receives about 150 replies, each, as requested, detailing a place and time of birth. Every respondent is then sent the identical horoscope, along with a questionnaire asking how accurate the horoscope had been. Ninety-four percent of the respondents (and 90 percent of their families and friends) reply that they were at least recognizable in the horoscope. However, the horoscope was drawn up for a French serial killer. If an astrologer can get this far without even meeting his subjects, think how well someone sensitive to human nuances and not overly scrupulous might do.

Why are we so easily taken in by fortune-tellers, psychic seers, palmists, tea-leaf, tarot, and yarrow readers, and their ilk? Of course, they note our posture, facial expressions, clothing, and answers to seemingly innocuous questions. Some of them are brilliant at it, and these are areas about which many scientists seem almost unconscious. There is also a computer network to which "professional" psychics subscribe, the details of their customers' lives available to their colleagues in an instant. A key tool is the so-called "cold read," a statement of opposing predispositions so tenuously balanced that anyone will recognize a grain of truth. Here's an example:

> At times you are extroverted, affable, sociable, while at other times you are introverted, wary, and reserved. You have found it unwise to be too frank in revealing yourself to others. You prefer a certain amount of change and variety, and become dissatisfied when hemmed in by restrictions and limitations. Disciplined and controlled on the outside, you tend to be worrisome and insecure on the inside. While you have some personality weaknesses, you are generally able to compensate for them. You have a great deal of unused capacity, which you have not turned to your advantage. You have a tendency to be critical of yourself. You have a strong need for other people to like you and for them to admire you.

Almost everyone finds this characterization recognizable, and many feel that it describes them perfectly. Small wonder: We are all human.

The list of "evidence" that some therapists think demonstrates repressed childhood sexual abuse (for example, in *The Courage to Heal*, by Ellen Bass and Laura Davis) is very long and prosaic: It includes sleep disorders, overeating, anorexia and bulimia, sexual dysfunction, vague anxieties, and even an inability to remember childhood sexual abuse. Another book, by the social worker E. Sue Blume, lists, among other telltale signs of forgotten incest: headaches, suspicion or its absence, excessive sexual passion or its absence, and adoring one's parents. Among diagnostic items for detecting "dysfunctional" families listed by Charles Whitfield, M.D., are "aches and pains," feeling "more alive" in a crisis, being anxious about "authority figures," and having "tried counseling or psychotherapy," yet feeling "that 'something' is wrong or missing." Like the cold read, if the list is long and broad enough, everyone will have "symptoms."

Skeptical scrutiny is not only the toolkit for rooting out bunkum and cruelty that prey on those least able to protect themselves and most in need of our compassion, people offered little other hope. It is also a timely reminder that mass rallies, radio and television, the print media, electronic marketing, and mail-order technology permit other kinds of lies to be injected into the body politic — to take advantage of the frustrated, the unwary, and the defenseless in a society riddled with political ills that are being treated ineffectively if at all.

Baloney, bamboozles, careless thinking, flimflam, and wishes disguised as facts are not restricted to parlor magic and ambiguous advice on matters of the heart. Unfortunately, they ripple through mainstream political, social, religious, and economic issues in every nation.

Chapter 14

ANTISCIENCE

There's no such thing as objective truth. We make our own truth. There's no such thing as objective reality. We make our own reality. There are spiritual, mystical, or inner ways of knowing that are superior to our ordinary ways of knowing. If an experience seems real, it is real. If an idea feels right to you, it is right. We are incapable of acquiring knowledge of the true nature of reality. Science itself is irrational or mystical. It's just another faith or belief system or myth, with no more justification than any other. It doesn't matter whether beliefs are true or not, as long as they're meaningful to you.

a summary of New Age beliefs, from
THEODORE SCHICK, JR., and LEWIS VAUGHN,
How to Think About Weird Things: Critical Thinking for a New Age
(Mountain View, CA: Mayfield Publishing Company, 1995)

If the established framework of science is plausibly in error (or arbitrary, or irrelevant, or unpatriotic, or impious, or mainly serving the interests of the powerful), then perhaps we can save ourselves the trouble of understanding what so many people think of as a complex, difficult, highly mathematical, and counterintuitive body of knowledge. Then all the scientists would have their comeuppance. Science envy could be transcended. Those who have pursued other paths to knowledge, those who have secretly harbored beliefs that science has scorned, could now have their place in the Sun.

The rate of change in science is responsible for some of the fire it draws. Just when we've finally understood something the scientists are talking about, they tell us it isn't any longer true. And even if it is, there's a slew of new things—things we never heard of, things difficult to believe, things with disquieting implications—that they claim to have discovered recently. Scientists can be perceived as toying with us, as wanting to overturn everything, as socially dangerous.

Edward U. Condon was a distinguished American physicist, a pioneer in quantum mechanics, a participant in the development of radar and nuclear weapons in World War II, research director of Corning Glass, director of the National Bureau of Standards, and president of the American Physical Society (as well as, late in his life, professor of physics at the University of Colorado, where he directed a controversial Air Force–funded scientific study of UFOs). He was one of the physicists whose loyalty to the United States was challenged by members of Congress—including Congressman Richard M. Nixon, who called for the revocation of his security clearance—in the late 1940s and early 1950s. The superpatriotic chairman of the House Committee on Un-American Activities (HCUA), Rep. J. Parnell Thomas, would call the physicist "Dr. Condom," the "weakest link" in American security, and—at one point—the "missing link." His view on Constitutional guarantees can be gleaned from the following response to a witness's lawyer: "The rights you have are the rights given you by this Committee. We will determine what rights you have and what rights you have not got before the Committee."

Albert Einstein publicly called on all those summoned before HCUA to refuse to cooperate. In 1948, President Harry Truman—at the Annual Meeting of the American Association for the Advancement of Science, and with Condon sitting beside him—denounced Rep. Thomas and HCUA on the grounds that vital scientific research "may be made impossible by the creation of an atmosphere in which no man feels safe against the public airing of unfounded rumors, gossip and vilification." He called HCUA's activities "the most un-American thing we have to contend with today. It is the climate of a totalitarian country."*

The playwright Arthur Miller wrote *The Crucible*, about the Salem Witch Trials, in this period. When the drama opened in Europe, Miller was denied a passport by the State Department on the grounds that it was not in the best interests of the United States for him to travel abroad. On opening night in Brussels the play was greeted with tumultuous applause, whereupon the U.S. Ambassador stood up and took a bow. Brought before HCUA, Miller was chastised for the suggestion that Congressional investigations might have something in common with witch trials; he replied, "The comparison is inevitable, sir." Thomas was shortly afterwards thrown in jail for fraud.

One summer in graduate school I was a student of Condon's. I remember vividly his account of being brought up before some loyalty review board:

"Dr. Condon, it says here that you have been at the forefront of a revolutionary movement in physics called"—and here the inquisitor read the words slowly and carefully—"quantum mechanics. It strikes this hearing that if you could be at the forefront of one revolutionary movement . . . you could be at the forefront of another."

---

* But Truman's responsibility for the witch-hunt atmosphere of the late 1940s and early 1950s is considerable. His 1947 Executive Order 9835 authorized inquiries into the opinions and associates of all federal employees, without the right to confront the accuser or even, in most cases, to know what the accusation was. Those found wanting were fired. His Attorney General, Tom Clark, established a list of "subversive" organizations so wide that at one time it included Consumer's Union.

Condon, quick on his feet, replied that the accusation was untrue. He was not a revolutionary in physics. He raised his right hand: "I believe in Archimedes' Principle, formulated in the third century B.C. I believe in Kepler's laws of planetary motion, discovered in the seventeenth century. I believe in Newton's laws. . . ." And on he went, invoking the illustrious names of Bernoulli, Fourier, Ampère, Boltzmann, and Maxwell. This physicist's catechism did not gain him much. The tribunal did not appreciate humor on so serious a matter. But the most they were able to pin on Condon, as I recall, was that in high school he had a job delivering a socialist newspaper door-to-door on his bicycle.

———

Imagine you seriously want to understand what quantum mechanics is about. There is a mathematical underpinning that you must first acquire, mastery of each mathematical subdiscipline leading you to the threshold of the next. In turn you must learn arithmetic, Euclidian geometry, high school algebra, differential and integral calculus, ordinary and partial differential equations, vector calculus, certain special functions of mathematical physics, matrix algebra, and group theory. For most physics students, this might occupy them from, say, third grade to early graduate school—roughly 15 years. Such a course of study does not actually involve learning any quantum mechanics, but merely establishing the mathematical framework required to approach it deeply.

The job of the popularizer of science, trying to get across some idea of quantum mechanics to a general audience that has not gone through these initiation rites, is daunting. Indeed, there are no successful popularizations of quantum mechanics in my opinion—partly for this reason. These mathematical complexities are compounded by the fact that quantum theory is so resolutely counterintuitive. Common sense is almost useless in approaching it. It's no good, Richard Feynman once said, asking why it *is* that way. No one knows why it is that way. That's just the way it is.

Now suppose we were to approach some obscure religion or New Age doctrine or shamanistic belief system skeptically. We have an open mind; we understand there's something interesting here; we introduce ourselves to the practitioner and ask for an intelligible summary. Instead we are told that it's intrinsically too difficult to be explained simply, that it's replete with "mysteries," but if we're willing to become acolytes for 15 years, at the end of that time we might begin to be prepared to consider the subject seriously. Most of us, I think, would say that we simply don't have the time; and many would suspect that the business about 15 years just to get to the threshold of understanding is evidence that the whole subject is a bamboozle: If it's too hard for us to understand, doesn't it follow that it's too hard for us to criticize knowledgeably? Then the bamboozle has free rein.

So how is shamanistic or theological or New Age doctrine different from quantum mechanics? The answer is that even if we cannot understand it, we can verify that quantum mechanics works. We can compare the quantitative predictions of quantum theory with the measured wavelengths of spectral lines of the chemical elements, the behavior of semiconductors and liquid helium, microprocessors, which kinds of molecules form from their constituent atoms, the existence and properties of white dwarf stars, what happens in masers and lasers, and which materials are susceptible to which kinds of magnetism. We don't have to understand the theory to see what it predicts. We don't have to be accomplished physicists to read what the experiments reveal. In every one of these instances—as in many others—the predictions of quantum mechanics are strikingly, and to high accuracy, confirmed.

But the shaman tells us that his doctrine is true because it too works—not on arcane matters of mathematical physics but on what really counts: He can cure people. Very well, then, let's accumulate the statistics on shamanistic cures, and see if they work better than placebos. If they do, let's willingly grant that there's something here—even if it's only that some illnesses are psychogenic, and can be cured or mitigated by the right attitudes and mental states. We can also compare the efficacy of alternative shamanistic systems.

Whether the shaman grasps why his cures work is another story. In quantum mechanics we have a purported understanding of Nature on the basis of which, step by step and quantitatively, we make predictions about what will happen if a certain experiment, never before attempted, is carried out. If the experiment bears out the prediction—especially if it does so numerically and precisely—we have confidence that we knew what we were doing. There are at best few examples with this character among shamans, priests, and New Age gurus.

Another important distinction was suggested in *Reason and Nature*, the 1931 book by Morris Cohen, a celebrated philosopher of science:

> To be sure, the vast majority of people who are untrained can accept the results of science only on authority. But there is obviously an important difference between an establishment that is open and invites every one to come, study its methods, and suggest improvement, and one that regards the questioning of its credentials as due to wickedness of heart, such as [Cardinal] Newman attributed to those who questioned the infallibility of the Bible. . . Rational science treats its credit notes as always redeemable on demand, while non-rational authoritarianism regards the demand for the redemption of its paper as a disloyal lack of faith.

The myths and folklore of many premodern cultures have explanatory or at least mnemonic value. In stories that everyone can appreciate and even witness, they encode the environment. Which constellations are rising or the orientation of the Milky Way on a given day of the year can be remembered by a story about lovers reunited or a canoe negotiating the sacred river. Since recognizing the sky is essential for planting and reaping and following the game, such stories have important practical value. They can also be helpful as psychological projective tests or as reassurances of humanity's place in the Universe. But that doesn't mean that the Milky Way really is a river or that a canoe really is traversing it before our eyes.

Quinine comes from an infusion of the bark of a particular tree from the Amazon rain forest. How did pre-modern people ever discover that a tea made from this tree, of all the plants in the forest, would relieve the symptoms of malaria? They must have tried every tree and every plant—roots, stems, bark, leaves—tried chewing on them, mashing them up, making an infusion. This constitutes a massive set of scientific experiments continuing over generations—experiments that moreover could not be duplicated today for reasons of medical ethics. Think of how many bark infusions from other trees must have been useless, or made the patient retch or even die. In such a case, the healer chalks these potential medicines off the list, and moves on to the next. The data of ethnopharmacology may not be systematically or even consciously acquired. By trial and error, though, and carefully remembering what worked, eventually they get there—using the molecular riches in the plant kingdom to accumulate a pharmacopoeia that works. Absolutely essential, life-saving information can be acquired from folk medicine and in no other way. We should be doing much more than we are to mine the treasures in such folk knowledge worldwide.

Likewise for, say, predicting the weather in a valley near the Orinoco: It is perfectly possible that pre-industrial peoples have noted over the millennia regularities, premonitory indications, cause-and-effect relationships at a particular geographic locale of which professors of meteorology and climatology in some distant university are wholly ignorant. But it does not follow that the shamans of such cultures are able to predict the weather in Paris or Tokyo, much less the global climate.

Certain kinds of folk knowledge are valid and priceless. Others are at best metaphors and codifiers. Ethnomedicine, yes; astrophysics, no. It is certainly true that all beliefs and all myths are worthy of a respectful hearing. It is not true that all folk beliefs are equally valid—if we're talking not about an internal mindset, but about understanding the external reality.

—

For centuries, science has been under a line of attack that, rather than pseudoscience, can be called antiscience. Science, and academic scholarship in general, the contention these days goes, is too subjective. Some even allege it's entirely subjective, as is, they say, history. History generally is written by the victors to justify their actions, to arouse patriotic fervor, and to suppress the legitimate claims of the vanquished. When no overwhelming victory takes place, each side writes self-promotional accounts of what *really* happened. English histories castigated the French, and vice versa; U.S. histories until very recently ignored the de facto policies of lebensraum and genocide toward Native Americans; Japanese histories of the events leading to World War II minimize Japanese atrocities, and suggest that their chief purpose was altruistically to free East Asia from European and American colonialism; Poland was invaded in 1939, Nazi historians asserted, because Poland, ruthless and unprovoked, attacked Germany; Soviet historians pretended that the Soviet troops that put down the Hungarian (1956) and Czech (1968) revolutions were invited in by general acclamation in the invaded nations rather than by Russian stooges; Belgian histories tend to gloss over the atrocities committed when the Congo was a private fiefdom of the King of Belgium; Chinese historians are strangely oblivious of the tens of millions of deaths caused by Mao Zedong's "Great Leap Forward"; that God condones and even advocates slavery was repeatedly argued from the pulpit and in the schools in Christian slave-holding societies, but Christian polities that have freed their slaves are mostly silent on the matter; as brilliant, widely read, and sober a historian as Edward Gibbon would not meet with Benjamin Franklin when they found themselves at the same English country inn — because of the late unpleasantness of the American Revolution. (Franklin then volunteered source material to Gibbon when he turned, as Franklin was sure he soon would, from the decline and fall of the Roman Empire to the decline and fall of the British Empire. Franklin was right about the British Empire, but his timetable was about two centuries early.)

These histories have traditionally been written by admired academic historians, often pillars of the establishment. Local dissent is given short shrift. Objectivity is sacrificed in the service of higher goals. From this doleful fact, some have gone so far as to conclude that there is no such thing as history, no possibility of reconstructing the actual events; that all we have are biased self-justifications; and that this conclusion stretches from history to all of knowledge, science included.

And yet who would deny that there were actual sequences of historical events, with real causal threads, even if our ability to reconstruct them in their full weave is limited, even if the signal is awash in an ocean of self-congratulatory noise? The danger of subjectivity and prejudice has been apparent from the beginning of history. Thucydides warned against it. Cicero wrote

The first law is that the historian shall never dare to set down what is false; the second, that he shall never dare to conceal the truth; the third, that there shall be no suspicion in his work of either favoritism or prejudice.

Lucian of Samosata, in *How History Should Be Written*, published in the year 170, urged "The historian should be fearless and incorruptible; a man of independence, loving frankness and truth."

It is the responsibility of those historians with integrity to try to reconstruct that actual sequence of events, however disappointing or alarming it may be. Historians learn to suppress their natural indignation about affronts to their nations and acknowledge, where appropriate, that their national leaders may have committed atrocious crimes. They may have to dodge outraged patriots as an occupational hazard. They recognize that accounts of events have passed through biased human filters, and that historians themselves have biases. Those who want to know what actually happened will become fully conversant with the views of historians in other, once adversary, nations. All that can be hoped for is a set of successive approximations: By slow steps, and through improving self-knowledge, our understanding of historical events improves.

Something similar is true in science. We have biases; we breathe in the prevailing prejudices from our surroundings like everyone else. Scientists have on occasion given aid and comfort to a variety of noxious doctrines (including the supposed "superiority" of one ethnic group or gender over another from measurements of brain size or skull bumps or IQ tests). Scientists are often reluctant to offend the rich and powerful. Occasionally, a few of them cheat and steal. Some worked—many without a trace of moral regret—for the Nazis. Scientists also exhibit biases connected with human chauvinisms and with our intellectual limitations. As I've discussed earlier, scientists are also responsible for deadly technologies—sometimes inventing them on purpose, sometimes being insufficiently cautious about unintended side-effects. But it is also scientists who, in most such cases, have blown the whistle alerting us to the danger.

Scientists make mistakes. Accordingly, it is the job of the scientist to recognize our weaknesses, to examine the widest range of opinions, to be ruthlessly self-critical. Science is a collective enterprise with the error-correction machinery often running smoothly. It has an overwhelming advantage over history, because in science we can do experiments. If you are unsure of the negotiations leading to the Treaty of Paris in 1814–1815, replaying the events is an unavailable option. You can only dig into old records. You cannot even ask questions of the participants. Every one of them is dead.

But for many questions in science, you can rerun the event as many times as you like, examine it in new ways, test a wide range of alternative hypotheses. When new tools are devised, you can perform the experiment again and see what emerges from your improved sensitivity. In those historical sciences where you cannot arrange a rerun, you can examine related cases and begin to recognize their common components. We can't make stars explode at our convenience, nor can we repeatedly evolve through many trials a mammal from its ancestors. But we can simulate some of the physics of supernova explosions in the laboratory, and we can compare in staggering detail the genetic instructions of mammals and reptiles.

The claim is also sometimes made that science is as arbitrary or irrational as all other claims to knowledge, or that reason itself is an illusion. The American revolutionary Ethan Allen—leader of the Green Mountain Boys in their capture of Fort Ticonderoga—had some words on this subject:

> Those who invalidate reason ought seriously to consider whether they argue against reason with or without reason; if with reason, then they establish the principle that they are laboring to dethrone: but if they argue without reason (which, in order to be consistent with themselves they must do), they are out of reach of rational conviction, nor do they deserve a rational argument.

The reader can judge the depth of this argument.

—

Anyone who witnesses the advance of science first-hand sees an intensely personal undertaking. There are always a few—driven by simple wonder and great integrity, or by frustration with the inadequacies of existing knowledge, or simply upset with themselves for their imagined inability to understand what everyone else can—who proceed to ask the devastating key questions. A few saintly personalities stand out amidst a roiling sea of jealousies, ambition, backbiting, suppression of dissent, and absurd conceits. In some fields, highly productive fields, such behavior is almost the norm.

I think all that social turmoil and human weakness aids the enterprise of science. There is an established framework in which any scientist can prove another wrong and make sure everyone else knows about it. Even when our motives are base, we keep stumbling on something new.

The American chemistry Nobel laureate Harold C. Urey once

confided to me that as he got older (he was then in his seventies), he experienced increasingly concerted efforts to prove him wrong. He described it as "the fastest gun in the West" syndrome: The young man who could outdraw the celebrated old gunslinger would inherit his reputation and the respect paid to him. It was annoying, he grumbled, but it did help direct the young whippersnappers into important areas of research that they would never have entered on their own.

Being human, scientists also sometimes engage in observational selection: they like to remember those cases when they've been right and forget when they've been wrong. But in many instances, what is "wrong" is partly right, or stimulates others to find out what's right. One of the most productive astrophysicists of our time has been Fred Hoyle, responsible for monumental contributions to our understanding of the evolution of stars, the synthesis of the chemical elements, cosmology, and much else. Sometimes he's succeeded by being right before anyone else even understood that there was something that needed explaining. Sometimes he's succeeded by being wrong—by being so provocative, by suggesting such outrageous alternatives that the observers and experimentalists feel obliged to check it out. The impassioned and concerted effort to "prove Fred wrong" has sometimes failed and sometimes succeeded. In almost every case, it has pushed forward the frontiers of knowledge. Even Hoyle at his most outrageous—for example, proposing that the influenza and HIV viruses are dropped down on Earth from comets, and that interstellar dust grains are bacteria—has led to significant advances in knowledge (although turning up nothing to support those particular notions).

It might be useful for scientists now and again to list some of their mistakes. It might play an instructive role in illuminating and demythologizing the process of science and in enlightening younger scientists. Even Johannes Kepler, Isaac Newton, Charles Darwin, Gregor Mendel, and Albert Einstein made serious mistakes. But the scientific enterprise arranges things so that teamwork prevails: What one of us, even the most brilliant among us, misses, another of us, even someone much less celebrated and capable, may detect and rectify.

For myself, I've tended in past books to recount some of the occasions when I've been right. Let me here mention a few of the cases where I've been wrong: At a time when no spacecraft had been to Venus, I thought at first that the atmospheric pressure was several times that on Earth, rather than many tens of times. I thought the clouds of Venus were made mainly of water, when they turn out to be only 25 percent water. I thought there might be plate tectonics on Mars, when close-up spacecraft observations now show hardly a hint of plate tectonics. I thought the highish infrared temperatures of Titan might be due to a sizable greenhouse effect there; instead, it turns out, it is caused by a stratospheric temperature inversion. Just before Iraq torched the Kuwaiti oil wells in January 1991, I warned that so much smoke might get so high as to disrupt agriculture in much of South Asia; as events transpired, it *was* pitch black at noon and the temperatures dropped 4–6°C over the Persian Gulf, but not much smoke reached stratospheric altitudes and Asia was spared. I did not sufficiently stress the uncertainty of the calculations.

Different scientists have different speculative styles, some being much more cautious than others. As long as new ideas are testable and scientists are not overly dogmatic, no harm is done; indeed, considerable progress can be made. In the first four instances I've just mentioned where I was wrong, I was trying to understand a distant world from a few clues in the absence of thorough spacecraft investigations. In the natural course of planetary exploration more data come in, and we find an army of old ideas plowed down by an armamentarium of new facts.

———

Postmodernists have criticized Kepler's astronomy because it emerged out of his medieval, monotheistic religious views; Darwin's evolutionary biology for being motivated by a wish to perpetuate the privileged social class from which he came, or to justify his supposed prior atheism; and so on. Some of these claims are just. Some are not. But why does it matter what biases and emotional predispositions scientists bring to their studies—so long as they are scrupulously honest and other people with different proclivities check their results? Presumably no one would argue that the conservative view on the sum of 14 and 27 differs from the liberal view, or that the mathematical function that is its own derivative is the exponential in the northern hemisphere but some other function in the southern. Any regular periodic function can be represented to arbitrary accuracy by a Fourier series in Muslim as well as in Hindu mathematics. Non-commutative algebras (where

A times B does not equal B times A) are as self-consistent and meaningful for speakers of Indo-European languages as for speakers of Finno-Ugric. Mathematics might be prized or ignored, but it is equally true everywhere—independent of ethnicity, culture, language, religion, ideology.

Towards the opposite extreme, there are questions such as whether abstract expressionism can be "great" art, or rap "great" music; whether it's more important to curb inflation or unemployment; whether French culture is superior to German culture; or whether prohibitions against murder should apply to the nation state. Here the questions are oversimple, or the dichotomies false, or the answers dependent on unspoken assumptions. Here local biases might very well determine the answers.

Where in this subjective continuum, from almost fully independent of cultural norms to almost wholly dependent on them, does science lie? Although issues of bias and cultural chauvinism certainly arise, and although its content is continually being refined, science is clearly much closer to mathematics than it is to fashion. The claim that its findings are in general arbitrary and biased is not merely tendentious, but specious.

The historians Joyce Appleby, Lynn Hunt, and Margaret Jacob (in *Telling the Truth About History*, 1994) criticize Isaac Newton: He is said to have rejected the philosophical position of Descartes because it might challenge conventional religion and lead to social chaos and atheism. Such criticisms amount only to the charge that scientists are human. How Newton was buffeted by the intellectual currents of his time is of course of interest to the historian of ideas; but it has little bearing on the truth of his propositions. For them to be generally accepted, they must convince atheists and theists alike. This is just what happened.

Appleby and her colleagues claim that "When Darwin formulated his theory of evolution, he was an atheist and a materialist," and suggest that evolution was a product of a purported atheist agenda. They have hopelessly confused cause and effect. Darwin was about to become a minister of the Church of England when the opportunity to sail on H.M.S. *Beagle* presented itself. His religious ideas, as he himself described them, were at the time highly conventional. He found every one of the Anglican Articles of Faith entirely believable.

Through his interrogation of Nature, through science, it slowly dawned on him that at least some of his religion was false. That's why he changed his religious views.

Appleby and her colleagues are appalled at Darwin's description of " 'the low morality of savages . . . their insufficient powers of reasoning . . . [their] weak power of self-command,' " and state that "Now many people are shocked by his racism." But there was no racism at all, as far as I can tell, in Darwin's comment. He was alluding to the inhabitants of Tierra del Fuego, suffering from grinding scarcity in the most barren and Antarctic province of Argentina. When he described a South American woman of African origin who threw herself to her death rather than submit to slavery, he noted that it was only prejudice that kept us from seeing her defiance in the same heroic light as we would a similar act by the proud matron of a noble Roman family. He was himself almost thrown off the *Beagle* by Captain FitzRoy for his militant opposition to the Captain's racism. Darwin was head and shoulders above most of his contemporaries in this regard.

But again, even if he was not, how does it affect the truth or falsity of natural selection? Thomas Jefferson and George Washington owned slaves; Albert Einstein and Mohandas Gandhi were imperfect husbands and fathers. The list goes on indefinitely. We are all flawed and creatures of our times. Is it fair to judge us by the unknown standards of the future? Some of the habits of our age will doubtless be considered barbaric by later generations—perhaps for insisting that small children and even infants sleep alone instead of with their parents; or exciting nationalist passions as a means of gaining popular approval and achieving high political office; or allowing bribery and corruption as a way of life; or keeping pets; or eating animals and jailing chimpanzees; or criminalizing the use of euphoriants by adults; or allowing our children to grow up ignorant.

Occasionally, in retrospect, someone stands out. In my book, the English-born American revolutionary Thomas Paine is one such. He was far ahead of his time. He courageously opposed monarchy, aristocracy, racism, slavery, superstition and sexism when all of these constituted the conventional wisdom. He was unswerving in his criticism of conventional religion. He wrote in *The Age of Reason:* "Whenever we read the obscene stories, the voluptuous debaucheries, the cruel and torturous executions, the unrelenting vindictiveness with which more than half the Bible is filled, it would be more consistent that we called it the word of a demon than the word of God. It . . . has served to corrupt and brutalize mankind." At the same time the book exhibited the deepest reverence for a Creator of the Universe whose existence Paine argued was apparent at a glance at the natural world. But condemning much of the Bible while embracing God seemed an impossible position to most of his contemporaries. Christian theologians concluded he was drunk, mad, or corrupt. The Jewish scholar David Levi forbade his co-religionists from even touching, much less reading, the book. Paine was made to suffer so much for his views (including being thrown into prison after the French Revolution for being too consistent in his opposition to tyranny), that he became an embittered old man.*

Yes, the Darwinian insight can be turned upside down and grotesquely misused: Voracious robber barons may explain their cutthroat practices by an appeal to Social Darwinism; Nazis and other racists may call on "survival of the fittest" to justify genocide. But Darwin did not make John D. Rockefeller or Adolf Hitler. Greed, the Industrial Revolution, the free enterprise system, and corruption of government by the monied are adequate to explain nineteenth-century capitalism. Ethnocentrism, xenophobia, social hierarchies, the long history of anti-Semitism in Germany, the Versailles Treaty, German child-rearing practices, inflation, and the Depression seem adequate to explain Hitler's rise to power. Very likely these or similar events would have transpired with or without Darwin. And modern Darwinism makes it abundantly clear that many less ruthless traits, some not always admired by robber barons and Führers—altruism, general intelligence, compassion—may be the key to survival.

If we could censor Darwin, what other kinds of knowledge could also be censored? Who would do the censoring? Who among us is wise enough to know which information and insights we can safely dispense with, and which will be necessary ten or a hundred or a thou-

---

* Paine was the author of the revolutionary pamphlet *Common Sense*. Published on January 10, 1776, it sold over half a million copies in the next few months and stirred many Americans to the cause of independence. He was the author of the three best-selling books of the eighteenth century. Later generations reviled him for his social and religious views. Theodore Roosevelt called him a "filthy little atheist"—despite his profound belief in God. He is probably the most illustrious American revolutionary uncommemorated by a monument in Washington, D.C.

sand years into the future? Surely we can exert some discretion on which kinds of machines and products it is safe to develop. We must in any case make such decisions, because we do not have the resources to pursue all possible technologies. But censoring knowledge, telling people what they must think, is the aperture to thought police, foolish and incompetent decision-making, and long-term decline.

Fervid ideologues and authoritarian regimes find it easy and natural to impose their views and suppress the alternatives. Nazi scientists, such as the Nobel laureate physicist Johannes Stark, distinguished fanciful, imaginary "Jewish science," including relativity and quantum mechanics, from realistic, practical "Aryan science." Another example: "A new era of the magical explanation of the world is rising," said Adolf Hitler, "an explanation based on will rather than knowledge. There is no truth, in either the moral or the scientific sense."

As he described it to me three decades later, in 1922 the American geneticist Hermann J. Muller flew from Berlin to Moscow in a light plane to witness the new Soviet society firsthand. He must have liked what he saw, because—after his discovery that radiation makes mutations (a discovery that would later win him a Nobel Prize)—he moved to Moscow to help establish modern genetics in the Soviet Union. But by the middle 1930s a charlatan named Trofim Lysenko had caught the notice and then the enthusiastic support of Stalin. Lysenko argued that genetics—which he called "Mendelism-Weissmanism-Morganism," after some of the founders of the field—had an unacceptable philosophical base, and that philosophically "correct" genetics, genetics that paid proper obeisance to communist dialectical materialism, would yield very different results. In particular, Lysenko's genetics would permit an additional crop of winter wheat—welcome news to a Soviet economy reeling from Stalin's forced collectivization of agriculture.

Lysenko's purported evidence was suspect, there were no experimental controls, and his broad conclusions flew in the face of an immense body of contradictory data. As Lysenko's power grew, Muller passionately argued that classical Mendelian genetics was in full harmony with dialectical materialism, while Lysenko, who believed in the inheritance of acquired characteristics and denied a material basis of heredity, was an "idealist," or worse. Muller was strongly supported by N. I. Vavilov, erstwhile president of the All-Union Academy of Agricultural Sciences.

In a 1936 address to the Academy of Agricultural Sciences, now presided over by Lysenko, Muller gave a stirring address that included these words:

> If the outstanding practitioners are going to support theories and opinions that are obviously absurd to everyone who knows even a little about genetics—such views as those recently put forward by President Lysenko and those who think as he does—then the choice before us will resemble the choice between witchcraft and medicine, between astrology and astronomy, between alchemy and chemistry.

In a country of arbitrary arrests and police terror, this speech displayed exemplary—many thought foolhardy—integrity and courage. In *The Vavilov Affair* (1984), the Soviet emigré historian Mark Popovsky describes these words as being accompanied by "thunderous applause from the whole hall" and "remembered by everyone still living who took part in the session."

Three months later, Muller was visited in Moscow by a Western geneticist who expressed astonishment at a widely circulated letter, signed by Muller, that condemned the prevalence of "Mendelism-Weissmanism-Morganism" in the West and that urged a boycott of the forthcoming International Congress of Genetics. Having never seen, much less signed, such a letter, an outraged Muller concluded that it was a forgery perpetrated by Lysenko. Muller promptly wrote an angry denunciation of Lysenko to *Pravda* and mailed a copy to Stalin.

The next day Vavilov came to Muller in a state of some agitation, informing him that he, Muller, had just volunteered to serve in the Spanish Civil War. The letter to *Pravda* had put Muller's life in danger. He left Moscow the next day, just evading, so he was later told, the NKVD, the secret police. Vavilov was not so lucky, and perished in 1943 in Siberia.

With the continuing support of Stalin and later of Khrushchev, Lysenko ruthlessly suppressed classical genetics. Soviet school biology texts in the early 1960s had as little about chromosomes and classical genetics as many American school biology texts have about evolution today. But no new crop of winter wheat grew; incantations of the phrase "dialectical materialism" went unheard by the DNA of domesticated plants; Soviet agriculture remained in the doldrums; and today, partly for this reason, Russia—world-class in many other sciences—is still almost hopelessly backward in molecular biology and genetic engineering. Two generations of modern biologists have been lost. Lysenkoism was not overthrown until 1964, in a series of debates and votes at the Soviet Academy of Sciences—one of the few institutions to maintain a degree of independence from the leaders of party and state—in which the nuclear physicist Andrei Sakharov played an outstanding role.

Americans tend to shake their heads in astonishment at the Soviet experience. The idea that some state-endorsed ideology or popular prejudice would hog-tie scientific progress seems unthinkable. For 200 years Americans have prided themselves on being a practical, pragmatic, nonideological people. And yet anthropological and psychological pseudoscience has flourished in the United States—on race, for example. Under the guise of "creationism," a serious effort continues to be made to prevent evolutionary theory—the most powerful integrating idea in all of biology, and essential for other sciences ranging from astronomy to anthropology—from being taught in the schools.

—

Science is different from many another human enterprise—not, of course, in its practitioners being influenced by the culture they grew up in, nor in sometimes being right and sometimes wrong (which are common to every human activity), but in its passion for framing testable hypotheses, in its search for definitive experiments that confirm or deny ideas, in the vigor of its substantive debate, and in its willingness to abandon ideas that have been found wanting. If we were not aware of our own limitations, though, if we were not seeking further data, if we were unwilling to perform controlled experiments, if we did not respect the evidence, we would have very little leverage in our quest for the truth. Through opportunism and timidity we might then be buffeted by every ideological breeze, with nothing of lasting value to hang on to.

Chapter 15

———

NEWTON'S SLEEP

May God keep us from single vision and Newton's sleep.

WILLIAM BLAKE,
from a poem included in a letter to Thomas Butts
(1802)

[I]gnorance more frequently begets confidence than
does knowledge: it is those who know little, and not those
who know much, who so positively assert that
this or that problem will never be solved by science.

CHARLES DARWIN,
Introduction, *The Descent of Man*
(1871)

B y "Newton's sleep," the poet, painter, and revolutionary William
Blake seems to have meant a tunnel vision in the perspective of
Newton's physics, as well as Newton's own (incomplete) disengage-
ment from mysticism. Blake thought the idea of atoms and particles of
light amusing, and Newton's influence on our species "satanic." A
common critique of science is that it is too narrow. Because of our
well-demonstrated fallibilities, it rules out of court, beyond serious dis-
course, a wide range of uplifting images, playful notions, earnest mysticism, and stupefying wonders. Without physical evidence, science
does not admit spirits, souls, angels, devils, or dharma bodies of the
Buddha. Or alien visitors.

The American psychologist Charles Tart, who believes the evidence for extrasensory perception is convincing, writes:

> An important factor in the current popularity of "New Age" ideas
> is a reaction against the dehumanizing, despiritualizing effects of
> *scientism*, the philosophical belief (masquerading as objective
> science and held with the emotional tenacity of born-again funda-
> mentalism) that we are *nothing but* material beings. To unthink-
> ingly embrace anything and everything labeled "spiritual" or
> "psychic" or "New Age" is, of course, foolish, for many of these
> ideas are factually wrong, however noble or inspiring they are. On
> the other hand, this New Age interest is a legitimate recognition of
> some of the realities of human nature: People have always had
> and continue to have experiences that seem to be "psychic" or
> "spiritual."

But why should "psychic" experiences challenge the idea that we
are made of matter and nothing but? There is very little doubt that, in
the everyday world, matter (and energy) exist. The evidence is all
around us. In contrast, as I've mentioned earlier, the evidence for

something non-material called "spirit" or "soul" is very much in doubt. Of course each of us has a rich internal life. Considering the stupendous complexity of matter, though, how could we possibly prove that our internal life is not wholly due to matter? Granted, there is much about human consciousness that we do not fully understand and cannot yet explain in terms of neurobiology. Humans have limitations, and no one knows this better than scientists. But a multitude of aspects of the natural world that were considered miraculous only a few generations ago are now thoroughly understood in terms of physics and chemistry. At least some of the mysteries of today will be comprehensively solved by our descendants. The fact that we cannot now produce a detailed understanding of, say, altered states of consciousness in terms of brain chemistry no more implies the existence of a "spirit world" than a sunflower following the Sun in its course across the sky was evidence of a literal miracle before we knew about phototropism and plant hormones.

And if the world does not in all respects correspond to our wishes, is this the fault of science, or of those who would impose their wishes on the world? All the mammals—and many other animals as well—experience emotions: fear, lust, hope, pain, love, hate, the need to be led. Humans may brood about the future more, but there is nothing in our emotions unique to us. On the other hand, no other species does science as much or as well as we. How then can science be "dehumanizing"?

Still, it seems so unfair: Some of us starve to death before we're out of infancy, while others—by an accident of birth—live out their lives in opulence and splendor. We can be born into an abusive family or a reviled ethnic group, or start out with some deformity; we go through life with the deck stacked against us, and then we die, and that's it? Nothing but a dreamless and endless sleep? Where's the justice in this? This is stark and brutal and heartless. Shouldn't we have a second chance on a level playing field? How much better if we were born again in circumstances that took account of how well we played our part in the last life, no matter how stacked against us the deck was then. Or if there were a time of judgment after we die, then—so long as we did well with the persona we were given in this life, and were humble and faithful and all the rest—we should be rewarded by living joyfully until the end of time in a permanent refuge from the agony and turmoil of the world. That's how it would be if the world were thought out, preplanned, fair. That's how it would be if those suffering from pain and torment were to receive the consolation they deserve.

So societies that teach contentment with our present station in life, in expectation of post-mortem reward, tend to inoculate themselves against revolution. Further, fear of death, which in some respects is adaptive in the evolutionary struggle for existence, is maladaptive in warfare. Those cultures that teach an afterlife of bliss for heroes—or even for those who just did what those in authority told them—might gain a competitive advantage.

Thus, the idea of a spiritual part of our nature that survives death, the notion of an afterlife, ought to be easy for religions and nations to sell. This is not an issue on which we might anticipate widespread skepticism. People will want to believe it, even if the evidence is meager to nil. True, brain lesions can make us lose major segments of our memory, or convert us from manic to placid, or vice versa; and changes in brain chemistry can convince us there's a massive conspiracy against us, or make us think we hear the Voice of God. But as compelling testimony as this provides that our personality, character, memory—if you will, soul—resides in the matter of the brain, it is easy not to focus on it, to find ways to evade the weight of the evidence.

And if there are powerful social institutions insisting that there *is* an afterlife, it should be no surprise that dissenters tend to be sparse, quiet, and resented. Some Eastern, Christian, and New Age religions, as well as Platonism, hold that the world is unreal, that suffering, death and matter itself are illusions; and that nothing really exists except "Mind." In contrast, the prevailing scientific view is that the mind is how we perceive what the brain does; i.e., it's a property of the hundred trillion neural connections in the brain.

There is a strangely waxing academic opinion, with roots in the 1960s, that holds all views to be equally arbitrary and "true" or "false" to be a delusion. Perhaps it is an attempt to turn the tables on scientists who have long argued that literary criticism, religion, aesthetics, and much of philosophy and ethics are mere subjective opinion, because they cannot be demonstrated like a theorem in Euclidean geometry nor put to experimental test.

There are people who want everything to be possible, to have their reality unconstrained. Our imagination and our needs require more, they feel, than the comparatively little that science teaches we may be reasonably sure of. Many New Age gurus—the actress Shirley MacLaine among them—go so far as to embrace solipsism, to assert that the only reality is their own thoughts. "I am God," they actually say. "I really think we are creating our own reality," MacLaine once told a skeptic. "I think I'm creating you right here."

If I dream of being reunited with a dead parent or child, who is to tell me that it didn't *really* happen? If I have a vision of myself floating in space looking down on the Earth, maybe I was really there; who are some scientists, who didn't even share the experience, to tell me that it's all in my head? If my religion teaches that it is the inalterable and inerrant word of God that the Universe is a few thousand years old, then scientists are being offensive and impious, as well as mistaken, when they claim it's a few billion.

Irritatingly, science claims to set limits on what we can do, even in principle. Who says we can't travel faster than light? They used to say that about sound, didn't they? Who's going to stop us, if we have really powerful instruments, from measuring the position and the momentum of an electron simultaneously? Why can't we, if we're very clever, build a perpetual motion machine "of the first kind" (one that generates more energy than is supplied to it), or a perpetual motion machine "of the second kind" (one that never runs down)? Who dares to set limits on human ingenuity?

In fact, Nature does. In fact, a fairly comprehensive and very brief statement of the laws of Nature, of how the Universe works, is contained in just such a list of prohibited acts. Tellingly, pseudoscience and superstition tend to recognize no constraints in Nature. Instead, "all things are possible." They promise a limitless production budget, however often their adherents have been disappointed and betrayed.

———

A related complaint is that science is too simple-minded, too "reductionist"; it naïvely imagines that in the final accounting there will be only a few laws of Nature—perhaps even rather simple ones—that explain everything, that the exquisite subtlety of the world, all the snow crystals, spiderweb latticework, spiral galaxies, and flashes of human insight can ultimately be "reduced" to such laws. Reductionism seems to pay insufficient respect to the complexity of the Universe. It appears to some as a curious hybrid of arrogance and intellectual laziness.

To Isaac Newton—who in the minds of critics of science personifies "single vision"—it looked like a clockwork Universe. Literally. The regular, predictable orbital motions of the planets around the Sun, or the Moon around the Earth, were described to high precision by essentially the same differential equation that predicts the swing of a pendulum or the oscillation of a spring. We have a tendency today to think we occupy some exalted vantage point, and to pity the poor Newtonians for having so limited a world view. But within certain reasonable limitations, the same harmonic equations that describe clockwork really do describe the motions of astronomical objects throughout the Universe. This is a profound, not a trivial parallelism.

Of course, there are no gears in the Solar System, and the component parts of the gravitational clockwork do not touch. Planets generally have more complicated motions than pendulums and springs. Also, the clockwork model breaks down in certain circumstances: Over very long periods of time, the gravitational tugs of distant worlds—tugs that might seem wholly insignificant over a few orbits—can build up, and some little world can go unexpectedly careening out of its accustomed course. However, something like chaotic motion is also known in pendulum clocks; if we displace the bob too far from the perpendicular, a wild and ugly motion ensues. But the Solar System keeps better time than any mechanical clock, and the whole idea of keeping time comes from the observed motion of the Sun and stars.

The astonishing fact is that similar mathematics applies so well to planets and to clocks. It needn't have been this way. We didn't impose it on the Universe. That's the way the Universe is. If this is reductionism, so be it.

Until the middle twentieth century, there had been a strong belief—among theologians, philosophers, and many biologists—that life was not "reducible" to the laws of physics and chemistry, that there was a "vital force," an "entelechy," a tao, a mana that made living things go. It "animated" life. It was impossible to see how mere atoms and molecules could account for the intricacy and elegance, the fitting of form to function, of a living thing. The world's religions were invoked: God or the gods breathed life, soul-stuff, into inanimate matter. The eighteenth-century chemist Joseph Priestley tried to find the "vital force." He weighed a mouse just before and just after it died. It weighed the same. All such attempts have failed. If there is soul-stuff, evidently it weighs nothing—that is, it is not made of matter.

Nevertheless, even biological materialists entertained reservations; perhaps, if not plant, animal, fungal and microbial souls, some still undiscovered principle of science was needed to understand life. For example, the British physiologist J. S. Haldane (father of J.B.S. Haldane) asked in 1932:

> What intelligible account can the mechanistic theory of life give of the . . . recovery from disease and injuries? Simply none at all, except that these phenomena are so complex and strange that as yet we cannot understand them. It is exactly the same with the closely related phenomena of reproduction. We cannot by any stretch of the imagination conceive a delicate and complex mechanism which is capable, like a living organism, of reproducing itself indefinitely often.

But only a few decades later and our knowledge of immunology and molecular biology have enormously clarified these once impenetrable mysteries.

I remember very well when the molecular structure of DNA and the nature of the genetic code were first elucidated in the 1950s and 1960s, how biologists who studied whole organisms accused the new proponents of molecular biology of reductionism. ("They'll never understand even a worm with their DNA.") Of course reducing everything to a "vital force" is no less reductionism. But it is now clear that all life on Earth, every single living thing, has its genetic information encoded in its nucleic acids and employs fundamentally the same codebook to implement the hereditary instructions. We have learned how to read the code. The same few dozen organic molecules are used over and over again in biology for the widest variety of functions. Genes bearing significant responsibility for cystic fibrosis and breast cancer have been identified. The 1.8 million rungs of the DNA ladder of the bacterium *Haemophilis influenzae*, comprising its 1743 genes, have been sequenced. The specific function of most of these genes is beautifully detailed—from the manufacture and folding of hundreds of complex molecules, to protection against heat and antibiotics, to increasing the mutation rate, to making identical copies of the bacterium. Much of the genomes of many other organisms (including the roundworm *Caenorhabditis elegans*) have now been mapped. Molecular biologists are busily recording the sequence of the three billion nucleotides that specify how to make a human being. In another decade or two, they'll be done. (Whether the benefits will ultimately exceed the risks seems by no means certain.)

The continuity between atomic physics, molecular chemistry, and that holy of holies, the nature of reproduction and heredity, has now been established. No new principle of science needed to be invoked. It looks as if there *are* a small number of simple facts that can be used to understand the enormous intricacy and variety of living things. (Molecular genetics also teaches that each organism has its own particularity.)

Reductionism is even better established in physics and chemistry. I will later describe the unexpected coalescence of our understanding of electricity, magnetism, light and relativity into a single framework. We've known for centuries that a handful of comparatively simple laws not only explains but quantitatively and accurately predicts a breathtaking variety of phenomena, not just on Earth but through the entire Universe.

We hear—for example from the theologian Langdon Gilkey in his *Nature, Reality and the Sacred*—that the notion of the laws of Nature being everywhere the same is simply a preconception imposed on the Universe by fallible scientists and their social milieu. He longs for other kinds of "knowledge," as valid in their contexts as science is in its. But the order of the Universe is not an assumption; it's an observed fact. We detect the light from distant quasars only because the laws of electromagnetism are the same ten billion light-years away as here. The spectra of those quasars are recognizable only because the same chemical elements are present there as here, and because the same laws of quantum mechanics apply. The motion of galaxies around one another follows familiar Newtonian gravity. Gravitational lenses and binary pulsar spin-downs reveal general relativity in the depths of space. We *could* have lived in a Universe with different laws in every province, but we do not. This fact cannot but elicit feelings of reverence and awe.

We might have lived in a Universe in which nothing could be understood by a few simple laws, in which Nature was complex beyond our abilities to understand, in which laws that apply on Earth are invalid on Mars, or in a distant quasar. But the evidence—not the preconceptions, the evidence—proves otherwise. Luckily for us, we live in a Universe in which much *can* be "reduced" to a small number of comparatively simple laws of Nature. Otherwise we might have lacked the intellectual capacity and grasp to comprehend the world.

Of course, we may make mistakes in applying a reductionist program to science. There may be aspects which, for all we know, are not reducible to a few comparatively simple laws. But in light of the findings in the last few centuries, it seems foolish to complain about reductionism. It is not a deficiency but one of the chief triumphs of science. And, it seems to me, its findings are perfectly consonant with many religions (although it does not *prove* their validity). Why should a few simple laws of Nature explain so much and hold sway throughout this vast Universe? Isn't this just what you might expect from a Creator of the Universe? Why should some religious people oppose the reductionist program in science, except out of some misplaced love of mysticism?

———

Attempts to reconcile religion and science have been on the religious agenda for centuries—at least for those who did not insist on Biblical and Qu'ranic literalism with no room for allegory or metaphor. The crowning achievements of Roman Catholic theology are the *Summa Theologica* and the *Summa Contra Gentiles* ("Against the Gentiles") of St. Thomas Aquinas. Out of the maelstrom of sophisticated Islamic philosophy that tumbled into Christendom in the twelfth and thirteenth centuries were the books of the ancient Greeks, especially Aristotle—works even on casual inspection of high accomplishment. Was this ancient learning compatible with God's Holy Word?* In the *Summa Theologica*, Aquinas set himself the task of reconciling 631 questions between Christian and classical sources. But how to do this where a clear dispute arises? It cannot be accomplished without some supervening organizing principle, some superior way to know the

---

* This was no dilemma for many others. "I believe; therefore I understand" said St. Anselm in the eleventh century.

world. Often, Aquinas appealed to common sense and the natural world—i.e., science used as an error-correcting device. With some contortion of both common sense and Nature, he managed to reconcile all 631 problems. (Although when push came to shove, the desired answer was simply assumed. Faith always got the nod over Reason.) Similar attempts at reconciliation permeate Talmudic and post-Talmudic Jewish literature and medieval Islamic philosophy.

But tenets at the heart of religion can be tested scientifically. This in itself makes some religious bureaucrats and believers wary of science. Is the Eucharist, as the Church teaches, in fact, and not just as productive metaphor, the flesh of Jesus Christ, or is it—chemically, microscopically, and in other ways—just a wafer handed to you by a priest? * Will the world be destroyed at the end of the 52-year Venus cycle unless humans are sacrificed to the gods? ** Does the occasional uncircumcised Jewish man fare worse than his co-religionists who abide by the ancient covenant in which God demands a piece of foreskin from every male worshiper? Are there humans populating innumerable other planets, as the Latter Day Saints teach? Were whites created from blacks by a mad scientist, as the Nation of Islam asserts? Would the Sun indeed not rise if the Hindu sacrificial rite is omitted (as we are assured would be the case in the *Satapatha Brahmana*)?

We can gain some insight into the human roots of prayer by examining those of unfamiliar religions and cultures. Here, for example, is what is written in a cuneiform inscription on a Babylonian cylinder seal from the second millennium B.C.:

Oh, Ninlil, Lady of the Lands, in your marriage bed, in the abode

---

* There was a time when the answer to this question was a matter of life or death. Miles Phillips was an English sailor, stranded in Spanish Mexico. He and his fellows were brought up before the Inquisition in the year 1574. They were asked "Whether we did not believe that the Host of bread which the priest did hold up over his head, and the wine that was in the chalice, was the very true and perfect body and blood of our Saviour Christ, Yea or No? To which," Phillips adds, "if we answered not 'Yea!' then there was no way but death."

** Since this Mesoamerican ritual has not really been practiced for five centuries, we have the perspective to reflect on the tens of thousands of willing and unwilling sacrifices to the Aztec and Mayan gods who reconciled themselves to their fates with the serene faith and confident knowledge that they were dying to save the Universe.

of your delight, intercede for me with Enlil, your beloved. [Signed] Mili-Shipak, Shatammu of Ninmah.

It's been a long time since there's been a Shatammu in Ninmah, or even a Ninmah. Despite the fact that Enlil and Ninlil were major gods—people all over the civilized Western world had prayed to them for two thousand years—was poor Mili-Shipak in fact praying to a phantom, to a societally condoned product of his imagination? And if so, what about us? Or is this blasphemy, a forbidden question—as doubtless it was among the worshipers of Enlil?

Does prayer work at all? Which ones?

There's a category of prayer in which God is begged to intervene in human history or just to right some real or imagined injustice or natural calamity—for example, when a bishop from the American West prays for God to intervene and end a devastating dry spell. Why is the prayer needed? Didn't God know of the drought? Was he unaware that it threatened the bishop's parishioners? What is implied here about the limitations of a supposedly omnipotent and omniscient deity? The bishop asked his followers to pray as well. Is God more likely to intervene when many pray for mercy or justice than when only a few do? Or consider the following request, printed in 1994 in *The Prayer and Action Weekly News: Iowa's Weekly Christian Information Source:*

> Can you join me in praying that God will burn down the Planned Parenthood in Des Moines in a manner no one can mistake for any human torching, which impartial investigators will have to attribute to miraculous (unexplainable) causes, and which Christians will have to attribute to the Hand of God?

We've discussed faith healing. What about longevity through prayer? The Victorian statistician Francis Galton argued that—other things being equal—British monarchs ought to be very long-lived, because millions of people all over the world daily intoned the heartfelt mantra "God Save the Queen" (or King). Yet, he showed, if anything, they don't live as long as other members of the wealthy and pampered aristocratic class. Tens of millions of people in concert publicly wished (although they did not exactly pray) that Mao Zedong would live "for ten thousand years." Nearly everyone in ancient Egypt exhorted the gods to let the Pharaoh live "forever." These collective prayers failed. Their failure constitutes data.

By making pronouncements that are, even if only in principle, testable, religions, however unwillingly, enter the arena of science. Religions can no longer make unchallenged assertions about reality—so long as they do not seize secular power, provided they cannot coerce belief. This, in turn, has infuriated some followers of some religions. Occasionally they threaten skeptics with the direst imaginable penalties. Consider the following high-stakes alternative by William Blake in his innocuously titled *Auguries of Innocence:*

> *He who shall teach the Child to Doubt*
> *The rotting Grave shall ne'er get out.*
> *He who respects the Infant's Faith*
> *Triumphs over Hell & Death*

Of course many religions—devoted to reverence, awe, ethics, ritual, community, family, charity, and political and economic justice—are in no way challenged, but rather uplifted, by the findings of science. There is no necessary conflict between science and religion. On one level, they share similar and consonant roles, and each needs the other. Open and vigorous debate, even the consecration of doubt, is a Christian tradition going back to John Milton's *Areopagitica* (1644). Some of mainstream Christianity and Judaism embraces and even anticipated at least a portion of the humility, self-criticism, reasoned debate, and questioning of received wisdom that the best of science offers. But other sects, sometimes called conservative or fundamentalist—and today they seem to be in the ascendant, with the mainstream religions almost inaudible and invisible—have chosen to make a stand on matters subject to disproof, and thus have something to fear from science.

The religious traditions are often so rich and multivariate that they offer ample opportunity for renewal and revision, again especially when their sacred books can be interpreted metaphorically and allegorically. There is thus a middle ground of confessing past errors—as the Roman Catholic Church did in its 1992 acknowledgment that Galileo was right after all, that the Earth does revolve around the Sun: three centuries late, but courageous and most welcome nonetheless.

Modern Roman Catholicism has no quarrel with the Big Bang, with a Universe 15 billion or so years old, with the first living things arising from prebiological molecules, or with humans evolving from apelike ancestors—although it has special opinions on "ensoulment." Most mainstream Protestant and Jewish faiths take the same sturdy position.

In theological discussion with religious leaders, I often ask what their response would be if a central tenet of their faith were disproved by science. When I put this question to the current, Fourteenth, Dalai Lama, he unhesitatingly replied as no conservative or fundamentalist religious leaders do: In such a case, he said, Tibetan Buddhism would have to change.

Even, I asked, if it's a *really* central tenet, like (I searched for an example) reincarnation?

Even then, he answered.

However—he added with a twinkle—it's going to be hard to disprove reincarnation.

Plainly, the Dalai Lama is right. Religious doctrine that is insulated from disproof has little reason to worry about the advance of science. The grand idea, common to many faiths, of a Creator of the Universe is one such doctrine—difficult alike to demonstrate or to dismiss.

Moses Maimonides, in his *Guide for the Perplexed*, held that God could be truly known only if there were free and open study of both physics and theology [I, 55]. What would happen if science demonstrated an infinitely old Universe? Then theology would have to be seriously revamped [II, 25]. Indeed, this is the one conceivable finding of science that could disprove a Creator—because an infinitely old universe would never have been created. It would have always been here.

There are other doctrines, interests, and concerns that also worry about what science will find out. Perhaps, they suggest, it's better not to know. If men and women turn out to have different hereditary propensities, won't this be used as an excuse for the former to suppress the latter? If there's a genetic component of violence, might this justify repression of one ethnic group by another, or even precautionary incarceration? If mental illness is just brain chemistry, doesn't this unravel our efforts to keep a grasp on reality or to be responsible for our actions? If we are not the special handiwork of the Creator of the Universe, if our basic moral laws are merely invented by fallible lawgivers, isn't our struggle to maintain an orderly society undermined?

I suggest that in every one of these cases, religious or secular, we are much better off if we know the best available approximation to the truth—and if we keep before us a keen apprehension of the errors our interest group or belief system has committed in the past. In every case the imagined dire consequences of the truth being generally known are exaggerated. And again, we are not wise enough to know which lies, or even which shadings of the facts, can competently serve some higher social purpose—especially in the long run.

Chapter 16

WHEN SCIENTISTS KNOW SIN

The mind of man—how far will it advance?
Where will its daring impudence find limits?
If human villainy and human life shall wax
in due proportion, if the son shall always grow
in wickedness past his father, the gods must
add another world to this that all the sinners
may have space enough.

EURIPIDES,
Hippolytus
(428 B.C.)

In a post-war meeting with President Harry S Truman, J. Robert Oppenheimer—the scientific director of the Manhattan nuclear weapons Project—mournfully commented that scientists had bloody hands; they had now known sin. Afterwards, Truman instructed his aides that he never wished to see Oppenheimer again. Sometimes scientists are castigated for doing evil, and sometimes for warning about the evil uses to which science may be put.

More often, science is taken to task because it and its products are said to be morally neutral, ethically ambiguous, as readily employed in the service of evil as of good. This is an old indictment. It goes back probably to the flaking of stone tools and the domestication of fire. Since technology has been with our ancestral line from before the first human, since we are a technological species, this problem is not so much one of science as of human nature. By this I don't mean that science has no responsibility for the misuse of its findings. It has profound responsibility, and the more powerful its products the greater its responsibility.

Like assault weapons and market derivatives, the technologies that allow us to alter the global environment that sustains us should mandate caution and prudence. Yes, it's the same old humans who have made it so far. Yes, we're developing new technologies as we always have. But when the weaknesses we've always had join forces with a capacity to do harm on an unprecedented planetary scale, something more is required of us—an emerging ethic that also must be established on an unprecedented planetary scale.

Sometimes scientists try to have it both ways: to take credit for those applications of science that enrich our lives, but to distance themselves from the instruments of death, intentional and inadvertent, that also trace back to scientific research. The Australian philosopher John Passmore writes in his book *Science and Its Critics*,

The Spanish Inquisition sought to avoid direct responsibility for the burning of heretics by handing them over to the secular arm; to burn them itself, it piously explained, would be wholly inconsistent with its Christian principles. Few of us would allow the Inquisition thus easily to wipe its hands clean of bloodshed; it knew quite well what would happen. Equally, where the technological application of scientific discoveries is clear and obvious—as when a scientist works on nerve gases—he cannot properly claim that such applications are "none of his business," merely on the ground that it is the military forces, not scientists, who use the gases to disable or kill. This is even more obvious when the scientist deliberately offers help to governments, in exchange for funds. If a scientist, or a philosopher, accepts funds from some such body as an office of naval research, then he is cheating if he knows his work will be useless to them and must take some responsibility for the outcome if he knows that it will be useful. He is subject, properly subject, to praise or blame in relation to any innovations which flow from his work.

An important case history is provided by the career of the Hungarian-born physicist Edward Teller. Teller was marked at a young age by the Bela Kuhn communist revolution in Hungary, in which the property of middle-class families like his was expropriated, and by losing part of his leg in a streetcar accident, leaving him in permanent pain. His early contributions ranged from quantum mechanical selection rules and solid state physics to cosmology. It was he who chauffeured the physicist Leo Szilard to the vacationing Albert Einstein on Long Island in July 1939—a meeting that led to the historic letter from Einstein to President Franklin Roosevelt urging, in view of both scientific and political events in Nazi Germany, that the United States develop a fission, or "atomic" bomb. Recruited to work on the Manhattan Project, Teller arrived at Los Alamos and promptly refused to cooperate—not because he was dismayed at what an atomic bomb might do, but just the opposite: because he wanted to work on a much more destructive weapon, the fusion, or thermonuclear, or hydrogen bomb. (While there is a practical upper limit on the yield or destructive energy of an atomic bomb, there is no such limit for a hydrogen bomb. But a hydrogen bomb needs an atomic bomb as trigger.)

After the fission bomb was invented, after Germany and Japan surrendered, after the war was over, Teller remained a persistent advocate of what was called "the Super," specifically intended to intimidate the Soviet Union. Concern about the rebuilding, toughened, and militarized Soviet Union under Stalin and the national paranoia in America called McCarthyism eased Teller's path. A substantial obstacle was offered, though, in the person of Oppenheimer, who had become the chairman of the General Advisory Committee to the post-war Atomic Energy Commission. Teller provided critical testimony at a government hearing, questioning Oppenheimer's loyalty to the United States. Teller's involvement is generally thought to have played a major role in the aftermath: Although Oppenheimer's loyalty was not exactly impugned by the review board, somehow his security clearance was denied, he was retired from the AEC, and Teller's way to the Super was greased.

The technique for making a thermonuclear weapon is generally attributed to Teller and the mathematician Stanislas Ulam. Hans Bethe, the Nobel laureate physicist who headed the Theoretical Division at the Manhattan Project and who played a major role in the development of both the atomic and the hydrogen bombs, attests that Teller's original suggestion was flawed, and that the work of many people was necessary to bring the thermonuclear weapon to reality. With fundamental technical contributions from a young physicist named Richard Garwin, the first U.S. thermonuclear "device" was exploded in 1952—it was too unwieldy to be carried by a missile or bomber; it just sat there where it was assembled and blew up. The first true hydrogen bomb was a Soviet invention exploded one year later. There has been debate on whether the Soviet Union would have developed a thermonuclear weapon if the United States had not, and whether a U.S. thermonuclear weapon was even needed to deter Soviet use of their hydrogen bomb—since the U.S. by then possessed a substantial arsenal of fission weapons. The preponderance of current evidence is that the USSR—even before it exploded its first fission bomb—had a workable design for a thermonuclear weapon. It was "the next logical step." But Soviet pursuit of fusion weapons was much aided by the knowledge, from espionage, that the Americans were working on them.

From my point of view, the consequences of global nuclear war became much more dangerous with the invention of the hydrogen bomb, because airbursts of thermonuclear weapons are much more capable of burning cities, generating vast amounts of smoke, cooling and darkening the Earth, and inducing global-scale nuclear winter. This was perhaps the most controversial scientific debate I've been involved in (from about 1983–1990). Much of the debate was politically driven. The strategic implications of nuclear winter were disquieting to those wedded to a policy of massive retaliation to deter a nuclear attack, or to those wishing to preserve the option of a massive first strike. In either case, the environmental consequences work the self-destruction of any nation launching large numbers of thermonuclear weapons even with no retaliation from the adversary. A major segment of the strategic policy of decades, and the reason for accumulating tens of thousands of nuclear weapons, suddenly became much less credible.

The global temperature declines predicted in the original (1983) nuclear winter scientific paper were 15–20°C; current estimates are 10–15°C. The two values are in good agreement considering the irreducible uncertainties in the calculations. Both temperature declines are much greater than the difference between current global temperatures and those of the last Ice Age. The long-term consequences of global thermonuclear war have been estimated by an international team of 200 scientists, who concluded that through nuclear winter the global civilization and most of the people on Earth—including those far from the northern mid-latitude target zone—would be at risk, mainly from starvation. If large-scale nuclear war ever occurs, with cities targeted, the effort of Edward Teller and his colleagues in the United States (and the counterpart team headed by Andrei Sakharov in the Soviet Union) might be responsible for lowering the curtain on the human future. The hydrogen bomb is by far the most horrific weapon ever invented.

When nuclear winter was discovered in 1983, Teller was quick to argue both (1) that the physics was mistaken, and (2) that the discovery had been made years earlier under his tutelage at the Lawrence Livermore National Laboratory. There is in fact no evidence for such a prior discovery, and considerable evidence that those in every nation charged to inform their national leaders of the effects of nuclear weapons had consistently overlooked nuclear winter. But if Teller is right, then it was unconscionable of him not to have disclosed the purported discovery to the affected parties—the citizens and leaders of his nation and the world. As in the Stanley Kubrick movie *Dr.*

Strangelove, classifying the ultimate weapon—so no one knows that it exists or what it can do—is the ultimate absurdity.

It seems to me impossible for any normal human being to be untroubled by helping to make such an invention, even putting nuclear winter aside. The stresses, conscious or unconscious, on those who take credit for the contrivance must be considerable. Whatever his actual contributions, Edward Teller has been widely described as the "father" of the hydrogen bomb. In an admiring 1954 article, *Life* magazine described his "almost fanatic determination" to build the hydrogen bomb. Much of his subsequent career can, I think, be understood as an attempt to justify what he begat. Teller has contended, not implausibly, that hydrogen bombs keep the peace, or at least prevent thermonuclear war, because the consequences of warfare between nuclear powers are now too dangerous. We haven't had a nuclear war yet, have we? But all such arguments assume that the nuclear-armed nations are and always will be, without exception, rational actors, and that bouts of anger and revenge and madness will never overtake their leaders (or military and secret police officers in charge of nuclear weapons). In the century of Hitler and Stalin, this seems ingenuous.

Teller has been a major force in preventing a comprehensive treaty banning nuclear weapons tests. He made it much more difficult to accomplish the 1963 Limited (above-ground) Test Ban Treaty. His argument that above-ground testing was essential to maintain and "improve" the nuclear arsenals, that ratifying the treaty would "give away the future safety of our country," has proven specious. He has also been a vigorous proponent of the safety and cost-effectiveness of fission power plants, claiming himself to be the only casualty of the Three Mile Island nuclear accident in Pennsylvania in 1979; he had a heart attack, he says, debating the issue.

Teller advocated exploding nuclear weapons from Alaska to South Africa, to dredge harbors and canals, to obliterate troublesome mountains, to do heavy earth-moving. When he proposed such a scheme to Queen Frederika of Greece, she is said to have responded, "Thank you, Dr. Teller, but Greece has enough quaint ruins already." Want to test Einstein's general relativity? Then explode a nuclear weapon on the far side of the Sun, Teller proposed. Want to understand the chemical composition of the Moon? Then fly a hydrogen bomb to the Moon, explode it, and examine the spectrum of the flash and fireball.

Also in the 1980s, Teller sold President Ronald Reagan the notion of Star Wars—called by them the "Strategic Defense Initiative," SDI. Reagan seems to have believed a highly imaginative story of Teller's that it was possible to build a desk-sized orbiting hydrogen-bomb-driven X-ray laser that would destroy 10,000 Soviet warheads in flight, and provide genuine protection for the citizens of the United States in case of global thermonuclear war.

It is claimed by apologists for the Reagan Administration that, whatever the exaggerations in capability, some of it intentional, SDI was responsible for the collapse of the Soviet Union. There is no serious evidence in support of this contention. Andrei Sakharov, Yevgeny Velikhov, Roald Sagdeev, and other scientists who advised President Mikhail Gorbachev made it clear that if the United States really went ahead with a Star Wars program, the safest and cheapest Soviet response would be merely to augment its existing arsenal of nuclear weapons and delivery systems. In this way Star Wars could have increased, not decreased, the peril of thermonuclear war. At any rate, Soviet expenditures on space-based defenses against American nuclear missiles were comparatively paltry—hardly of a magnitude to trigger a collapse of the Soviet economy. The fall of the USSR has much more to do with the failure of the command economy, growing awareness of the standard of living in the West, widespread disaffection from a moribund Communist ideology, and—although he did not intend such an outcome—Gorbachev's promotion of *glasnost*, or openness.

Ten thousand American scientists and engineers publicly pledged they would not work on Star Wars or accept money from the SDI organization. This provides an example of widespread and courageous non-cooperation by scientists (at some conceivable personal cost) with a democratic government that had, temporarily at least, lost its way.

Teller has also advocated the development of burrowing nuclear warheads—so underground command centers and deeply buried shelters for the leadership (and their families) of an adversary nation might be dug down to and wiped out; and 0.1-kiloton nuclear warheads that would saturate an enemy country, obliterating its infrastructure "without a single casualty": Civilians would be alerted in advance. Nuclear war would be humane.

As I write, Edward Teller—still vigorous and retaining considerable intellectual powers into his late eighties—has mounted a campaign, with his counterparts in the former Soviet nuclear weapons establishment, to develop and explode new generations of high-yield thermonuclear weapons in space, in order to destroy or deflect asteroids that might be on collision trajectories with the Earth. I worry that premature experimentation with the orbits of nearby asteroids may involve extreme dangers for our species.

Dr. Teller and I have met privately. We've debated at scientific meetings, in the national media, and in a closed rump session of Congress. We've had strong disagreements, especially on Star Wars, nuclear winter, and asteroid defense. Perhaps all this has hopelessly colored my view of him. Although he has always been a fervent anticommunist and technophile, as I look back over his life it seems to me I see something more in his desperate attempt to justify the hydrogen bomb: Its effects aren't as bad as you might think. It can be used to defend the world from other hydrogen bombs, for science, for civil engineering, to protect the population of the United States against an enemy's thermonuclear weapons, to wage war humanely, to save the planet from random hazards from space. Somehow, somewhere, he wants to believe, thermonuclear weapons, and he, will be acknowledged by the human species as its savior and not its destroyer.

When scientific research provides fallible nations and political leaders with formidable, indeed awesome powers, many dangers present themselves: One is that some of the scientists involved may lose all but a superficial semblance of objectivity. As always, power tends to corrupt. In this circumstance, the institution of secrecy is especially pernicious, and the checks and balances of a democracy become especially valuable. (Teller, who has flourished in the secrecy culture, has also repeatedly attacked it.) The CIA Inspector General commented in 1995 that "absolute secrecy corrupts absolutely." The most open and vigorous debate is often the only protection against the most perilous misuse of technology. The critical piece of the counterargument may be something obvious—that many scientists or even lay people could come up with provided there were no penalties for speaking out. Or it might be something more subtle, something that would be noted by an obscure graduate student in some locale remote from Washington, D.C.—who, if the arguments were closely held and highly secret, would never have the opportunity to address the issue.

What realm of human endeavor is not morally ambiguous? Even folk institutions that purport to give us advice on behavior and ethics seem fraught with contradictions. Consider aphorisms: Haste makes waste. Yes, but a stitch in time saves nine. Better safe than sorry; but nothing ventured, nothing gained. Where there's smoke, there's fire; but you can't tell a book by its cover. A penny saved is a penny earned; but you can't take it with you. He who hesitates is lost; but fools rush in where angels fear to tread. Two heads are better than one; but too many cooks spoil the broth. There was a time when people planned or justified their actions on the basis of such contradictory platitudes. What is the moral responsibility of the aphorist? Or the Sun-sign astrologer, the Tarot card reader, the tabloid prophet?

Or consider the mainstream religions. We are enjoined in Micah to do justly and love mercy; in Exodus we are forbidden to commit murder; in Leviticus we are commanded to love our neighbor as ourselves; and in the Gospels we are urged to love our enemies. Yet think of the rivers of blood spilled by fervent followers of the books in which these well-meaning exhortations are embedded.

In Joshua and in the second half of Numbers is celebrated the mass murder of men, women, children, down to the domestic animals in city after city across the whole land of Canaan. Jericho is obliterated in a *kherem*, a "holy war." The only justification offered for this slaughter is the mass murderers' claim that, in exchange for circumcising their sons and adopting a particular set of rituals, their ancestors were long before promised that this land was their land. Not a hint of self-reproach, not a muttering of patriarchal or divine disquiet at these campaigns of extermination can be dug out of holy scripture. Instead, Joshua "destroyed all that breathed, as the Lord God of Israel commanded" (Joshua 10:40). And these events are not incidental, but central to the main narrative thrust of the Old Testament. Similar stories of mass murder (and in the case of the Amalekites, genocide) can be found in the books of Saul, Esther, and elsewhere in the Bible, with hardly a pang of moral doubt. It was all, of course, troubling to liberal theologians of a later age.

It is properly said that the Devil can "quote Scripture to his purpose." The Bible is full of so many stories of contradictory moral purpose that every generation can find scriptural justification for nearly any action it proposes—from incest, slavery, and mass murder to the most refined love, courage, and self-sacrifice. And this moral multiple personality disorder is hardly restricted to Judaism and Christianity. You can find it deep within Islam, the Hindu tradition, indeed nearly all the world's religions. Perhaps then it is not so much scientists as people who are morally ambiguous.

It is the particular task of scientists, I believe, to alert the public to possible dangers, especially those emanating from science or foreseeable through the use of science. Such a mission is, you might say, prophetic. Clearly the warnings need to be judicious and not more flamboyant than the dangers require; but if we must make errors, given the stakes, they should be on the side of safety.

Among the !Kung San hunter-gatherers of the Kalahari Desert, when two men, perhaps testosterone-inflamed, would begin to argue, the women would reach for their poison arrows and put the weapons out of harm's way. Today our poison arrows can destroy the global civilization and just possibly annihilate our species. The price of moral ambiguity is now too high. For this reason—and not because of its approach to knowledge—the ethical responsibility of scientists must also be high, extraordinarily high, unprecedentedly high. I wish graduate science programs explicitly and systematically raised these questions with fledgling scientists and engineers. And sometimes I wonder whether in our society, too, the women—and the children—will eventually put the poison arrows out of harm's way.

Chapter 17

———

THE
MARRIAGE OF
SKEPTICISM
AND WONDER

Nothing is too wonderful to be true.

Remark attributed to
MICHAEL FARADAY
(1791–1867)

Insight, untested and unsupported, is an insufficient guarantee of truth.

BERTRAND RUSSELL,
Mysticism and Logic
(1929)

When we are asked to swear in American courts of law—that we will tell "the truth, the whole truth, and nothing but the truth"—we are being asked the impossible. It is simply beyond our powers. Our memories are fallible; even scientific truth is merely an approximation; and we are ignorant about nearly all of the Universe. Nevertheless, a life may depend on our testimony. To swear to tell the truth, the whole truth, and nothing but the truth *to the limit of our abilities* is a fair request. Without the qualifying phrase, though, it's simply out of touch. But such a qualification, however consonant with human reality, is unacceptable to any legal system. If everyone tells the truth only to a degree determined by individual judgment, then incriminating or awkward facts might be withheld, events shaded, culpability hidden, responsibility evaded, and justice denied. So the law strives for an impossible standard of accuracy, and we do the best we can.

In the jury selection process, the court needs to be reassured that the verdict will be based on evidence. It makes heroic efforts to weed out bias. It is aware of human imperfection. Does the potential juror personally know the district attorney, or the prosecutor, or the defense attorney? What about the judge or the other jurors? Has she formed an opinion about this case not from the facts laid out in court but from pre-trial publicity? Will she assign evidence from police officers greater or lesser weight than evidence from witnesses for the defense? Is she biased against the defendant's ethnic group? Does the potential juror live in the neighborhood where the crimes were committed, and might that influence her judgment? Does she have a scientific background about matters on which expert witnesses will testify? (This is often a count against her.) Are any of her relatives or close family members employed in law enforcement or criminal law? Has she herself ever had any run-ins with police that might influence her judgment in the trial? Was any close friend or relative ever arrested on a similar charge?

The American system of jurisprudence recognizes a wide range of factors, predispositions, prejudices, and experiences that might cloud our judgment, or affect our objectivity—sometimes even without our knowing it. It goes to great, perhaps even extravagant, lengths to safe-guard the process of judgment in a criminal trial from the human weaknesses of those who must decide on innocence or guilt. Even then, of course, the process sometimes fails.

Why would we settle for anything less when interrogating the nat-ural world, or when attempting to decide on vital matters of politics, economics, religion, and ethics?

—

If it is to be applied consistently, science imposes, in exchange for its manifold gifts, a certain onerous burden: We are enjoined, no matter how uncomfortable it might be, to consider *ourselves* and our cultural institutions scientifically—not to accept uncritically whatever we're told; to surmount as best we can our hopes, conceits, and unexamined beliefs; to view ourselves as we really are. Can we conscientiously and courageously follow planetary motion or bacterial genetics wherever the search may lead, but declare the origin of matter or human behav-ior off-limits? Because its explanatory power is so great, once you get the hang of scientific reasoning you're eager to apply it everywhere. However, in the course of looking deeply within ourselves, we may challenge notions that give comfort before the terrors of the world. I'm aware that some of the discussion in, say, the preceding chapter may have such a character.

When anthropologists survey the thousands of distinct cultures and ethnicities that comprise the human family, they are struck by how few features there are that are givens, always present no matter how exotic the society. There are, for example, cultures—the Ik of Uganda is one—where all Ten Commandments seem to be systematically, insti-tutionally ignored. There are societies that abandon their old and their newborn, that eat their enemies, that use seashells or pigs or young women for money. But they all have a strong incest taboo, they all use technology, and almost all believe in a supernatural world of gods and spirits—often connected with the natural environment they inhabit and the well-being of the plants and animals they eat. (The ones with a supreme god who lives in the sky tend to be the most ferocious—torturing their enemies for example. But this is a statistical correlation only; the causal link has not been established, although speculations naturally present themselves.)

In every such society, there is a cherished world of myth and metaphor which co-exists with the workaday world. Efforts to reconcile the two are made, and any rough edges at the joints tend to be off-limits and ignored. We compartmentalize. Some scientists do this too, effortlessly stepping between the skeptical world of science and the credulous world of religious belief without skipping a beat. Of course, the greater the mismatch between these two worlds, the more difficult it is to be comfortable, with untroubled conscience, with both.

In a life short and uncertain, it seems heartless to do anything that might deprive people of the consolation of faith when science cannot remedy their anguish. Those who cannot bear the burden of science are free to ignore its precepts. But we cannot have science in bits and pieces, applying it where we feel safe and ignoring it where we feel threatened—again, because we are not wise enough to do so. Except by sealing the brain off into separate airtight compartments, how is it possible to fly in airplanes, listen to the radio or take antibiotics while holding that the Earth is around 10,000 years old or that all Sagittarians are gregarious and affable?

Have I ever heard a skeptic wax superior and contemptuous? Certainly. I've even sometimes heard, to my retrospective dismay, that unpleasant tone in my own voice. There are human imperfections on both sides of this issue. Even when it's applied sensitively, scientific skepticism may come across as arrogant, dogmatic, heartless, and dismissive of the feelings and deeply held beliefs of others. And, it must be said, some scientists and dedicated skeptics apply this tool as a blunt instrument, with little finesse. Sometimes it looks as if the skeptical conclusion came first, that contentions were dismissed before, not after, the evidence was examined. All of us cherish our beliefs. They are, to a degree, self-defining. When someone comes along who challenges our belief system as insufficiently well-based—or who, like Socrates, merely asks embarrassing questions that we haven't thought of, or demonstrates that we've swept key underlying assumptions under the rug—it becomes much more than a search for knowledge. It feels like a personal assault.

The scientist who first proposed to consecrate doubt as a prime virtue of the inquiring mind made it clear that it was a tool and not an end in itself. René Descartes wrote,

> I did not imitate the skeptics who doubt only for doubting's sake, and pretend to be always undecided; on the contrary, my whole intention was to arrive at a certainty, and to dig away the drift and the sand until I reached the rock or the clay beneath.

In the way that skepticism is sometimes applied to issues of public concern, there *is* a tendency to belittle, to condescend, to ignore the fact that, deluded or not, supporters of superstition and pseudoscience are human beings with real feelings, who, like the skeptics, are trying to figure out how the world works and what our role in it might be. Their motives are in many cases consonant with science. If their culture has not given them all the tools they need to pursue this great quest, let us temper our criticism with kindness. None of us comes fully equipped.

Clearly there are limits to the uses of skepticism. There is some cost-benefit analysis which must be applied, and if the comfort, consolation and hope delivered by mysticism and superstition is high, and the dangers of belief comparatively low, should we not keep our misgivings to ourselves? But the issue is tricky. Imagine that you enter a big-city taxicab and the moment you get settled in, the driver begins a harangue about the supposed iniquities and inferiorities of another ethnic group. Is your best course to keep quiet, bearing in mind that silence conveys assent? Or is it your moral responsibility to argue with him, to express outrage, even to leave the cab—because you know that every silent assent will encourage him next time, and every vigorous dissent will cause him next time to think twice? Likewise, if we offer too much silent assent about mysticism and superstition—even when it seems to be doing a little good—we abet a general climate in which skepticism is considered impolite, science tiresome, and rigorous thinking somehow stuffy and inappropriate. Figuring out a prudent balance takes wisdom.

The Committee for the Scientific Investigation of Claims of the Paranormal is an organization of scientists, academics, magicians, and others dedicated to skeptical scrutiny of emerging or full-blown pseudo-sciences. It was founded by the University of Buffalo philosopher Paul Kurtz in 1976. I've been affiliated with it since its beginning. Its acronym, CSICOP, is pronounced "sci-cop"—as if it's an organization of scientists performing a police function. Those wounded by CSICOP's analyses sometimes make just such a complaint: It's hostile to every new idea, they say, will go to absurd lengths in its knee-jerk debunking, is a vigilante organization, a New Inquisition, and so on.

CSICOP *is* imperfect. In certain cases such a critique is to some degree justified. But from my point of view CSICOP serves an important social function—as a well-known organization to which media can apply when they wish to hear the other side of the story, especially when some amazing claim of pseudoscience is adjudged newsworthy. It used to be (and for much of the global news media it still is) that every levitating guru, visiting alien, channeler, and faith healer, when covered by the media, would be treated nonsubstantively and uncritically. There would be no institutional memory at the television studio or newspaper or magazine about other, similar claims previously shown to be scams and bamboozles. CSICOP represents a counterbalance, although not yet nearly a loud enough voice, to the pseudoscience gullibility that seems second nature to so much of the media.

One of my favorite cartoons shows a fortune-teller scrutinizing the mark's palm and gravely concluding, "You are very gullible." CSICOP publishes a bimonthly periodical called *The Skeptical Inquirer*. On the day it arrives, I take it home from the office and pore through its pages, wondering what new misunderstandings will be revealed. There's always another bamboozle that I never thought of. Crop circles! Aliens have come and made perfect circles and mathematical messages in wheat! . . . Who would have thought it? So unlikely an artistic medium. Or they've come and eviscerated cows—on a large scale, systematically. Farmers are furious. At first, I'm impressed by the inventiveness of the stories. But then, on more sober reflection, it always strikes me how dull and routine these accounts are; what a compilation of unimaginative, stale ideas, chauvinisms, hopes, and fears dressed up as facts. The contentions, from this point of view, are suspect on their face. That's all they can conceive the extraterrestrials doing . . . making circles in wheat? What a failure of the imagination! With every issue, another facet of pseudoscience is revealed and criticized.

And yet, the chief deficiency I see in the skeptical movement is in its polarization: Us vs. Them—the sense that *we* have a monopoly on the truth; that those other people who believe in all these stupid doctrines are morons; that if you're sensible, you'll listen to us; and if not, you're beyond redemption. This is unconstructive. It does not get the message across. It condemns the skeptics to permanent minority status; whereas, a compassionate approach that from the beginning acknowledges the human roots of pseudoscience and superstition might be much more widely accepted.

If we understand this, then of course we feel the uncertainty and pain of the abductees, or those who dare not leave home without consulting their horoscopes, or those who pin their hopes on crystals from Atlantis. And such compassion for kindred spirits in a common quest also works to make science and the scientific method less off-putting, especially to the young.

Many pseudoscientific and New Age belief systems emerge out of dissatisfaction with conventional values and perspectives—and are therefore themselves a kind of skepticism. (The same is true of the origins of most religions.) David Hess (in *Science and the New Age*) argues that

the world of paranormal beliefs and practices cannot be reduced to cranks, crackpots, and charlatans. A large number of sincere people are exploring alternative approaches to questions of personal meaning, spirituality, healing, and paranormal experience in general. To the skeptic, their quest may ultimately rest on a delusion, but debunking is hardly likely to be an effective rhetorical device for their rationalist project of getting [people] to recognize what appears to the skeptic as mistaken or magical thinking. . . . [T]he skeptic might take a clue from cultural anthropology and develop a more sophisticated skepticism by understanding alternative belief systems from the perspective of the people who hold them and by situating these beliefs in their historical, social, and cultural contexts. As a result, the world of the paranormal may appear less as a silly turn toward irrationalism and more as an idiom through which segments of society express their conflicts, dilemmas, and identities. . .

To the extent that skeptics have a psychological or sociological theory of New Age beliefs, it tends to be very simplistic: paranormal beliefs are "comforting" to people who cannot handle the reality of an atheistic universe, or their beliefs are the product of an irresponsible media that is not encouraging the public to think critically. . .

But Hess's just criticism promptly deteriorates into complaints that parapsychologists "have had their careers ruined by skeptical colleagues," and that skeptics exhibit "a kind of religious zeal to defend the materialistic and atheistic world view that smacks of what has been called 'scientific fundamentalism' or 'irrational rationalism.' "

This is a common but to me deeply mysterious—indeed, occult—complaint. Again, we know a great deal about the existence and properties of matter. If a given phenomenon can already be plausibly understood in terms of matter and energy, why should we hypothesize that something else—something for which there is as yet no other good evidence—is responsible? Yet the complaint persists: Skeptics won't accept that there's an invisible fire-breathing dragon in my garage because they're all atheistic materialists.

In *Science in the New Age*, skepticism is discussed, but it is not understood, and it is certainly not practiced. All sorts of paranormal claims are quoted, skeptics are "deconstructed," but you can never learn from reading it that there are ways to decide whether New Age and parapsychological claims to knowledge are promising or false. It's all, as in many postmodernist texts, a matter of how strongly people feel and what their biases may be.

Robert Anton Wilson (in *The New Inquisition: Irrational Rationalism and the Citadel of Science* [Phoenix: Falcon Press, 1986]) describes skeptics as the "New Inquisition." But to my knowledge no skeptic compels belief. Indeed, on most TV documentaries and talk shows, skeptics get short shrift and almost no air time. All that's happening is that some doctrines and methods are being criticized—at the worst, ridiculed—in magazines like *The Skeptical Inquirer* with circulations of a few tens of thousands. New Agers are not much, as in earlier times, being called up before criminal tribunals, nor whipped for having visions, and they are certainly not being burned at the stake. Why fear a little criticism? Aren't they interested to see how well their beliefs hold up against the best counterarguments the skeptics can muster?

—

Perhaps one percent of the time, someone who has an idea that smells, feels, and looks indistinguishable from the usual run of pseudoscience will turn out to be right. Maybe some undiscovered reptile left over from the Cretaceous period will indeed be found in Loch Ness or the Congo Republic; or we will find artifacts of an advanced, non-human species elsewhere in the Solar System. At the time of writing there are three claims in the ESP field which, in my opinion, deserve serious study: (1) that by thought alone humans can (barely) affect random number generators in computers; (2) that people under mild sensory deprivation can receive thoughts or images "projected" at them; and (3) that young children sometimes report the details of a previous life, which upon checking turn out to be accurate and which they could not have known about in any other way than reincarnation. I pick these claims not because I think they're likely to be valid (I don't), but as examples of contentions that *might* be true. The last three have at least some, although still dubious, experimental support. Of course, I could be wrong.

In the middle 1970s an astronomer I admire put together a modest manifesto called "Objections to Astrology" and asked me to endorse it. I struggled with his wording, and in the end found myself unable to sign—not because I thought astrology has any validity whatever, but because I felt (and still feel) that the tone of the statement was authoritarian. It criticized astrology for having origins shrouded in superstition. But this is true as well for religion, chemistry, medicine, and astronomy, to mention only four. The issue is not what faltering and rudimentary knowledge astrology came from, but what is its present validity. Then there was speculation on the psychological motivations of those who believe in astrology. These motivations—for example, the feeling of powerlessness in a complex, troublesome and unpredictable world—might explain why astrology is not generally given the skeptical scrutiny it deserves, but is quite peripheral to whether it works.

The statement stressed that we can think of no mechanism by which astrology could work. This is certainly a relevant point but by itself it's unconvincing. No mechanism was known for continental drift (now subsumed in plate tectonics) when it was proposed by Alfred Wegener in the first quarter of the twentieth century to explain a range of puzzling data in geology and paleontology. (Ore-bearing veins of rocks and fossils seemed to run continuously from Eastern South America to West Africa; were the two continents once touching and the Atlantic Ocean new to our planet?) The notion was roundly dismissed by all the great geophysicists, who were certain that continents were fixed, not floating on anything, and therefore unable to "drift." Instead, the key twentieth-century idea in geophysics turns out to be plate tectonics; we now understand that continental plates do indeed float and "drift" (or better, are carried by a kind of conveyor belt driven by the great heat engine of the Earth's interior), and all those great geophysicists were simply wrong. Objections to pseudoscience on the grounds of unavailable mechanism can be mistaken—although if the contentions violate well-established laws of physics, such objections of course carry great weight.

Many valid criticisms of astrology can be formulated in a few sentences: for example, its acceptance of precession of the equinoxes in announcing an "Age of Aquarius" and its rejection of precession of the equinoxes in casting horoscopes; its neglect of atmospheric refraction; its list of supposedly significant celestial objects that is mainly limited to naked eye objects known to Ptolemy in the second century, and that ignores an enormous variety of new astronomical objects discovered since (where is the astrology of near-Earth asteroids?); inconsistent requirements for detailed information on the time as compared to the latitude and longitude of birth; the failure of astrology to pass the identical-twin test; the major differences in horoscopes cast from the same birth information by different astrologers; and the absence of demonstrated correlation between horoscopes and such psychological tests as the Minnesota Multiphasic Personality Inventory.

What I would have signed is a statement describing and refuting the principal tenets of astrological belief. Such a statement would have been far more persuasive than what was actually circulated and published. But astrology, which has been with us for four thousand years or more, today seems more popular than ever. At least a quarter of all Americans, according to opinion polls, "believe" in astrology. A third think Sun-sign astrology is "scientific." The fraction of schoolchildren believing in astrology rose from 40 percent to 59 percent between 1978 and 1984. There are perhaps ten times more astrologers than astronomers in the United States. In France there are more astrologers than Roman Catholic clergy. No stuffy dismissal by a gaggle of scientists makes contact with the social needs that astrology—no matter how invalid it is—addresses, and science does not.

—

As I've tried to stress, at the heart of science is an essential balance between two seemingly contradictory attitudes—an openness to new ideas, no matter how bizarre or counterintuitive, and the most ruthlessly skeptical scrutiny of all ideas, old and new. This is how deep truths are winnowed from deep nonsense. The collective enterprise of creative thinking *and* skeptical thinking, working together, keeps the field on track. Those two seemingly contradictory attitudes are, though, in some tension.

Consider this claim: As I walk along, time—as measured by my wristwatch or my aging process—slows down. Also, I shrink in the direction of motion. Also, I get more massive. Who has ever witnessed such a thing? It's easy to dismiss it out of hand. Here's another: Matter and antimatter are all the time, throughout the Universe, being created from nothing. Here's a third: Once in a *very* great while, your car will spontaneously ooze through the brick wall of your garage and be found the next morning on the street. They're all absurd! But the first is a statement of special relativity, and the other two are consequences of quantum mechanics (vacuum fluctuations and barrier tunneling,* they're called). Like it or not, that's the way the world is. If you insist it's ridiculous, you'll be forever closed to some of the major findings on the rules that govern the Universe.

If you're only skeptical, then no new ideas make it through to you. You never learn anything. You become a crotchety misanthrope convinced that nonsense is ruling the world. (There is, of course, much data to support you.) Since major discoveries at the borderlines of science are rare, experience will tend to confirm your grumpiness. But every now and then a new idea turns out to be on the mark, valid and wonderful. If you're too resolutely and uncompromisingly skeptical,

---

* The average waiting time per stochastic ooze is *much* longer than the age of the Universe since the Big Bang. But, however improbable, in principle it might happen tomorrow.

you're going to miss (or resent) the transforming discoveries in science, and either way you will be obstructing understanding and progress. Mere skepticism is not enough.

At the same time, science requires the most vigorous and uncompromising skepticism, because the vast majority of ideas are simply wrong, and the only way to winnow the wheat from the chaff is by critical experiment and analysis. If you're open to the point of gullibility and have not a microgram of skeptical sense in you, then you cannot distinguish the promising ideas from the worthless ones. Uncritically accepting every proffered notion, idea, and hypothesis is tantamount to knowing nothing. Ideas contradict one another; only through skeptical scrutiny can we decide among them. Some ideas really are better than others.

The judicious mix of these two modes of thought is central to the success of science. Good scientists do both. On their own, talking to themselves, they churn up many new ideas, and criticize them systematically. Most of the ideas never make it to the outside world. Only those that pass a rigorous self-filtration make it out to be criticized by the rest of the scientific community.

Because of this dogged mutual criticism and self-criticism, and the proper reliance on experiment as the arbiter between contending hypotheses, many scientists tend to be diffident about describing their own sense of wonder at the dawning of a wild surmise. This is a pity, because these rare exultant moments demystify and humanize the scientific endeavor.

No one can be entirely open or completely skeptical. We all must draw the line somewhere.* An ancient Chinese proverb advises, "Better to be too credulous than too skeptical," but this is from an extremely conservative society in which stability was much more prized than freedom and where the rulers had a powerful vested interest in not being challenged. Most scientists, I believe, would say, "Better to be too skeptical than too credulous." But neither is easy. Responsible, thoroughgoing, rigorous skepticism requires a hardnosed habit of thought that takes practice and training to master. Credulity—I think a better word here is "openness" or "wonder"—does not come easily

---

* And in some cases skepticism would be simply silly, as for example in learning to spell.

either. If we really are to be open to counterintuitive ideas in physics or social organization or anything else, we must grasp those ideas. It means nothing to be open to a proposition we don't understand.

Both skepticism and wonder are skills that need honing and practice. Their harmonious marriage within the mind of every schoolchild ought to be a principal goal of public education. I'd love to see such a domestic felicity portrayed in the media, television especially: a community of people really working the mix—full of wonder, generously open to every notion, dismissing nothing except for good reason, but at the same time, and as second nature, demanding stringent standards of evidence—and these standards applied with at least as much rigor to what they hold dear as to what they are tempted to reject with impunity.

Chapter 18

———

THE WIND
MAKES DUST

[T]he wind makes dust because it intends to blow, taking away our footprints.

Specimens of Bushmen Folklore,
W. H. I. Bleek and L. C. Lloyd, collectors,
L. C. Lloyd, editor (1911)

[E]very time a savage tracks his game he employs a minuteness of observation, and an accuracy of inductive and deductive reasoning which, applied to other matters, would assure some reputation as a man of science . . . [T]he intellectual labour of a "good hunter or warrior" considerably exceeds that of an ordinary Englishman.

THOMAS H. HUXLEY,
Collected Essays, Volume II, Darwiniana:
Essays (London: Macmillan, 1907),
pp. 175–6 [from "Mr. Darwin's Critics" (1871)]

W hy should so many people find science hard to learn and hard to teach? I've tried to suggest some of the reasons—its precision, its counterintuitive and disquieting aspects, its prospects of misuse, its independence of authority, and so on. But is there something deeper? Alan Cromer is a physics professor at Northeastern University in Boston who was surprised to find so many students unable to grasp the most elementary concepts in his physics class. In *Uncommon Sense: The Heretical Nature of Science* (1993), Cromer proposes that science is difficult because it's new. We, a species that's a few hundred thousand years old, discovered the method of science only a few centuries ago, he says. Like writing, which is only a few millennia old, we haven't gotten the hang of it yet—or at least not without very serious and attentive study.

Except for an unlikely concatenation of historical events, he suggests, we would never have invented science:

> This hostility to science, in the face of its obvious triumphs and benefits, is . . . evidence that it is something outside the mainstream of human development, perhaps a fluke.

Chinese civilization invented movable type, gunpowder, the rocket, the magnetic compass, the seismograph, and systematic observations and chronicles of the heavens. Indian mathematicians invented the zero, the key to comfortable arithmetic and therefore to quantitative science. Aztec civilization developed a far better calendar than that of the European civilization that inundated and destroyed it; they were better able, and for longer periods into the future, to predict where the planets would be. But none of these civilizations, Cromer argues, had developed the skeptical, inquiring, experimental method of science. All of that came out of ancient Greece:

The development of objective thinking by the Greeks appears to have required a number of specific cultural factors. First was the assembly, where men first learned to persuade one another by means of rational debate. Second was a maritime economy that prevented isolation and parochialism. Third was the existence of a widespread Greek-speaking world around which travelers and scholars could wander. Fourth was the existence of an independent merchant class that could hire its own teachers. Fifth was the *Iliad* and the *Odyssey*, literary masterpieces that are themselves the epitome of liberal rational thinking. Sixth was a literary religion not dominated by priests. And seventh was the persistence of these factors for 1,000 years.

That all these factors came together in one great civilization is quite fortuitous; it didn't happen twice.

I'm sympathetic to part of this thesis. The ancient Ionians were the first we know of to argue systematically that laws and forces of Nature, rather than gods, are responsible for the order and even the existence of the world. As Lucretius summarized their views, "Nature free at once and rid of her haughty lords is seen to do all things spontaneously of herself without the meddling of the gods." Except for the first week of introductory philosophy courses, though, the names and notions of the early Ionians are almost never mentioned in our society. Those who dismiss the gods tend to be forgotten. We are not anxious to preserve the memory of such skeptics, much less their ideas. Heroes who try to explain the world in terms of matter and energy may have arisen many times in many cultures, only to be obliterated by the priests and philosophers in charge of the conventional wisdom—as the Ionian approach was almost wholly lost after the time of Plato and Aristotle. With many cultures and many experiments of this sort, it may be that only on rare occasions does the idea take root.

Plants and animals were domesticated and civilization began only ten or twelve thousand years ago. The Ionian experiment is 2,500 years old. It was almost entirely expunged. We can see steps towards science in ancient China, India, and elsewhere, even though faltering, incomplete, and bearing less fruit. But suppose the Ionians had never existed, and Greek science and mathematics never flourished. Is it possible that never again in the history of the human species would science have emerged? Or, given many cultures and many alternative historical skeins, isn't it likely that the right combination of factors would come into play somewhere else, sooner or later—in the islands of Indonesia, say, or in the Caribbean on the outskirts of a Mesoamerican civilization untouched by conquistadores, or in Norse colonies on the shores of the Black Sea?

The impediment to scientific thinking is not, I think, the difficulty of the subject. Complex intellectual feats have been mainstays even of oppressed cultures. Shamans, magicians, and theologians are highly skilled in their intricate and arcane arts. No, the impediment is political and hierarchical. In those cultures lacking unfamiliar challenges, external or internal, where fundamental change is unneeded, novel ideas need not be encouraged. Indeed, heresies can be declared dangerous; thinking can be rigidified; and sanctions against impermissible ideas can be enforced—all without much harm. But under varied and changing environmental or biological or political circumstances, simply copying the old ways no longer works. Then, a premium awaits those who, instead of blandly following tradition, or trying to foist their preferences onto the physical or social Universe, are open to what the Universe teaches. Each society must decide where in the continuum between openness and rigidity safety lies.

Greek mathematics was a brilliant step forward. Greek science, on the other hand—its first steps rudimentary and often uninformed by experiment—was riddled with error. Despite the fact that we cannot see in pitch darkness, they believed that vision depends on a kind of radar that emanates from the eye, bounces off what we're seeing, and returns to the eye. (Nevertheless, they made substantial progress in optics.) Despite the obvious resemblance of children to their mothers, they believed that heredity was carried by semen alone, the woman a mere passive receptacle. They believed that the horizontal motion of a thrown rock somehow lifts it up, so that it takes longer to reach the ground than a rock dropped from the same height at the same moment. Enamored of simple geometry, they believed the circle to be "perfect"; despite the "Man in the Moon" and sunspots (occasionally visible to the naked eye at sunset), they held the heavens also to be "perfect"; therefore, planetary orbits had to be circular.

Being freed from superstition isn't enough for science to grow. One must also have the idea of interrogating Nature, of doing experiments. There were some brilliant examples—Eratosthenes' measurement of the Earth's diameter, say, or Empedocles' clepsydra experiment demonstrating the material nature of air. But in a society in which manual labor is demeaned and thought fit only for slaves, as in the classical Græco-Roman world, the experimental method does not thrive. Science requires us to be freed of gross superstition and gross injustice both. Often, superstition and injustice are imposed by the same ecclesiastical and secular authorities, working hand in glove. It is no surprise that political revolutions, skepticism about religion, and the rise of science might go together. Liberation from superstition is a necessary but not a sufficient condition for science.

At the same time, it is undeniable that central figures in the transition from medieval superstition to modern science were profoundly influenced by the idea of one Supreme God who created the Universe and established not only commandments that humans must live by, but laws that Nature itself must abide by. The seventeenth-century German astronomer Johannes Kepler, without whom Newtonian physics might not have come to be, described his pursuit of science as a wish to know the mind of God. In our own time, leading scientists, including Albert Einstein and Stephen Hawking, have described their quest in nearly identical terms. The philosopher Alfred North Whitehead and the historian of Chinese technology Joseph Needham have also suggested that what was lacking in the development of science in non-Western cultures was monotheism.

And yet, I think there is strong contrary evidence to this whole thesis, calling out to us from across the millennia. . .

———

The small hunting party follows the trail of hoofprints and other spoor. They pause for a moment by a stand of trees. Squatting on their heels, they examine the evidence more carefully. The trail they've been following has been crossed by another. Quickly they agree on which animals are responsible, how many of them, what ages and sexes, whether any are injured, how fast they're traveling, how long ago they passed, whether any other hunters are in pursuit, whether the party can overtake the game, and if so, how long it will take. The decision made, they flick their hands over the trail they will follow, make a quiet sound between their teeth like the wind, and off they lope. Despite their bows and poi-

son arrows, they continue at championship marathon racing form for hours. Almost always they've read the message in the ground correctly. The wildebeests or elands or okapis are where they thought, in the numbers and condition they estimated. The hunt is successful. Meat is carried back to the temporary camp. Everyone feasts.

This more or less typical hunting vignette comes from the !Kung San people of the Kalahari Desert, in the Republics of Botswana and Namibia, who are now, tragically, on the verge of extinction. But for decades they and their way of life were studied by anthropologists. The !Kung San may be typical of the hunter-gatherer mode of existence in which we humans spent most of our time—until ten thousand years ago, when plants and animals were domesticated and the human condition began to change, perhaps forever. They were trackers of such legendary prowess that they were enlisted by the apartheid South African army to hunt down human prey in the wars against the "frontline states." This encounter with the white South African military in several different ways accelerated the destruction of the !Kung-San way of life—that had, in any case, been deteriorating bit by bit over the centuries from every contact with European civilization.

How did they do it? How could they tell so much from a glance? Saying they're keen observers explains nothing. What actually did they do? According to anthropologist Richard Lee:

They scrutinized the shape of the depressions. The footprints of a fast-moving animal display a more elongated symmetry. A slightly lame animal favors the afflicted foot, puts less weight on it, and leaves a fainter imprint. A heavier animal leaves a deeper and broader hollow. The correlation functions are in the heads of the hunters.

In the course of the day, the footprints erode a little. The walls of the depression tend to crumble. Windblown sand accumulates on the floor of the hollow. Perhaps bits of leaf, twigs, or grass are blown into it. The longer you wait, the more erosion there is.

This method is essentially identical to what planetary astronomers use in analyzing craters left by impacting worldlets: other things being equal, the shallower the crater, the older it is. Craters with slumped walls, with modest depth-to-diameter ratios, with fine particles accumulated in their interiors tend to be more ancient—because they had to be around long enough for these erosive processes to come into play.

The sources of degradation may differ from world to world, or desert to desert, or epoch to epoch. But if you know what they are you can determine a great deal from how crisp or blurred the crater is. If insect or other animal tracks are superposed on the hoofprints, this also argues against their freshness. The subsurface moisture content of the soil and the rate at which it dries out after being exposed by a hoof determine how crumbly the crater walls are. All these matters are closely studied by the !Kung.

The galloping herd hates the hot Sun. The animals will use whatever shade they can find. They will alter course to take brief advantage of the shade from a stand of trees. But where the shadow *is* depends on the time of day, because the Sun is moving across the sky. In the morning, as the Sun is rising in the east, shadows are cast west of the trees. Later in the afternoon, as the Sun is setting toward the west, shadows are cast to the east. From the swerve of the tracks, it's possible to tell how long ago the animals passed. This calculation will be different in different seasons of the year. So the hunters must carry in their heads a kind of astronomical calendar predicting the apparent solar motion.

To me, all of these formidable forensic tracking skills are science in action.

Not only are hunter-gatherers expert in the tracks of other animals; they also know human tracks very well. Every member of the band is recognizable by his or her footprints; they are as familiar as their faces. Laurens van der Post recounts,

> [M]any miles from home and separated from the rest, Nxou and I, on the track of a wounded buck, suddenly found another set of prints and spoor joining our own. He gave a deep grunt of satisfaction and said it was Bauxhau's foot-marks made not many minutes before. He declared Bauxhau was running fast and that we would soon see him and the animal. We topped the dune in front of us and there was Bauxhau, already skinning the animal.

Or Richard Lee, also among the !Kung San, relates how when briefly examining some tracks a hunter commented, "Oh, look, Tunu is here with his brother-in-law. But where is his son?"

Is this really science? Does every tracker in the course of his training sit on his haunches for hours, following the slow degradation of an eland hoofprint? When the anthropologist asks this question, the answer given is that hunters have always used such methods. They observed their fathers and other accomplished hunters during their apprenticeships. They learned by imitation. The general principles were passed down from generation to generation. The local variations—wind speed, soil moisture—are updated as needed in each generation, or seasonally, or day-by-day.

But modern scientists do just the same. Every time we try to judge the age of a crater on the Moon or Mercury or Triton by its degree of erosion, we do not perform the calculation from scratch. We dust off a certain scientific paper and read the tried-and-true numbers that have been set down perhaps as much as a generation earlier. Physicists do not derive Maxwell's equations or quantum mechanics from scratch. They try to understand the principles and the mathematics, they observe its utility, they note how Nature follows these rules, and they take these sciences to heart, making them their own.

Yet someone had to figure out all these tracking protocols for the first time, perhaps some paleolithic genius, or more likely a succession of geniuses in widely separated times and places. There is no hint in the !Kung tracking protocols of magical methods—examining the stars the night before or the entrails of an animal, or casting dice, or interpreting dreams, or conjuring demons, or any of the myriad other spurious claims to knowledge that humans have intermittently entertained. Here there's a specific, well-defined question: Which way did the prey go and what are its characteristics? You need a precise answer that magic and divination simply do not provide—or at least not often enough to stave off starvation. Instead hunter-gatherers—who are not very superstitious in their everyday life, except during trance dances around the fire and under the influence of mild euphoriants—are practical, workaday, motivated, social, and often very cheerful. They employ skills winnowed from past successes and failures.

Scientific thinking has almost certainly been with us from the beginning. You can even see it in chimpanzees when tracking on patrol of the frontiers of their territory, or when preparing a reed to insert into the termite mound to extract a modest but much-needed source of protein. The development of tracking skills delivers a powerful evolutionary selective advantage. Those groups unable to figure it out get less protein and leave fewer offspring. Those with a scientific bent, those able to patiently observe, those with a penchant for figuring out acquire more food, especially more protein, and live in more varied habitats; they and their hereditary lines prosper. The same is true, for instance, of Polynesian seafaring skills. A scientific bent brings tangible rewards.

The other principal food-garnering activity of pre-agrarian societies is foraging. To forage, you must know the properties of many plants, and you must certainly be able to distinguish one from another. Botanists and anthropologists have repeatedly found that all over the world hunter-gatherer peoples have distinguished the various plant species with the precision of Western taxonomists. They have mentally mapped their territory with the finesse of cartographers. Again, all this is a precondition for survival.

So the claim that, just as children are not developmentally ready for certain concepts in mathematics or logic, so "primitive" peoples are not intellectually able to grasp science and technology, is nonsense. This vestige of colonialism and racism is belied by the everyday activities of people living with no fixed abode and almost no possessions, the few remaining hunter-gatherers—the custodians of our deep past.

Of Cromer's criteria for "objective thinking," we can certainly find in hunter-gatherer peoples vigorous and substantive debate, direct participatory democracy, wide-ranging travel, no priests, and the persistence of these factors not for 1,000 but for 300,000 years or more. By his criteria hunter-gatherers *ought* to have science. I think they do. Or did.

What Ionia and ancient Greece provided is not so much inventions or technology or engineering, but the idea of systematic inquiry, the notion that laws of Nature, rather than capricious gods, govern the world. Water, air, earth, and fire all had their turn as candidate "explanations" of the nature and origin of the world. Each such explanation—identified with a different pre-Socratic philosopher—was deeply flawed in its details. But the mode of explanation, an alternative to divine intervention, was productive and new. Likewise, in the history of ancient Greece, we can see nearly *all* significant events driven by the caprice of the gods in Homer, only a few events in Herodotus, and essentially none at all in Thucydides. In a few hundred years, history passed from god-driven to human-driven.

Something akin to laws of Nature was once glimpsed in a determinedly polytheistic society, in which some scholars toyed with a form of atheism. This approach of the pre-Socratics was, beginning in about the fourth century B.C., quenched by Plato, Aristotle, and then Christian theologians. If the skein of historical causality had been different—if the brilliant guesses of the atomists on the nature of matter, the plurality of worlds, the vastness of space and time had been treasured and built upon, if the innovative technology of Archimedes had been taught and emulated, if the notion of invariable laws of Nature that humans must seek out and understand had been widely propagated— I wonder what kind of world we would live in now.

I don't think science is hard to teach because humans aren't ready for it, or because it arose only through a fluke, or because, by and large, we don't have the brainpower to grapple with it. Instead, the enormous zest for science that I see in first-graders and the lesson from the remnant hunter-gatherers both speak eloquently: A proclivity for science is embedded deeply within us, in all times, places and cultures. It has been the means for our survival. It is our birthright. When, through indifference, inattention, incompetence, or fear of skepticism, we discourage children from science, we are disenfranchising them, taking from them the tools needed to manage their future.

Chapter 19

NO SUCH
THING
AS A DUMB
QUESTION

So we keep asking, over and over,
Until a handful of earth
Stops our mouths—
But is that an answer?

HEINRICH HEINE,
"Lazarus" (1854)

In East Africa, in the records of the rocks dating back to about two million years ago, you can find a sequence of worked tools that our ancestors designed and executed. Their lives depended on making and using these tools. This was, of course, Early Stone Age technology. Over time, specially fashioned stones were used for stabbing, chipping, flaking, cutting, carving. Although there are many ways of making stone tools, what is remarkable is that in a given site for enormous periods of time the tools were made in the same way—which means that there must have been educational institutions hundreds of thousands of years ago, even if it was mainly an apprenticeship system. While it's easy to exaggerate the similarities, it's also easy to imagine the equivalent of professors and students in loincloths, laboratory courses, examinations, failing grades, graduation ceremonies, and postgraduate education.

When the training is unchanged for immense periods of time, traditions are passed on intact to the next generation. But when what needs to be learned changes quickly, especially in the course of a single generation, it becomes much harder to know what to teach and how to teach it. Then, students complain about relevance; respect for their elders diminishes. Teachers despair at how educational standards have deteriorated, and how lackadaisical students have become. In a world in transition, students and teachers both need to teach themselves one essential skill—learning how to learn.

——

Except for children (who don't know enough not to ask the important questions), few of us spend much time wondering why Nature is the way it is; where the Cosmos came from, or whether it was always here; if time will one day flow backward, and effects precede causes; or whether there are ultimate limits to what humans can know. There are even children, and I have met some of them, who want to know what a

black hole looks like; what is the smallest piece of matter; why we remember the past and not the future; and why there *is* a Universe.

Every now and then, I'm lucky enough to teach a kindergarten or
first-grade class. Many of these children are natural-born scientists—
although heavy on the wonder side and light on skepticism. They're
curious, intellectually vigorous. Provocative and insightful questions
bubble out of them. They exhibit enormous enthusiasm. I'm asked
follow-up questions. They've never heard of the notion of a "dumb
question."

But when I talk to high school seniors, I find something different. They memorize "facts." By and large, though, the joy of discovery,
the life behind those facts, has gone out of them. They've lost much of
the wonder, and gained very little skepticism. They're worried about
asking "dumb" questions; they're willing to accept inadequate answers; they don't pose follow-up questions; the room is awash with
sidelong glances to judge, second-by-second, the approval of their
peers. They come to class with their questions written out on pieces of
paper, which they surreptitiously examine, waiting their turn and
oblivious of whatever discussion their peers are at this moment
engaged in.

Something has happened between first and twelfth grade, and it's
not just puberty. I'd guess that it's partly peer pressure *not* to excel
(except in sports); partly that the society teaches short-term gratifica-
tion; partly the impression that science or mathematics won't buy
you a sports car; partly that so little is expected of students; and partly
that there are few rewards or role models for intelligent discussion of
science and technology—or even for learning for its own sake. Those
few who remain interested are vilified as "nerds" or "geeks" or "grinds."

But there's something else: I find many adults are put off when
young children pose scientific questions. Why is the Moon round? the
children ask. Why is grass green? What is a dream? How deep can you
dig a hole? When is the world's birthday? Why do we have toes? Too
many teachers and parents answer with irritation or ridicule, or quickly
move on to something else: "What did you expect the Moon to be,
square?" Children soon recognize that somehow this kind of question
annoys the grown-ups. A few more experiences like it, and another
child has been lost to science. Why adults should pretend to omni-
science before 6-year-olds , I can't for the life of me understand. What's

more than four and a half centuries after Copernicus, most people on Earth still think, in their heart of hearts, that our planet sits immobile at the center of the Universe, and that we are profoundly "special."

These are typical questions in "scientific literacy." The results are appalling. But what do they measure? The memorization of authoritative pronouncements. What they *should* be asking is *how we know*— that antibiotics discriminate between microbes, that electrons are "smaller" than atoms, that the Sun is a star which the Earth orbits once a year. Such questions are a much truer measure of public understanding of science, and the results of such tests would doubtless be more disheartening still.

If you accept the literal truth of every word of the Bible, then the Earth must be flat. The same is true for the Qu'ran. Pronouncing the Earth round then means you're an atheist. In 1993, the supreme religious authority of Saudi Arabia, Sheik Abdel-Aziz Ibn Baaz, issued an edict, or fatwa, declaring that the world is flat. Anyone of the round persuasion does not believe in God and should be punished. Among many ironies, the lucid evidence that the Earth is a sphere, accumulated by the second century Græco-Egyptian astronomer Claudius Ptolemaeus, was transmitted to the West by astronomers who were Muslim and Arab. In the ninth century, they named Ptolemy's book in which the sphericity of the Earth is demonstrated, the *Almagest*, "The Greatest."

I meet many people offended by evolution, who passionately prefer to be the personal handicraft of God than to arise by blind physical and chemical forces over aeons from slime. They also tend to be less than assiduous in exposing themselves to the evidence. Evidence has little to do with it: What they wish to be true, they believe *is* true. Only 9 percent of Americans accept the central finding of modern biology that human beings (and all the other species) have slowly evolved by natural processes from a succession of more ancient beings with no divine intervention needed along the way. (When asked merely if they accept evolution, 45 percent of Americans say yes. The figure is 70 percent in China.) When the movie *Jurassic Park* was shown in Israel, it was condemned by some Orthodox rabbis because it accepted evolution and because it taught that dinosaurs lived a hundred million years ago—when, as is plainly stated at every Rosh Hashonah and every Jewish wedding ceremony, the Universe is less than 6,000 years old. The clearest evidence of our evolution can be found in our genes. But evolution is still being fought, ironically by those whose own DNA proclaims it — in the schools, in the courts, in textbook publishing houses, and on the question of just how much pain we can inflict on other animals without crossing some ethical threshold.

During the Great Depression, teachers enjoyed job security, good salaries, respectability. Teaching was an admired profession, partly because learning was widely recognized as the road out of poverty. Little of that is true today. And so science (and other) teaching is too often incompetently or uninspiringly done, its practitioners, astonishingly, having little or no training in their subjects, impatient with the method and in a hurry to get to the findings of science — and sometimes themselves unable to distinguish science from pseudoscience. Those who do have the training often get higher-paying jobs elsewhere.

Children need hands-on experience with the experimental method rather than just reading about science in a book. We can be told about oxidation of wax as the explanation of the candle flame. But we have a much more vivid sense of what's going on if we witness the candle burning briefly in a bell jar until the carbon dioxide produced by the burning surrounds the wick, blocks access to oxygen, and the flame flickers and dies. We can be taught about mitochondria in cells, how they mediate the oxidation of food like the flame burning the wax, but it's another thing altogether to see them under the microscope. We may be told that oxygen is necessary for the life of some organisms and not others. But we begin to really understand when we test the proposition in a bell jar fully depleted of oxygen. What does oxygen do for *us*? Why do we die without it? Where does the oxygen in the air come from? How secure is the supply?

Experiment and the scientific method can be taught in many matters other than science. Daniel Kunitz is a friend of mine from college. He's spent his life as an innovative junior and senior high school social sciences teacher. Want the students to understand the Constitution of the United States? You could have them read it, Article by Article, and en discuss it in class — but, sadly, this will put most of them to sleep. you could try the Kunitz method: You forbid the students to read Constitution. Instead, you assign them, two for each state, to at- a Constitutional Convention. You brief each of the thirteen

Those in America with the most favorable view of science tend to be young, well-to-do, college-educated white males. But three-quarters of new American workers in the next decade will be women, non-whites, and immigrants. Failing to rouse their enthusiasm—to say nothing of discriminating against them—isn't only unjust, it's also stupid and self-defeating. It deprives the economy of desperately needed skilled workers.

African-American and Hispanic students are doing significantly better in standardized science tests now than in the late 1960s, but they're the only ones who are. The average math gap between white and black U.S. high school graduates is still huge—two to three grade levels; but the gap between white U.S. high school graduates and those in, say, Japan, Canada, Great Britain, or Finland is more than twice as large (with the U.S. students behind). If you're poorly motivated and poorly educated, you won't know much—no mystery there. Suburban African-Americans with college-educated parents do just as well in college as suburban whites with college-educated parents. According to some statistics, enrolling a poor child in a Head Start program doubles his or her chances to be employed later in life; one who completes an Upward Bound program is four times as likely to get a college education. If we're serious, we know what to do.

What about college and university? There are obvious steps to take: improved status based on teaching success, and promotions of teachers based on the performance of their students in standardized, double-blind tests; salaries for teachers that approach what they could get in industry; more scholarships, fellowships, and laboratory equipment; imaginative, inspiring curricula and textbooks in which the leading faculty members play a major role; laboratory courses required of everyone to graduate; and special attention paid to those traditionally steered away from science. We should also encourage the best academic scientists to spend more time on public education—textbooks, lectures, newspaper and magazine articles, TV appearances. And a mandatory freshman or sophomore course in skeptical thinking and the methods of science might be worth trying.

———

The mystic William Blake stared at the Sun and saw angels there, while others, more worldly, "perceived only an object of about the size and colour of a golden guinea." Did Blake really see angels in the Sun, or was it some perceptual or cognitive error? I know of no photograph of the Sun that shows anything of the sort. Did Blake see what the camera and the telescope cannot? Or does the explanation lie much more inside Blake's head than outside? And is not the truth of the Sun's nature as revealed by modern science far more wonderful: no mere angels or gold coin, but an enormous sphere into which a million Earths could be packed, in the core of which the hidden nuclei of atoms are being jammed together, hydrogen transfigured into helium, the energy latent in hydrogen for billions of years released, the Earth and other planets warmed and lit thereby, and the same process repeated four hundred billion times elsewhere in the Milky Way galaxy?

The blueprints, detailed instructions, and job orders for building you from scratch would fill about 1,000 encyclopedia volumes if written out in English. Yet every cell in your body has a set of these encyclopedias. A quasar is so far away that the light we see from it began its intergalactic voyage before the Earth was formed. Every person on Earth is descended from the same not-quite-human ancestors in East Africa a few million years ago, making us all cousins.

Whenever I think about any of these discoveries, I feel a tingle of exhilaration. My heart races. I can't help it. Science is an astonishment and a delight. Every time a spacecraft flies by a new world, I find myself amazed. Planetary scientists ask themselves: "Oh, is *that* the way it is? Why didn't we think of that?" But nature is *always* more subtle, more intricate, more elegant than what we are able to imagine. Given our manifest human limitations, what is surprising is that we have been able to penetrate so far into the secrets of Nature.

Nearly every scientist has experienced, in a moment of discovery or sudden understanding, a reverential astonishment. Science—pure science, science not for any practical application but for its own sake—is a deeply emotional matter for those who practice it, as well as for those nonscientists who every now and then dip in to see what's been discovered lately.

And, as in a detective story, it's a joy to frame key questions, to work through alternative explanations, and maybe even to advance the process of scientific discovery. Consider these examples, some very simple, some not, chosen more or less at random:

- Could there be an undiscovered integer between 6 and 7?
- Could there be an undiscovered chemical element between

atomic number 6 (which is carbon) and atomic number 7 (which is nitrogen)?

• Yes, the new preservative causes cancer in rats. But what if you have to give a person, who weighs much more than a rat, a pound a day of the stuff to induce cancer? In that case, maybe the new preservative isn't all that dangerous. Might the benefit of having food preserved for long periods outweigh the small additional risk of cancer? Who decides? What data do they need to make a prudent decision?

• In a 3.8-billion-year-old rock, you find a ratio of carbon isotopes typical of living things today, and different from inorganic sediments. Do you deduce abundant life on Earth 3.8-billion years ago? Or could the chemical remains of more modern organisms have infiltrated into the rock? Or is there a way for isotopes to separate in the rock apart from biological processes?

• Sensitive measurements of electrical currents in the human brain show that when certain memories or mental processes occur, particular regions of the brain go into action. Can our thoughts, memories, and passions all be generated by particular circuitry of the brain neurons? Might it ever be possible to simulate such circuitry in a robot? Would it ever be feasible to insert new circuits or alter old ones in the brain in such a way as to change opinions, memories, emotions, logical deductions? Is such tampering wildly dangerous?

• Your theory of the origin of the Solar System predicts many flat disks of gas and dust all over the Milky Way galaxy. You look through the telescope and you find flat disks everywhere. You happily conclude that your theory is confirmed. But it turns out the disks you sighted were spiral galaxies far beyond the Milky Way, and much too big to be nascent solar systems. Should you abandon your theory? Or should you look for a different kind of disk? Or is this just an expression of your unwillingness to abandon a discredited hypothesis?

• A growing cancer sends out an all-points bulletin to the cells lining adjacent blood vessels: "We need blood," the message says. The endothelial cells obligingly build blood vessel bridges to supply the cancer cells with blood. How does this come about? Can the message be intercepted or canceled?

• You mix violet, blue, green, yellow, orange, and red paints and make a murky brown. Then you mix light of the same colors and you get white. What's going on?

• In the genes of humans and many other animals there are long, repetitive sequences of hereditary information (called "nonsense"). Some of these sequences cause genetic diseases. Could it be that segments of the DNA are rogue nucleic acids, reproducing on their own, in business for themselves, disdaining the well-being of the organism they inhabit?

• Many animals behave strangely just before an earthquake. What do they know that seismologists don't?

• The ancient Aztec and the ancient Greek words for "God" are nearly the same. Is this evidence of some contact or commonality between the two civilizations, or should we expect occasional such coincidences between two wholly unrelated languages merely by chance? Or could, as Plato thought in the *Cratylus*, certain words be built into us from birth?

• The Second Law of Thermodynamics states that in the Universe as a whole, disorder increases as time goes on. (Of course, locally worlds and life and intelligence can emerge, at the cost of a decrease in order elsewhere in the Universe.) But if we live in a Universe in which the present Big Bang expansion will slow, stop, and be replaced by a contraction, might the Second Law then be reversed? Can effects precede causes?

• The human body uses concentrated hydrochloric acid in the stomach to dissolve food and aid digestion. Why doesn't the hydrochloric acid dissolve the stomach?

• The oldest stars seem to be, at the time I'm writing, older than the Universe. Like the claim that an acquaintance has children older than she is, you don't have to know very much to recognize that someone has made a mistake. Who?

• The technology now exists to move individual atoms around, so long and complex messages can be written on an ultra-microscopic scale. It is also possible to make machines the size of molecules. Rudimentary examples of both these "nano-technologies" are now well-demonstrated. Where does this take us in another few decades?

• In several different laboratories, complex molecules have been found that under suitable conditions make copies of themselves in the test tube. Some of these molecules are, like DNA and RNA, built out of nucleotides; others are not. Some use enzymes to hasten the pace of the chemistry; others do not. Sometimes there is a mistake in copying;

from that point forward the mistake is copied in successive generations of molecules. Thus there get to be slightly different species of self-replicating molecules, some of which reproduce faster or more efficiently than others. These preferentially thrive. As time goes on, the molecules in the test tube become more and more efficient. We are beginning to witness the evolution of molecules. How much insight does this provide about the origin of life?

· Why is ordinary ice white, but pure glacial ice blue?

· Life has been found miles below the surface of the Earth. How deep does it go?

· The Dogon people in the Republic of Mali are said by a French anthropologist to have a legend that the star Sirius has an extremely dense companion star. Sirius in fact does have such a companion, although it requires fairly sophisticated astronomy to detect it. So (1) did the Dogon people descend from a forgotten civilization that had large optical telescopes and theoretical astrophysics? Or, (2) were they instructed by extraterrestrials? Or, (3) did the Dogon hear about the white dwarf companion of Sirius from a visiting European? Or, (4) was the French anthropologist mistaken and the Dogon in fact never had such a legend?

———

Why should it be hard for scientists to get science across? Some scientists—including some very good ones—tell me they'd love to popularize, but feel they lack talent in this area. Knowing and explaining, they say, are not the same thing. What's the secret?

There's only one, I think: Don't talk to the general audience as you would to your scientific colleagues. There are terms that convey your meaning instantly and accurately to fellow experts. You may parse these phrases every day in your professional work. But they do no more than mystify an audience of nonspecialists. Use the simplest possible language. Above all, remember how it was before you yourself grasped whatever it is you're explaining. Remember the misunderstandings that you almost fell into, and note them explicitly. Keep firmly in mind that there was a time when you didn't understand any of this either. Recapitulate the first steps that led you from ignorance to knowledge. Never forget that native intelligence is widely distributed in our species. Indeed, it is the secret of our success.

The effort involved is slight, the benefits great. Among the potential pitfalls are oversimplification, the need to be sparing with qualifications (and quantifications), inadequate credit given to the many scientists involved, and insufficient distinctions drawn between helpful analogy and reality. Doubtless, compromises must be made.

The more you make such presentations, the clearer it is which approaches work and which do not. There is a natural selection of metaphors, images, analogies, anecdotes. After a while you find that you can get almost anywhere you want to go, walking on consumer-tested stepping-stones. You can then fine-tune your presentations for the needs of a given audience.

Like some editors and television producers, some scientists believe the public is too ignorant or too stupid to understand science, that the enterprise of popularization is fundamentally a lost cause, or even that it's tantamount to fraternization, if not outright cohabitation, with the enemy. Among the many criticisms that could be made of this judgment—along with its insufferable arrogance and its neglect of a host of examples of highly successful science popularizations—is that it is self-confirming. And also, for the scientists involved, self-defeating:

Large-scale government support for science is fairly new, dating back only to World War II—although patronage of a few scientists by the rich and powerful is much older. With the end of the Cold War, the national-defense trump card that provided support for all sorts of fundamental science became virtually unplayable. Only partly for this reason, most scientists, I think, are now comfortable with the idea of popularizing science. (Since nearly all support for science comes from the public coffers, it would be an odd flirtation with suicide for scientists to oppose competent popularization.) What the public understands and appreciates, it is more likely to support. I don't mean writing articles for *Scientific American*, say, that are read by science enthusiasts and scientists in other fields. I'm not just talking about teaching introductory courses for undergraduates. I'm talking about efforts to communicate the substance and approach of science in newspapers, magazines, on radio and television, in lectures for the general public, and in elementary, middle, and high school textbooks.

Of course there are judgment calls to be made in popularizing. It's important neither to mystify nor to patronize. In attempting to prod public interest, scientists have on occasion gone too far—for example, in drawing unjustified religious conclusions. Astronomer George Smoot described his discovery of small irregularities in the radio radiation left over from the Big Bang as "seeing God face-to-face." Physics Nobel laureate Leon Lederman described the Higgs boson, a hypothetical building block of matter, as "the God particle," and so titled a book. (In my opinion, they're all God particles.) If the Higgs boson doesn't exist, is the God hypothesis disproved? Physicist Frank Tipler proposes that computers in the remote future will prove the existence of God and work our bodily resurrection.

Periodicals and television can strike sparks as they give us a glimpse of science, and this is very important. But—apart from apprenticeship or well-structured classes and seminars—the best way to popularize science is through textbooks, popular books, CD-ROMs, and laser disks. You can mull things over, go at your own pace, revisit the hard parts, compare texts, dig deep. It has to be done right, though, and in the schools especially it generally isn't. There, as the philosopher John Passmore comments, science is often presented

> as a matter of learning principles and applying them by routine procedures. It is learned from textbooks, not by reading the works of great scientists or even the day-to-day contributions to the scientific literature. . . . The beginning scientist, unlike the beginning humanist, does not have an immediate contact with genius. Indeed . . . school courses can attract quite the wrong sort of person into science—unimaginative boys and girls who *like* routine.

I hold that popularization of science is successful if, at first, it does no more than spark the sense of wonder. To do that, it is sufficient to provide a glimpse of the findings of science without thoroughly explaining how those findings were achieved. It is easier to portray the destination than the journey. But, where possible, popularizers should try to chronicle some of the mistakes, false starts, dead ends, and apparently hopeless confusion along the way. At least every now and then, we should provide the evidence and let the reader draw his or her own conclusion. This converts obedient assimilation of new knowledge into personal discovery. When you make the finding yourself—even if you're the last person on Earth to see the light—you never forget it.

As a youngster, I was inspired by the popular science books and articles of George Gamow, James Jeans, Arthur Eddington, J.B.S. Haldane, Julian Huxley, Rachel Carson, and Arthur C. Clarke—all of them trained in, and most of them leading practitioners of science. The popularity of well-written, well-explained, deeply imaginative books on science that touch our hearts as well as our minds seems greater in the last twenty years than ever before, and the number and disciplinary diversity of scientists writing these books is likewise unprecedented. Among the best contemporary scientist-popularizers, I think of Stephen Jay Gould, E. O. Wilson, Lewis Thomas, and Richard Dawkins in biology; Steven Weinberg, Alan Lightman, and Kip Thorne in physics; Roald Hoffmann in chemistry; and the early works of Fred Hoyle in astronomy. Isaac Asimov wrote capably on everything. (And while requiring calculus, the most consistently exciting, provocative, and inspiring science popularization of the last few decades seems to me to be Volume I of Richard Feynman's *Introductory Lectures on Physics*.) Nevertheless, current efforts are clearly nowhere near commensurate with the public good. And, of course, if we can't read, we can't benefit from such works, no matter how inspiring they are.

I want us to rescue Mr. "Buckley" and the millions like him. I also want us to stop turning out leaden, incurious, uncritical, and unimaginative high school seniors. Our species needs, and deserves, a citizenry with minds wide awake and a basic understanding of how the world works.

Science, I maintain, is an absolutely essential tool for any society with a hope of surviving well into the next century with its fundamental values intact—not just science as engaged in by its practitioners, but science understood and embraced by the entire human community. And if the scientists will not bring this about, who will?

Chapter 20

———

HOUSE ON FIRE*

* Written with Ann Druyan.

The Lord [Buddha] replied to the Venerable Sariputra:

"In some village, city, market town, country district, province, kingdom, or capital there lived a householder, old, advanced in years, decrepit, weak in health and strength, but rich, wealthy, and well-to-do. His house was a large one, both extensive and high, and it was old, having been built a long time ago. It was inhabited by many living beings, some two, three, four, or five hundred. It had one single door only. It was thatched with straw, its terraces had fallen down, its foundations were rotten, its walls, matting-screens, and plaster were in an advanced state of decay. Suddenly a great blaze of fire broke out, and the house started burning on all sides. And that man had many young sons, five, or ten, or twenty, and he himself got out of the house.

"When that man saw his own house ablaze all around with that great mass of fire, he became afraid and trembled, his mind became agitated, and he thought to himself: 'I, it is true, have been competent enough to run out of the door, and to escape from my burning house, quickly and safely, without being touched or scorched by that great mass of fire. But what about my sons, my young boys, my little sons? There, in this burning house, they play, sport, and amuse themselves with all sorts of games. They do not know that this dwelling is afire, they do not understand it, do not perceive it, pay no attention to it, and so they feel no agitation. Though threatened by this great [fire], though in such close contact with so much ill, they pay no attention to their danger, and make no efforts to get out.' "

from *The Saddharmapundarika*, in *Buddhist Scriptures*,
Edward Conze, ed. (Harmondsworth, Middlesex, England:
Penguin Books, 1959)

One of the reasons it's so interesting to write for *Parade* magazine is feedback. With eighty million readers you can really sample the opinion of the citizens of the United States. You can understand how people think, what their anxieties and hopes are, and even perhaps where we have lost our way.

An abbreviated version of the preceding chapter, emphasizing the performance of students and teachers, was published in *Parade*. I was flooded with mail. Some people denied there was a problem; others said that Americans were losing cutting-edge intelligence and know-how. Some thought there were easy solutions; others, that the problems were too deeply ingrained to fix. Many opinions were a surprise to me.

A tenth-grade teacher in Minnesota handed out copies of the article and asked his students to tell me what they thought. Here's what some American high school students wrote (spelling, grammar, and punctuation as in the original letters):

• Not a Americans are stupid We just rank lower in school big deal.
• Maybe that's good that we are not as smart as the other countries. So then we can just import all of our products and then we don't have to spend all of our money on the parts for the goods.
• And if other countries are doing better, what does it matter, their most likely going to come over the U.S. anyway?
• Our society is doing just fine with what discoveries we are making. It's going slowly, but the cure for cancer is coming right along.
• The U.S. has its own learning system and it may not be as advanced as theirs, but it is just as good. Otherwise I think your article is a very educating one.
• Not one kid in this school likes science. I really didn't understand the point of the article. I thought that it was very boreing. I'm just not into anything like that.

- I am studying to be a lawyer and frankly I do agree with my parents when they say I have an attitude problem toward science.
- It's true that some American kids don't try, but we could be smarter than any other country if we wanted to.
- Instead of homework, kids will watch TV. I have to agree that I do it. I have cut it down from about 4 hrs. a day.
- I don't believe it is the school systems fault, I think the whole country is brought up with not enough emphasis on school. I know my mom would rather be watching me play basketball or soccer, instead of helping me with an assignment. Most of the kids I know could care less about making sure there doing there work right.
- I don't think American kids are stupid. It just they don't study hard enough because most of kids work. . . Lots of people said that Asian people are smarter than American and they are good at everything, but that's not true. They are not good at sports. They don't have time to play sports.
- I'm in sports myself, and I feel that the other kids on my team push to you to excel more in that sport than in school.
- If we want to rank first, we could go to school all day and not have any social life.
- I can see why a lot of science teachers would get mad at you for insulting there job.
- Maybe if the teachers could be more exciting, the children will want to learn. . . If science is made to be fun, kids will want to learn. To accomplish this, it needs to be started early on, not just taught as facts and figures.
- I really find it hard to believe those facts about the U.S. in science. If we are so far behind, how come Michael Gorbachev came to Minnesota and Montana to Control Data to see how we run are computers and thing?
- Around 33 hours for fifth graders! In my opinion thats too much thats almost as many hours as a full job practically. So instead of homework we can be making money.
- When you put down how far behind we are in science and math, why don't you try tell us this in a little nicer manner? . . . Have a little pride in your country and its capabilities.
- I think your facts were inconclusive and the evidence very flimsy. All in all, you raised a good point.

All in all, these students don't think there's much of a problem; and if there is, not much can be done about it. Many also complained that the lectures, classroom discussions, and homework were "boring." Especially for an MTV generation beset by attention deficit disorders in various degrees of severity, it *is* boring. But spending three or four grades practicing once again the addition, subtraction, multiplication, and division of fractions would bore anyone — and the tragedy is that, say, elementary probability theory is within reach of these students. Likewise for the forms of plants and animals presented without evolution; history presented as wars, dates, and kings without the role of obedience to authority, greed, incompetence, and ignorance; English without new words entering the language and old words disappearing; and chemistry without where the elements come from. The means of awakening these students are at hand and ignored. Since most school children emerge with only a tiny fraction of what they've been taught permanently engraved in their long-term memories, isn't it essential to infect them with consumer-tested topics that aren't boring . . . and a zest for learning?

Most adults who wrote thought there's a substantial problem. I received letters from parents about inquisitive children willing to work hard, passionate about science but with no adequate community or school resources to satisfy their interests. Other letters told of parents who knew nothing about science sacrificing their own comfort so their children could have science books, microscopes, telescopes, computers, or chemistry sets; of parents teaching their children that hard work will get them out of poverty; of a grandmother bringing tea to a student up late at night still doing homework; of peer pressure not to do well in school because "it makes the other kids look bad."

Here's a sampling — not an opinion poll, but representative commentary — of other responses by parents:

- Do parents understand that you can't be a full human being if you're ignorant? Are there books at home? How about a magnifying glass? Encyclopedia? Do they encourage children to learn?
- Parents have to teach patience and perseverance. The most important gift they can give their children is the ethos of hard work, but they

can't just talk about it. The kids who learn to work hard are the ones who see their parents work hard and never give up.
• My child is fascinated by science, but she doesn't get any in school or on TV.
• My child is identified as gifted, but the school has no program for science enrichment. The guidance counselor told me to send her to a private school, but we can't afford a private school.
• There's enormous peer pressure; shy children don't want to "stand out" by doing well in science. When my daughter reached 13 and 14, her life-long interest in science seemed to disappear.

———

Parents also had much to say about teachers, and some of the comments by teachers echoed the parents. For example, people complained that teachers are trained *how* to teach but not *what* to teach; that a large number of physics and chemistry teachers have no degree in physics or chemistry and are "uncomfortable and incompetent" in teaching science; that teachers themselves have too much science and math anxiety; that they resist being asked questions, or they answer, "It's in the book. Look it up." Some complained that the biology teacher was a "Creationist"; some complained that he wasn't. Among other comments by or about teachers:

• We are breeding a collection of half-wits.
• It's easier to memorize than to think. Kids have to be taught to think.
• The teachers and curricula are "dumbing down" to the lowest common denominator.
• Why is the basketball coach teaching chemistry?
• Teachers are required to spend much too much time on discipline and on "social curricula." There's no incentive to use our own judgment. The "brass" are always looking over our shoulders.
• Abandon tenure in schools and colleges. Get rid of the deadwood. Leave hiring and firing to principals, deans, and superintendents.
• My joy in teaching was repeatedly thwarted by militaristic-type principals.
• Teachers should be rewarded on the basis of performance—especially student performance on standardized, nationwide tests, and im-

provements in student performance on such tests from one year to the next.

• Teachers are stifling our children's minds by telling them they're not "smart" enough—for example, for a career in physics. Why not give the students a chance to take the course?

• My son was promoted even though he's reading two grade levels behind the rest of his class. The reason given was social, not educational. He'll never catch up unless he's left back.

• Science should be required in all school (and especially high school) curricula. It should be carefully coordinated with the math courses the students are taking at the same time.

• Most homework is "busy work" rather than something that makes you think.

• I think Diane Ravitch [*New Republic*, March 6, 1989] tells it like it is: "As a female student at Hunter High School in New York City recently explained, 'I make straight As, but I never talk about it. . . It's cool to do really badly. If you are interested in school and you show it, you're a nerd' . . . The popular culture—through television, movies, magazines, and videos—incessantly drums in the message to young women that it is better to be popular, sexy, and 'cool' than to be intelligent, accomplished, and outspoken. . . In 1986 researchers found a similar anti-academic ethos among both high school and female students in Washington, D.C. They noted that able students faced strong peer pressure not to succeed in school. If they did well in their studies, they might be accused of 'acting white.' "

• Schools could easily give much more recognition and rewards to kids who are outstanding in science and math. Why don't they? Why not special jackets with school letters? Announcements in assembly and the school newspaper and the local press? Local industry and social organizations to give special awards? This costs very little and could overcome peer pressure not to excel.

• Headstart is the single most effective . . . program for improving children's understanding of science and everything else.

———

There were also many passionate, highly controversial opinions expressed which, at the very least, give a sense of how deeply people feel about the subject. Here's a smattering:

- All the smart kids are looking for the fast buck these days, so they become lawyers, not scientists.
- I don't want you to improve education. Then there'd be nobody to drive the cabs.
- The problem in science education is that God isn't sufficiently honored.
- The fundamentalist teaching that science is "humanism" and is to be mistrusted is the reason nobody understands science. Religions are afraid of the skeptical thinking at the heart of science. Students are brainwashed not to accept scientific thinking long before they get to college.
- Science has discredited itself. It works for politicians. It makes weapons, it lies about marijuana "hazards," it ignores about the dangers of agent orange, etc.
- The public schools don't work. Abandon them. Let's have private schools only.
- We have let the advocates of permissiveness, fuzzy thinking, and rampant socialism destroy what was once a great educational system.
- The school system has enough money. The problem is that the white males, usually coaches, who run the schools would never (and I mean never) hire an intellectual. . . They care more about the football team than the curriculum and hire only submediocre, flag-waving, God-loving automatons to teach. What kind of students can emerge from schools that oppress, punish, and neglect logical thinking?
- Release schools from the stranglehold of the ACLU [American Civil Liberties Union], NEA [National Education Association], and others engaged in the breakdown of the discipline and competence in the schools.
- I'm afraid you have no understanding of the country in which you live. The people are incredibly ignorant and fearful. They will not tolerate listening to any [new] idea. . . Don't you get it? The system survives only because it has an ignorant God-fearing population. There's a reason lots of [educated people] are unemployed.
- I'm sometimes required to explain technological issues to Congressional staffers. Believe me, there's a problem in science education in this country.

—

There is no single solution to the problem of illiteracy in science—or math, history, English, geography, and many of the other skills which our society needs more of. The responsibilities are broadly shared—parents, the voting public, local school boards, the media, teachers, administrators, federal, state, and local governments, plus, of course, the students themselves. At every level teachers complain that the problem lies in earlier grades. And first-grade teachers can with justice despair of teaching children with learning deficits because of malnutrition, or no books in the home, or a culture of violence in which the leisure to think is unavailable.

I know very well from my own experience how much a child can benefit from parents who have a little learning, and are able to pass it on. Even small improvements in the education, communication skills, and passion for learning in one generation might work much larger improvements in the next. I think of this every time I hear a complaint that school and collegiate "standards" are falling, or that a bachelor's degree doesn't "mean" what it once did.

Dorothy Rich, an innovative teacher from Yonkers, New York, believes that far more important than specific academic subjects is the honing of key skills which she lists as: "confidence, perseverance, caring, teamwork, common sense and problem-solving." To which I'd add skeptical thinking and an aptitude for wonder.

At the same time, children with special abilities and skills need to be nourished and encouraged. They are a national treasure. Challenging programs for the "gifted" are sometimes decried as "elitism." Why aren't intensive practice sessions for varsity football, baseball, and basketball players and interschool competition deemed elitism? After all, only the most gifted athletes participate. There is a self-defeating double standard at work here, nationwide.

—

The problems in public education in science and other subjects run so deep that it's easy to despair and conclude that they can never be fixed. And yet, there are institutions hidden away in big cities and small towns that provide reason for hope, places that strike the spark, awaken slumbering curiosities, and ignite the scientist that lives in all of us:

• The enormous metallic iron meteorite in front of you is as full of holes as a Swiss cheese. Gingerly you reach out to touch it. It feels smooth and cold. The thought occurs to you that this is a piece of another world. How did it get to Earth? What happened in space to make it so beat up? . . .

• The display shows maps of eighteenth-century London, and the spread of a horrifying cholera epidemic. People in one house got it from people in neighboring houses. By running the wave of infection back, you can see where it started. It's like being a detective. And when you pinpoint the origin you find it's a place with open sewers. It occurs to you that there's a life and death reason why modern cities have adequate sanitation. You think of all those cities and towns and villages in the world that don't. You get to thinking maybe there's a simpler, cheaper way to do it. . .

• You're crawling through a long, utterly black tunnel. There are sudden turns, ups and downs. You go through a forest of feathery things, beady things, big solid round things. You imagine what it must be like to be blind. You think about how little we rely on our sense of touch. In the dark and the quiet, you're alone with your thoughts. Somehow the experience is exhilarating. . .

• You examine a detailed reconstruction of a procession of priests climbing up one of the great ziggurats of Sumer, or a gorgeously painted tomb in the Valley of the Kings in ancient Egypt, or a house in ancient Rome, or a full-scale turn-of-the-century street in small-town America. You think of all those civilizations, so different from yours, how if you'd been born into them you would have thought them completely natural, how you'd consider *our* society—if you had somehow been told of it—as weird . . .

• You squeeze the eyedropper, and a drop of pond water drips out onto the microscope stage. You look at the projected image. The drop is full of life—strange beings swimming, crawling, tumbling; high dramas of pursuit and escape, triumph and tragedy. This is a world populated by beings far more exotic than in any science fiction movie. . .

• Seated in the theater, you find yourself inside the head of an eleven-year-old boy. You look out through his eyes. You encounter his typical daily crises: bullies, authoritarian adults, crushes on girls. You hear the voice inside his head. You witness his neurological and hormonal responses to his social environment. And you get to wonder how *you* work on the inside . . .

• Following the simple instructions, you type in the commands. What will the Earth look like if we continue to burn coal, oil, and gas, and double the amount of carbon dioxide in the atmosphere? How much hotter will it be? How much polar ice will melt? How much higher will the oceans be? *Why* are we pouring so much carbon dioxide into the atmosphere? What if we put five times more carbon dioxide into the atmosphere? Also, how could anybody know what the future climate will be like? It gets you thinking . . .

In my childhood, I was taken to the American Museum of Natural History in New York City. I was transfixed by the dioramas—lifelike representations of animals and their habitats all over the world. Penguins on the dimly lit Antarctic ice; okapi in the bright African veldt; a family of gorillas, the male beating his chest, in a shaded forest glade; an American grizzly bear standing on his hind legs, ten or twelve feet tall, and staring me right in the eye. These were three-dimensional freeze-frames captured by some genie of the lamp. Did the grizzly move just then? Did the gorilla blink? Might the genie return, lift the spell and permit this gorgeous array of living things to go on with their lives as, jaws agape, I watch?

Kids have an irresistible urge to touch. Back in those days, the most commonly heard two words in museums were "don't touch." Decades ago there was almost nothing "hands-on" in museums of science or natural history, not even a simulated tidal pool in which you could pick up a crab and inspect it. The closest thing to an interactive exhibit that I knew was the scales in the Hayden Planetarium, one for each planet. Weighing a mere forty pounds on Earth, there was something reassuring in the thought that if only you lived on Jupiter, you would weigh a hundred pounds. But sadly, on the Moon you would weigh only seven pounds; on the Moon it seemed you would hardly be there at all.

Today, children are encouraged to touch, to poke, to run through a branched contingency tree of questions and answers via computer, or to make funny noises and see what the sound waves look like. Even kids who don't get everything out of the exhibit, or who don't even get the point of the exhibit, usually extract something valuable. You go to these museums and you're struck by the wide-eyed looks of wonder, by kids racing from exhibit to exhibit, by the triumphant smiles of discovery. They're wildly popular. Almost as many of us go to them each year as attend professional baseball, basketball and football games combined.

These exhibits do not replace instruction in school or at home, but they awaken and excite. A great science museum inspires a child to read a book, or take a course, or return to the museum again to engage in a process of discovery—and, most important, to learn the method of scientific thinking.

Another glorious feature of many modern science museums is a movie theater showing IMAX or OMNIMAX films. In some cases the screen is ten stories tall and wraps around you. The Smithsonian's National Air and Space Museum, the most popular museum on Earth, has premiered in its Langley Theater some of the best of these films. *To Fly* brings a catch to my throat even after five or six viewings. I've seen religious leaders of many denominations witness *Blue Planet* and be converted on the spot to the need to protect the Earth's environment.

Not every exhibit and science museum is exemplary. A few still are commercials for firms that have contributed money to promote their products—how an automobile engine works or the "cleanliness" of one fossil fuel as compared to another. Too many museums that claim to be about science are really about technology and medicine. Too many biology exhibits are still afraid to mention the key idea of modern biology: evolution. Beings "develop" or "emerge," but never evolve. The absence of humans from the deep fossil record is underplayed. We are shown nothing of the anatomical and DNA near-identity between humans and chimps or gorillas. Nothing is displayed on complex organic molecules in space and on other worlds, nor about experiments showing the stuff of life forming in enormous numbers in the known atmospheres of other worlds and the presumptive atmosphere of the early Earth. A notable exception: the Natural History Museum of the Smithsonian Institution once had an unforgettable exhibit on evolution. It began with two roaches in a modern kitchen with open cereal boxes and other food. Left alone for a few weeks, the place was crowded with roaches, buckets of them everywhere, competing for the little food now available, and the long-term hereditary advantage that a slightly better adapted roach might have over its competitors became crystal clear. Also, too, many planetaria are still devoted to picking out constellations rather than traveling to other worlds, and depicting the evolution of galaxies, stars, and planets; they also have an insectlike projector always visible which robs the sky of its reality.

Perhaps the grandest museum exhibit can't be seen. It has no home: George Awad is one of the leading architectural model makers in America, specializing in skyscrapers. He is also a dedicated student of astronomy who has made a spectacular model of the Universe. Starting with a prosaic scene on Earth, and following a scheme proposed by the designers Charles and Ray Eames, he goes progressively by factors of ten to show us the whole Earth, the Solar System, the Milky Way and the Universe. Every astronomical body is meticulously detailed. You can lose yourself in them. It's one of the best tools I know of to explain the scale and nature of the Universe to children. Isaac Asimov described it as "the most imaginative representation of the universe that I have ever seen, or could have conceived of. I could have wandered through it for hours, seeing something new at every turn that I hadn't observed before." Versions of it ought to be available throughout the country—for stirring the imagination, for inspiration, and for teaching. But instead, Mr. Awad cannot give this exhibit to any major science museum in the country. No one is willing to devote to it the floor space needed. As I write, it still sits forlornly, crated in storage.

—

The population of my town, Ithaca, New York, doubles to a grand total of about 50,000 when Cornell University and Ithaca College are in session. Ethnically diverse, surrounded by farmland, it has suffered, like so much of the Northeast, the decline of its nineteenth century manufacturing base. Half the children at Beverly J. Martin Elementary School, which our daughter attended, live below the poverty line. Those are the kids that two volunteer science teachers, Debbie Levin and Ilma Levine, worried about most. It didn't seem right that for some, the children of Cornell faculty, say, even the sky wasn't the limit. For others there was no access to the liberating power of science education. Starting in the 1960s, they made regular trips to the school, dragging their portable library cart, laden with household chemicals and other familiar items to convey something of the magic of science. They dreamed of creating a place for kids to go, where they could get a personal, hands-on feel for science.

In 1983 Levin and Levine placed a small ad in our local paper inviting the community to discuss the idea. Fifty people showed up. From that group came the first board of directors of the Sciencenter. Within a year they secured exhibition space in the first floor of an un-rented office building. When the owner found a paying tenant, the tadpoles and litmus paper were packed up again and carted off to a vacant storefront.

Moves to other storefronts followed until an Ithacan named Bob Leathers, an architect world-renowned for designing innovative community-built playgrounds, drew up and donated the plans for a permanent Sciencenter. Gifts from local firms provided enough money to purchase an abandoned lot from the city and then hire an executive director, Charles Trautmann, a Cornell civil engineer. He and Leathers traveled to the annual meeting of the National Association of Homebuilders in Atlanta. Trautmann relates how they told the story "of a community eager to take responsibility for the education of its youth and secured donations of many key items such as windows, skylights and lumber."

Before they could start building, some of the old pumphouse on the site had to be torn down. Members of a Cornell fraternity were enlisted. With hardhats and sledge hammers, they demolished the place joyfully. "This is the kind of thing," they said, "we usually get into trouble for doing." In two days, they carted away 200 tons of rubble.

What followed were images straight out of an America that many of us fear has vanished. In the tradition of pioneer barn raising, members of the community—bricklayers, doctors, carpenters, university professors, plumbers, farmers, the very young, and the very old—all rolled up their sleeves to build the Sciencenter.

"The continuous seven-days-a-week schedule was maintained," says Trautmann, "so that anyone would be able to help anytime. Everyone was given a job. Experienced volunteers built stairs, laid carpet and tile, and trimmed windows. Others painted, nailed, and carried supplies." Some 2,200 townspeople donated more than 40,000 hours. Roughly 10 percent of the construction work was performed by people convicted of minor offenses; they preferred to do something for the community than to sit idle in jail. Ten months later, Ithaca had the only community-built science museum in the world.

Among the seventy-five interactive exhibits emphasizing both the processes and principles of science are: the Magicam, a microscope that visitors can use to view on a color monitor and then photograph any object at 40 times magnification; the world's only public connection to the satellite-based National Lightning Detection Network; a 6 x 9–foot walk-in camera; a fossil pit seeded with local shale where visitors hunt for fossils from 380 million years ago and keep their finds; an eight-foot-long boa constrictor named "Spot"; and a dazzling array of other experiments, computers, and activities.

Levin and Levine can still be found there, full-time volunteers teaching the citizens and scientists of the future. The DeWitt Wallace–Reader's Digest Fund supports and extends their dream of reaching kids who would ordinarily be denied their scientific birthright. Through the Fund's nationwide Youth-ALIVE program, Ithaca teenagers receive intensive mentoring to develop their science, conflict resolution, and employment skills.

Levin and Levine thought science should belong to everyone. Their community agreed and made a commitment to realize that dream. In the Sciencenter's first year, 55,000 people came from all 50 states and 60 countries. Not bad for a small town. It makes you wonder what else we could do if we worked together for a better future for our kids.

Chapter 21

THE PATH TO
FREEDOM*

* Written with Ann Druyan.

We must not believe the many, who say that only free people ought to be educated, but we should rather believe the philosophers who say that only the educated are free.

EPICTETUS,
Roman philosopher and former slave,
Discourses

Frederick Bailey was a slave. As a boy in Maryland in the 1820s, he had no mother or father to look after him. ("It is a common custom," he later wrote, "to part children from their mothers . . . before the child has reached its twelfth month.") He was one of countless millions of slave children whose realistic prospects for a hopeful life were nil.

What Bailey witnessed and experienced in his growing up marked him forever: "I have often been awakened at the dawn of day by the most heart-rending shrieks of an own aunt of mine, whom [the overseer] used to tie up to a joist, and whip upon her naked back till she was literally covered with blood . . . From the rising till the going down of the sun he was cursing, raving, cutting, and slashing among the slaves of the field . . . He seemed to take pleasure in manifesting his fiendish barbarity."

The slaves had drummed into them, from plantation and pulpit alike, from courthouse and statehouse, the notion that they were hereditary inferiors, that God *intended* them for their misery. The Holy Bible, as countless passages confirmed, condoned slavery. In these ways the "peculiar institution" maintained itself despite its monstrous nature—something even its practitioners must have glimpsed.

There was a most revealing rule: Slaves were to remain illiterate. In the antebellum South, whites who taught a slave to read were severely punished. "[To] make a contented slave," Bailey later wrote, "it is necessary to make a thoughtless one. It is necessary to darken his moral and mental vision, and, as far as possible, to annihilate the power of reason." This is why the slaveholders must control what slaves hear and see and think. This is why reading and critical thinking are dangerous, indeed subversive, in an unjust society.

So now picture Frederick Bailey in 1828—a 10-year-old African-American child, enslaved, with no legal rights of any kind, long since torn from his mother's arms, sold away from the tattered remnants of his extended family as if he were a calf or a pony, conveyed to an unknown household in the strange city of Baltimore, and condemned to a life of drudgery with no prospect of reprieve.

Bailey was sent to work for Capt. Hugh Auld and his wife, Sophia, moving from plantation to urban bustle, from field work to housework. In this new environment, he came every day upon letters, books, and people who could read. He discovered what he called "this mystery" of reading: There was a connection between the letters on the page and the movement of the reader's lips, a nearly one-to-one correlation between the black squiggles and the sounds uttered. Surreptitiously, he studied from young Tommy Auld's *Webster's Spelling Book*. He memorized the letters of the alphabet. He tried to understand the sounds they stood for. Eventually, he asked Sophia Auld to help him learn. Impressed with the intelligence and dedication of the boy, and perhaps ignorant of the prohibitions, she complied.

By the time Frederick was spelling words of three and four letters, Captain Auld discovered what was going on. Furious, he ordered Sophia to stop. In Frederick's presence he explained:

> A nigger should know nothing but to obey his master—to do as he is told to do. Learning would *spoil* the best nigger in the world. Now, if you teach that nigger how to read, there would be no keeping him. It would forever unfit him to be a slave.

Auld chastised Sophia in this way as if Frederick Bailey were not there in the room with them, or as if he were a block of wood.

But Auld had revealed to Bailey the great secret: "I now understood . . . the white man's power to enslave the black man. From that moment, I understood the pathway from slavery to freedom."

Without further help from the now reticent and intimidated Sophia Auld, Frederick found ways to continue learning how to read, including buttonholing white schoolchildren on the streets. Then he began teaching his fellow slaves: "Their minds had been starved . . . They had been shut up in mental darkness. I taught them, because it was the delight of my soul."

With his knowledge of reading playing a key role in his escape, Bailey fled to New England, where slavery was illegal and black people were free. He changed his name to Frederick Douglass (after a character in Walter Scott's *The Lady of the Lake*), eluded the bounty hunters who tracked down escaped slaves, and became one of the greatest orators, writers, and political leaders in American history. All his life, he understood that literacy had been the way out.

—

For 99 percent of the tenure of humans on earth, nobody could read or write. The great invention had not yet been made. Except for first-hand experience, almost everything we knew was passed on by word of mouth. As in the children's game "Telephone," over tens and hundreds of generations, information would slowly be distorted and lost.

Books changed all that. Books, purchasable at low cost, permit us to interrogate the past with high accuracy; to tap the wisdom of our species; to understand the point of view of others, and not just those in power; to contemplate—with the best teachers—the insights, painfully extracted from Nature, of the greatest minds that ever were, drawn from the entire planet and from all of our history. They allow people long dead to talk inside our heads. Books can accompany us everywhere. Books are patient where we are slow to understand, allow us to go over the hard parts as many times as we wish, and are never critical of our lapses. Books are key to understanding the world and participating in a democratic society.

By some standards, African-Americans have made enormous strides in literacy since Emancipation. In 1860, it is estimated, only about 5 percent of African-Americans could read and write. By 1890, 39 percent were judged literate—by the U.S. census; and by 1969, 96 percent. Between 1940 and 1992, the fraction of African-Americans who had completed high school soared from 7 percent to 82 percent. But fair questions can be asked about the quality of that education, and the standards of literacy tested. These questions apply to every ethnic group.

A national survey done for the U.S. Department of Education paints a picture of a country with more than 40 million barely literate adults. Other estimates are much worse. The literacy of young adults has slipped dramatically in the last decade. Only 3 to 4 percent of the population scores at the highest of five reading levels (essentially everybody in this group has gone to college). The vast majority have no idea how bad their reading is. Only 4 percent of those at the highest reading level are in poverty, but 43 percent of those at the lowest reading level are. Although it's not the only factor, of course, in general the better you read, the more you make—an average of about $12,000 a year at the lowest of these reading levels, and about $34,000 a year at the highest. It looks to be a necessary if not a sufficient condition for making money. And you're much more likely to be in prison if you're illiterate or barely literate. (In evaluating these facts, we must be careful not to improperly deduce causation from correlation.)

Also, marginally literate poorer people tend not to understand ballot initiatives that might help them and their children, and in stunningly disproportionate numbers fail to vote at all. This works to undermine democracy at its roots.

If Frederick Douglass as an enslaved child could teach himself into literacy and greatness, why should anyone in our more enlightened day and age remain unable to read? Well, it's not that simple—in part because few of us are as brilliant and courageous as Frederick Douglass, but for other important reasons as well:

If you grow up in a household where there are books, where you are read to, where parents, siblings, aunts, uncles, and cousins read for their own pleasure, naturally you learn to read. If no one close to you takes joy in reading, where is the evidence that it's worth the effort? If the quality of education available to you is inadequate, if you're taught rote memorization rather than how to think, if the content of what you're first given to read comes from a nearly alien culture, literacy can be a rocky road.

You have to internalize, so they're second nature, dozens of upper- and lower-case letters, symbols, and punctuation marks; memorize thousands of dumb spellings on a word-by-word basis; and conform to a range of rigid and arbitrary rules of grammar. If you're preoccupied by the absence of basic family support or dropped into a roiling sea of anger, neglect, exploitation, danger, and self-hatred, you might well conclude that reading takes too much work and just isn't worth the trouble. If you're repeatedly given the message that you're too stupid to learn (or, the functional equivalent, too cool to learn), and if there's no one there to contradict it, you might very well buy this pernicious advice. There are always some children—like Frederick Bailey—who beat the odds. Too many don't.

But, beyond all this, there's a particularly insidious way in which, if you're poor, you may have another strike against you in your effort to read—and even to think.

Ann Druyan and I come from families that knew grinding poverty. But our parents were passionate readers. One of our grandmothers learned to read because her father, a subsistence farmer, traded a sack of onions to an itinerant teacher. She read for the next hundred years. Our parents had personal hygiene and the germ theory of disease drummed into them by the New York public schools. They followed prescriptions on childhood nutrition recommended by the U.S. Department of Agriculture as if they had been handed down from Mount Sinai. Our government book on children's health had been repeatedly taped together as its pages fell out. The corners were tattered. Key advice was underlined. It was consulted in every medical crisis. For a while, my parents gave up smoking—one of the few pleasures available to them in the Depression years—so their infant could have vitamin and mineral supplements. Ann and I were very lucky.

Recent research shows that many children without enough to eat wind up with diminished capacity to understand and learn ("cognitive impairment"). Children don't have to be starving for this to happen. Even mild undernourishment—the kind most common among poor people in America—can do it. This can happen before the baby is born (if the mother isn't eating enough), in infancy, or in childhood. When there isn't enough food, the body has to decide how to invest the limited foodstuffs available. Survival comes first. Growth comes second. In this nutritional triage, the body seems obliged to rank learning last. Better to be stupid and alive, it judges, than smart and dead.

Instead of showing an enthusiasm, a zest for learning—as most healthy youngsters do—the undernourished child becomes bored, apathetic, unresponsive. More severe malnutrition leads to lower birth weights and, in its most extreme forms, smaller brains. However, even a child who looks perfectly healthy but has not enough iron, say, suffers an immediate decline in the ability to concentrate. Iron-deficiency anemia may affect as much as a quarter of all low-income children in America; it attacks the child's attention span and memory, and may have consequences reaching well into adulthood.

What once was considered relatively mild undernutrition is now understood to be potentially associated with lifelong cognitive impairment. Children who are undernourished even on a short-term basis have a diminished capacity to learn. And millions of American children go hungry every week. Lead poisoning, which is endemic in inner cities, also causes serious learning deficits. By many criteria, the prevalence of poverty in America has been steadily increasing since the early 1980s. Almost a quarter of American children now live in poverty—the highest rate of childhood poverty in the industrialized world. According to one estimate, between 1980 and 1985 alone more American infants and children died of preventable disease, malnutrition and other consequences of dire poverty than all American battle deaths during the Vietnam War.

Some programs wisely instituted on the Federal or state level in America deal with malnutrition. The Special Supplemental Food Program for Women, Infants, and Children (WIC), school breakfast and lunch programs, the Summer Food Service Program—all have been shown to work, although they do not get to all the people who need them. So rich a country is well able to provide enough food for all its children.

Some deleterious effects of undernutrition can be undone; iron-repletion therapy, for example, can repair some consequences of iron-deficiency anemia. But not all of the damage is reversible. Dyslexia—various disorders that impair reading skills—may affect 15 percent of us or more, rich and poor alike. Its causes (whether biological, psychological, or environmental) are often undetermined. But methods now exist to help many with dyslexia to learn to read.

No one should be unable to learn to read because education is unavailable. But there are many schools in America in which reading is taught as a tedious and reluctant excursion into the hieroglyphics of an unknown civilization, and many classrooms in which not a single book can be found. Sadly, the demand for adult literacy classes far outweighs the supply. High-quality early education programs such as Head Start can be enormously successful in preparing children for reading. But Head Start reaches only a third to a quarter of eligible preschoolers, many of its programs have been enfeebled by cuts in funding, and it and the nutrition programs I mentioned are under renewed Congressional attack as I write.

Head Start is criticized in a 1994 book called *The Bell Curve* by

Richard J. Herrnstein and Charles Murray. Their argument has been characterized by Gerald Coles of the University of Rochester:

> First, inadequately fund a program for poor children, then deny whatever success is achieved in the face of overwhelming obstacles, and finally conclude that the program must be eliminated because the children are intellectually inferior.

The book, which received surprisingly respectful attention from the media, concludes that there is an irreducible hereditary gap between blacks and whites—about 10 or 15 points on IQ tests. In a review, the psychologist Leon J. Kamin concludes that "[t]he authors repeatedly fail to distinguish between correlation and causation"—one of the fallacies in our baloney detection kit.

The National Center for Family Literacy, based in Louisville, Kentucky, has been implementing programs aimed at low-income families to teach both children and their parents to read. It works like this: The child, 3 to 4 years old, attends school three days a week along with a parent, or possibly a grandparent or guardian. While the grown-up spends the morning learning basic academic skills, the child is in a preschool class. Parent and child meet for lunch and then "learn how to learn together" for the rest of the afternoon.

A follow-up study of 14 such programs in three states revealed: (1) Although all of the children had been designated as being at risk for school failure as preschoolers, only 10 percent were still rated at risk by their current elementary school teachers. (2) More than 90 percent were considered by their current elementary school teachers as motivated to learn. (3) *Not one* of the children had to repeat any grade in elementary school.

The growth of the parents was no less dramatic. When asked to describe how their lives had changed as a result of the family literacy program, typical responses described improved self-confidence (nearly every participant) and self-control, passing high-school equivalency exams, admission to college, new jobs, and much better relations with their children. The children are described as more attentive to parents, eager to learn and—in some cases for the first time—hopeful about the future. Such programs could also be used in later grades for teaching mathematics, science, and much else.

—

Tyrants and autocrats have always understood that literacy, learning, books and newspapers are potentially dangerous. They can put independent and even rebellious ideas in the heads of their subjects. The British Royal Governor of the Colony of Virginia wrote in 1671:

I thank God there are no free schools nor printing; and I hope we shall not have [them] these [next] hundred years; for learning has brought disobedience, and heresy, and sects into the world, and printing has divulged them and libels against the best government. God keep us from both!

But the American colonists, understanding where liberty lies, would have none of this.

In its early years, the United States boasted one of the highest—perhaps the highest—literacy rates in the world. (Of course, slaves and women didn't count in those days.) As early as 1635, there had been public schools in Massachusetts, and by 1647, compulsory education in all townships there of more than 50 "households." By the next century and a half, educational democracy had spread all over the country. Political theorists came from abroad to witness this national wonder: vast numbers of ordinary working people who could read and write. The American devotion to education for all propelled discovery and invention, a vigorous democratic process, and an upward mobility that pumped the nation's economic vitality.

Today, the United States is not the world leader in literacy. Many of those judged literate are unable to read and understand very simple material—much less a sixth-grade textbook, an instruction manual, a bus schedule, a mortgage statement, or a ballot initiative. And the sixth-grade textbooks of today are much less challenging than those of a few decades ago, while the literacy requirements at the workplace have become more demanding than ever before.

The gears of poverty, ignorance, hopelessness, and low self-esteem mesh to create a kind of perpetual failure machine that grinds down dreams from generation to generation. We all bear the cost of keeping it running. Illiteracy is its linchpin.

Even if we hardened our hearts to the shame and misery experienced by the victims, the cost of illiteracy to everyone else is severe—the cost in medical expenses and hospitalization, the cost in crime and prisons, the cost in special education, the cost in lost productivity and in potentially brilliant minds who could help solve the dilemmas besetting us.

Frederick Douglass taught that literacy is the path from slavery to freedom. There are many kinds of slavery and many kinds of freedom. But reading is still the path.

Frederick Douglass
After the Escape

When he was barely twenty, he ran away to freedom. Settling in New Bedford with his bride, Anna Murray, he worked as a common laborer. Four years later Douglass was invited to address a meeting. By that time, in the North, it was not unusual to hear the great orators of the day—the white ones, that is—railing against slavery. But even many of those opposed to slavery thought of the slaves themselves as somehow less than human. On the night of August 16, 1841, on the small island of Nantucket, the members of the mostly Quaker Massachusetts Anti-Slavery Society leaned forward in their chairs to hear something new: a voice raised in opposition to slavery by someone who knew it from bitter personal experience.

His very appearance and demeanor destroyed the then-prevalent myth of the "natural servility" of African-Americans. By all accounts his eloquent analysis of the evils of slavery was one of the most brilliant debuts in American oratorical history. William Lloyd Garrison, the leading abolitionist of the day, sat in the front row. When Douglass finished his speech, Garrison rose, turned to the stunned audience, and challenged them with a shouted question: "Have we been listening to a thing, a chattel personal, or a man?"

"A man! A man!" the audience roared back as one voice.

"Shall such a man be held a slave in a Christian land?" called out Garrison.

"No! No!" shouted the audience.

And even louder, Garrison asked: "Shall such a man ever be sent back to bondage from the free soil of Old Massachusetts?"

And now the crowd was on its feet, crying out "No! No! No!"

He never did return to slavery. Instead, as an author, editor, and publisher of journals, as a speaker in America and abroad, and as the first African-American to occupy a high advisory position in the U.S. government, he spent the rest of his life fighting for human rights. During the Civil War, he was a consultant to President Lincoln. Douglass successfully advocated the arming of ex-slaves to fight for the North, Federal retaliation against Confederate prisoners-of-war for Confederate summary execution of captured African-American soldiers, and freedom for the slaves as a principal objective of the war.

Many of his opinions were scathing, and ill-designed to win him friends in high places:

> I assert most unhesitatingly, that the religion of the South is a mere covering for the most horrid crimes—a justifier of the most appalling barbarity, a sanctifier of the most hateful frauds, and a dark shelter under which the darkest, foulest, grossest, and most infernal deeds of slaveholders find the strongest protection. Were I to be again reduced to the chains of slavery, next to that enslavement, I should regard being the slave of a religious master the greatest calamity that could befall me. . . I . . . . hate the corrupt, slaveholding, women-whipping, cradle-plundering, partial and hypocritical Christianity of this land.

Compared to some of the religiously inspired racist rhetoric of that time and later, Douglass's comments do not seem hyperbolic. "Slavery is of God" they used to say in antebellum times. As one of many loathsome *post*–Civil War examples, Charles Carroll's *The Negro a Beast* (St. Louis: American Book and Bible House) taught its pious readers that "the Bible and Divine Revelation, as well as reason, all teach that the Negro is not human." More recently, some racists still reject the plain testimony written in the DNA that all the races are not only human but nearly indistinguishable with appeals to the Bible as an "impregnable bulwark" against even examining the evidence.

It is worth noting, though, that much of the abolitionist ferment arose out of Christian, especially Quaker, communities of the North; that the traditional black Southern Christian churches played a key role in the historic American civil rights struggle of the 1960s; and that many of its leaders—most notably Martin Luther King, Jr.—were ministers ordained in those churches.

Douglass addressed the white community in these words:

> [Slavery] fetters your progress, it is the enemy of improvement; the deadly foe of education; it fosters pride; it breeds indolence; it promotes vice; it shelters crime; it is a curse of the earth that supports it, and yet you cling to it as if it were the sheet anchor of all your hopes.

In 1843, on a speaking tour of Ireland shortly before the potato famine, he was moved by the dire poverty there to write home to Garrison: "I see much here to remind me of my former condition, and I confess I should be ashamed to lift my voice against American slavery, but that I know the cause of humanity is one the world over." He was outspoken in opposition to the policy of extermination of the Native Americans. And in 1848, at the Seneca Falls Convention, when Elizabeth Cady Stanton* had the nerve to call for an effort to secure the vote for women, he was the only man of any ethnic group to stand in support.

On the night of February 20, 1895—more than thirty years after Emancipation—following an appearance at a women's rights rally with Susan B. Anthony, he collapsed and died.

---

* Years later, she wrote of the Bible in words reminiscent of Douglass's: "I know of no other books that so fully teach the subjection and degradation of women."

Chapter 22

SIGNIFICANCE
JUNKIES

We also know how cruel the truth often is,
and we wonder whether delusion is not more consoling.

HENRI POINCARÉ
(1854–1912)

I hope no one will consider me unduly cynical if I assert that a good first-order model of how commercial and public television programming work is simply this: Money is everything. In prime time, a single rating point difference is worth millions of dollars in advertising. Especially since the early 1980s, television has become almost entirely profit-motivated. You can see this, say, in the decline of network news and news specials, or in the pathetic evasions that the major networks offered to circumvent a Federal Communications Commission mandate that they improve the level of children's programming. (For example, educational virtues were asserted for a cartoon series that systematically misrepresents the technology and lifestyles of our Pleistocene ancestors, and that portrays dinosaurs as pets.) As I write, public television in America is in real danger of losing government support and the content of commercial programming is in the course of a steep, long-term dumbing down.

In this perspective, fighting for more real science on television seems naïve and forlorn. But owners of networks and television producers have children and grandchildren about whose future they rightly worry. They must feel some responsibility for the future of their nation. There is evidence that science programming can be successful, and that people hunger for more of it. I remain hopeful that sooner or later we'll see real science skillfully and appealingly presented as regular fare on major network television worldwide.

———

Baseball and soccer have Aztec antecedents. Football is a thinly disguised re-enactment of hunting; we played it before we were human. Lacrosse is an ancient Native American game, and hockey is related to it. But basketball is new. We've been making movies longer than we've been playing basketball.

At first, they didn't think to make a hole in the peach basket so the ball could be retrieved without climbing a flight of stairs. But in the brief time since then, the game has evolved. In the hands mainly of African-American players, basketball has become—at its best—the paramount synthesis in sport of intelligence, precision, courage, audacity, anticipation, artifice, teamwork, elegance, and grace.

Five-foot-three-inch Muggsy Bogues negotiates a forest of giants; Michael Jordan sails in from some outer darkness beyond the free-throw line; Larry Bird threads a precise, no-look pass; Kareem Abdul-Jabbar unleashes a skyhook. This is not fundamentally a contact sport like football. It's a game of finesse. The full-court press, passes out of the double-team, the pick-and-roll, cutting off the passing lanes, a tip-in from a high-flying forward soaring from out of nowhere all constitute a coordination of intellect and athleticism, a harmony of mind and body. It's not surprising that the game has caught fire.

Ever since National Basketball Association games became a television staple, it's seemed to me that it could be used to teach science and mathematics. To appreciate a free-throw average of 0.926, you must know something about converting fractions into decimals. A lay-up is Newton's first law of motion in action. Every shot represents the launching of a basketball on a parabolic arc, a curve determined by the same gravitational physics that specifies the flight of a ballistic missile, or the Earth orbiting the Sun, or a spacecraft on its rendezvous with some distant world. The center of mass of the player's body during a slam dunk is briefly in orbit about the center of the Earth.

To get the ball in the basket, you must loft it at exactly the right speed; a 1 percent error and gravity will make you look bad. Three-point shooters, whether they know it or not, compensate for aerodynamic drag. Each successive bounce of a dropped basketball is nearer to the ground because of the Second Law of Thermodynamics. Daryl Dawkins or Shaquille O'Neal shattering a backboard is an opportunity for teaching—among some other things—the propagation of shock waves. A spin shot off the glass from under the backboard goes in because of the conservation of angular momentum. It's an infraction of the rules to touch the basketball in "the cylinder" above the basket; we're now talking about a key mathematical idea: generating n-dimensional objects by moving $(n-1)$-dimensional objects.

In classroom, newspapers, and television, why aren't we using sports to teach science?

When I was growing up, my father would bring home a daily paper and consume (often with great gusto) the baseball box scores. There they were, to me dry as dust, with obscure abbreviations (W, SS, K, W-L, AB, RBI), but they spoke to him. Newspapers everywhere printed them. I figured maybe they weren't too hard for me. Eventually I too got caught up in the world of baseball statistics. (I know it helped me in learning decimals, and I still cringe a little when I hear, usually at the very beginning of the baseball season, that someone's "batting a thousand." But 1.000 is not 1,000. The lucky player is batting one.)

Or take a look at the financial pages. Any introductory material? Explanatory footnotes? Definitions of abbreviations? Almost none. It's sink or swim. Look at those acres of statistics! Yet people voluntarily read the stuff. It's not beyond their ability. It's only a matter of motivation. Why can't we do the same with math, science, and technology?

———

In every sport the players seem to perform in streaks. In basketball it's called the hot hand. You can do no wrong. I remember a playoff game in which Michael Jordan, not ordinarily a superb long-range shooter, was effortlessly making so many consecutive three-point baskets from all over the floor that he shrugged his shoulders in amazement at himself. In contrast, there are times when you're cold, when nothing goes in. When a player is in the groove he seems to be tapping into some mysterious power, and when ice-cold he's under some kind of jinx or spell. But this is magical, not scientific thinking.

Streakiness, far from being remarkable, is expected, even for random events. What *would* be amazing would be no streaks. If I flip a penny 10 times in a row, I might get this sequence of heads and tails: H H H T H T H H H H . Eight heads out of 10, and four in a row! Was I exercising some psychokinetic control over my penny? Was I in a heads groove? It looks much too regular to be due to chance.

But then I remember that I was flipping before and after I got this run of heads, that it's embedded in a much longer and less interesting sequence: H H T H T T H H H T H T H H H H T H T T H T H T T. If I'm permitted to pay attention to some results and ignore others, I'll always be able to "prove" there's something exceptional about my streak. This is one of the fallacies in the baloney detection kit, the enumeration of favorable circumstances. We remember the hits and forget the misses. If your ordinary field goal shooting percentage is 50 percent and you can't improve your statistics by an effort of will, you're exactly as likely to have a hot hand in basketball as I am in coin-flipping. As often as I get eight out of ten heads, you'll get eight out of ten baskets. Basketball can teach something about probability and statistics, as well as critical thinking.

An investigation by my colleague Tom Gilovich, professor of psychology at Cornell, shows persuasively that our ordinary understanding of the basketball streak is a misperception. Gilovich studied whether shots made by NBA players tend to cluster more than you'd expect by chance. After making one or two or three baskets, players were no more likely to succeed than after a missed basket. This was true for the great and the near-great, not only for field goals but for free throws—where there's no hand in your face. (Of course some attenuation of shooting streaks can be attributed to increased attention by the defense to the player with the "hot hand.") In baseball, there's the related but contrary myth that someone batting below his average is "due" to make a hit. This is no more true than that a few heads in a row makes the chance of flipping tails next time anything other than 50 percent. If there are streaks beyond what you'd expect statistically, they're hard to find.

But somehow this doesn't satisfy. It doesn't feel true. Ask the players, or the coaches, or the fans. We seek meaning, even in random numbers. We're significance junkies. When the celebrated coach Red Auerbach heard of Gilovich's study, his response was: "Who is this guy? So he makes a study. I couldn't care less." And you know exactly how he feels. But if basketball streaks don't show up more often than sequences of heads or tails, there's nothing magical about them. Does this reduce players to mere marionettes, manipulated by the laws of chance? Certainly not. Their average shooting percentages are a true reflection of their personal skills. This is only about the frequency and duration of streaks.

Of course, it's much more fun to think that the gods have touched the player who's on a streak and scorned the one with a cold hand. So what? What's the harm of a little mystification? It sure beats boring statistical analyses. In basketball, in sports, no harm. But as a habitual way of thinking, it gets us into trouble in some of the other games we like to play.

—

"Scientist, yes; mad, no" giggles the mad scientist on *Gilligan's Island* as he adjusts the electronic device that permits him to control the minds of others for his own nefarious purpose.

"I'm sorry, Dr. Nerdnik, the people of Earth will not appreciate being shrunk to 3 inches high, even if it *will* save room and energy. . ." The cartoon superhero is patiently explaining an ethical dilemma to the typical scientist portrayed on Saturday-morning children's television. Many of these so-called scientists—judging from the programs I've seen (and plausible inference about ones I haven't, such as the *Mad Scientist's 'Toon Club*)—are moral cripples driven by a lust for power or endowed with a spectacular insensitivity to the feelings of others. The message conveyed to the moppet audience is that science is dangerous and scientists worse than weird: They're crazed.

The applications of science, of course, *can* be dangerous, and, as I've tried to stress, virtually every major technological advance in the history of the human species—back to the invention of stone tools and the domestication of fire—has been ethically ambiguous. These advances can be used by ignorant or evil people for dangerous purposes or by wise and good people for the benefit of the human species. But only one side of the ambiguity ever seems to be presented in these offerings to our children.

Where in all these programs are the joys of science? The delights in discovering how the universe is put together? The exhilaration in knowing a deep thing well? What about the crucial contributions that science and technology have made to human welfare—or the billions of lives saved or made possible by medical and agricultural technology? (In fairness, though, I should mention that the Professor in *Gilligan's Island* often used his knowledge of science to solve practical problems for the castaways.)

We live in a complex age where many of the problems we face can, whatever their origins, only have solutions that involve a deep understanding of science and technology. Modern society desperately needs the finest minds available to devise solutions to these problems. I do not think that many gifted youngsters will be encouraged toward a career in science or engineering by watching Saturday-morning television—or much of the rest of the available American video menu.

Over the years, a profusion of credulous, uncritical TV series and "specials"—on ESP, channeling, the Bermuda Triangle, UFOs, ancient astronauts, Big-Foot, and the like—have been spawned. The style-setting series *In Search Of*. . . begins with a disclaimer disavowing any responsibility to present a balanced view of the subject. You can see a thirst for wonder here untempered by even rudimentary scientific skepticism. Pretty much whatever anyone says on camera is true. The idea that there might be alternative explanations to be decided among by the weight of evidence never surfaces. The same is true of *Sightings* and *Unsolved Mysteries*—in which, as the very title suggests, prosaic solutions are unwelcome—and innumerable other clones.

In Search of. . . frequently takes an intrinsically interesting subject and systematically distorts the evidence. If there is a mundane scientific explanation and one which requires the most extravagant paranormal or psychic explanation, you can be sure which will be highlighted. An almost random example: An author is presented who argues that a major planet lies beyond Pluto. His evidence is cylinder seals from ancient Sumer, carved long before the invention of the telescope. His views are increasingly accepted by professional astronomers, he says. Not a word is mentioned of the failure of astronomers—studying the motions of Neptune, Pluto, and the four spacecraft beyond—to find a trace of the alleged planet.

The graphics are indiscriminate. When an offscreen narrator is talking about dinosaurs, we see a woolly mammoth. The narrator describes a hovercraft; the screen shows a shuttle liftoff. We hear about lakes and flood plains, but are shown mountains. It doesn't matter. The visuals are as indifferent to the facts as is the voice-over.

A series called *The X Files*, which pays lip service to skeptical examination of the paranormal, is skewed heavily towards the reality of alien abductions, strange powers and government complicity in covering up just about everything interesting. Almost never does the paranormal claim turn out to be a hoax or a psychological aberration or a misunderstanding of the natural world. Much closer to reality, as well as a much greater public service, would be an adult series (*Scooby Doo* does it for children) in which paranormal claims are systematically investigated and every case is found to be explicable in prosaic terms. The dramatic tension would lie in uncovering how misapprehension and hoax could generate apparently genuine paranormal phenomena.

Perhaps one of the investigators would always be disappointed, hoping that *next* time an unambiguously paranormal case will survive skeptical scrutiny.

Other shortcomings are evident in television science fiction programming. *Star Trek*, for example, despite its charm and strong international and interspecies perspective, often ignores the most elementary scientific facts. The idea that Mr. Spock could be a cross between a human being and a life-form independently evolved on the planet Vulcan is genetically far less probable than a successful cross of a man and an artichoke. The idea does, however, provide a precedent in popular culture for the extraterrestrial/human hybrids that later became so central a component of the alien abduction story. There must be dozens of alien species on the various *Star Trek* TV series and movies. Almost all we spend any time with are minor variants of humans. This is driven by economic necessity, costing only an actor and a latex mask, but it flies in the face of the stochastic nature of the evolutionary process. If there are aliens, almost all of them I think will look devastatingly less human than Klingons and Romulans (and be at wildly different levels of technology). *Star Trek* doesn't come to grips with evolution.

In many TV programs and films, even the casual science—the throwaway lines that are not essential to a plot already innocent of science—is done incompetently. It costs very little to hire a graduate student to read the script for scientific accuracy. But, so far as I can tell, this is almost never done. As a result we have such howlers as "parsec" mentioned as a unit of speed instead of distance in the—in many other ways exemplary—film *Star Wars*. If such things were done with a modicum of care, they might even improve the plot; certainly, they might help convey a little science to a mass audience.

There's a great deal of pseudoscience for the gullible on TV, a fair amount of medicine and technology, but hardly any science—especially on the big commercial networks, whose executives tend to think that science programming means ratings declines and lost profits, and nothing else matters. There are network employees with the title "Science Correspondent," and an occasional news feature said to be devoted to science. But we almost never hear any science from them, just medicine and technology. In all the networks, I doubt if there's a single employee whose job it is to read each week's issue of *Nature* or *Science* to see if anything newsworthy has been discovered. When the

Nobel Prizes in science are announced each fall, there's a superb news "hook" for science: a chance to explain what the prizes were given for. But, almost always, all we hear is something like ". . . may one day lead to a cure for cancer. Today in Belgrade . . ."

How much science is there on the radio or television talk shows, or on those dreary Sunday morning programs in which middle-aged white people sit around agreeing with each other? When is the last time you heard an intelligent comment on science by a President of the United States? Why in all America is there no TV drama that has as its hero someone devoted to figuring out how the Universe works? When a highly publicized murder trial has everyone casually mentioning DNA testing, where are the prime-time network specials devoted to nucleic acids and heredity? I can't even recall seeing an accurate and comprehensible description on television of how *television* works.

By far the most effective means of raising interest in science is television. But this enormously powerful medium is doing close to nothing to convey the joys and methods of science, while its "mad scientist" engine continues to huff and puff away.

In American polls in the early 1990s, two-thirds of all adults had no idea what the "information superhighway" was; 42 percent didn't know where Japan is; and 38 percent were ignorant of the term "holocaust." But the proportion was in the high 90s who had heard of the Menendez, Bobbitt, and O. J. Simpson criminal cases; 99 percent had heard that the singer Michael Jackson had allegedly sexually molested a boy. The United States may be the best-entertained nation on Earth, but a steep price is being paid.

Surveys in Canada and in the United States in the same period show that television viewers wish there were more science programming. In North America, often there's a good science program in the *Nova* series of the Public Broadcasting System, and occasionally on the Discovery or Learning channels, or the Canadian Broadcasting Company. Bill Nye's "The Science Guy" programs for young children on PBS are fast paced, feature arresting graphics, range over many realms of science, and sometimes even illuminate the process of discovery. But the depth of public interest in science engrossingly and accurately presented—to say nothing of the immense good that would result from better public understanding of science—is not yet reflected in network programming.

—

How could we put more science on television? Here are some possibilities:

• The wonders and methods of science routinely presented on news and talk programs. There's real human drama in the process of discovery.
• A series called "Solved Mysteries," in which tremulous speculations have rational resolutions, including puzzling cases in forensic medicine and epidemiology.
• Ring My Bells Again: A series in which we relive the media and the public falling hook, line and sinker for a coordinated government lie. The first two episodes might be the Gulf of Tonkin "incident" and the systematic irradiation of unsuspecting and unprotected American civilians and military personnel in the alleged requirements of "national defense" following 1945.
• A separate series on fundamental misunderstandings and mistakes made by famous scientists, national leaders, and religious figures.
• Regular exposés of pernicious pseudoscience, and audience-participation "how-to" programs: how to bend spoons, read minds, appear to foretell the future, perform psychic surgery, do cold reads, and press the TV viewers' personal buttons. How we're bamboozled: Learn by doing.
• A state-of-the-art computer graphics facility to prepare in advance scientific visuals for a wide range of news contingencies.
• A set of inexpensive televised debates, each perhaps an hour long, with a computer graphics budget for each side provided by the producers, rigorous standards of evidence required by the moderator, and the widest range of topics broached. They could address issues where the scientific evidence is overwhelming, as on the matter of the shape of the Earth; controversial matters where the answer is less clear, such as the survival of one's personality after death, or abortion, or animal rights, or genetic engineering; or any of the presumptive pseudosciences mentioned in this book.

There is a pressing national need for more public knowledge of science. Television cannot provide it all by itself. But if we want to make short-term improvements in the understanding of science, television is the place to start.

Chapter 23

MAXWELL
AND
THE NERDS

Why should we subsidize intellectual curiosity?

RONALD REAGAN,
campaign speech, 1980

There is nothing which can better deserve
our patronage than the promotion of science
and literature. Knowledge is in every country
the surest basis of public happiness.

GEORGE WASHINGTON,
address to Congress, January 8, 1790

Stereotypes abound. Ethnic groups are stereotyped, the citizens of other nations and religions are stereotyped, the genders and sexual preferences are stereotyped, people born in various times of the year are stereotyped (Sun-sign astrology), and occupations are stereotyped. The most generous interpretation ascribes it to a kind of intellectual laziness: Instead of judging people on their individual merits and deficits, we concentrate on one or two bits of information about them, and then place them in a small number of previously constructed pigeonholes.

This saves the trouble of thinking, at the price in many cases of committing a profound injustice. It also shields the stereotyper from contact with the enormous variety of people, the multiplicity of ways of being human. Even if stereotyping were valid on average, it is bound to fail in many individual cases: Human variation runs to bell-type curves. There's an average value of any quality, and smaller numbers of people running off in both extremes.

Some stereotyping is the result of not controlling the variables, of forgetting what other factors might be in play. For example, it used to be that there were almost no women in science. Many male scientists were vehement: This proved that women lacked the ability to do science. Temperamentally, it didn't fit them, it was too difficult, it required a kind of intelligence that women don't have, they're too emotional to be objective, can you think of any great women theoretical physicists? . . . and so on. Since then the barriers have come tumbling down. Today women populate most of the subdisciplines of science. In my own fields of astronomy and planetary studies, women have recently burst upon the scene, making discovery after discovery, and providing a desperately needed breath of fresh air.

So what data were they missing—all those famous male scientists of the 1950s and '60s and earlier who had pronounced so authoritatively on the intellectual deficiencies of women? Plainly, the society was preventing women from entering science, and then criticizing them for it, confusing cause and effect:

You want to be an astronomer, young woman? Sorry.

Why can't you? Because you're unsuited.

How do we know you're unsuited? Because women have never been astronomers.

Put so baldly, the case sounds absurd. But the contrivances of bias can be subtle. The despised group is rejected by spurious arguments, sometimes done with such confidence and contempt that many of us, including some of the victims themselves, fail to recognize it as self-serving sleight of hand.

Casual observers of meetings of skeptics, and those who glance at the list of CSICOP Fellows, have noted a great preponderance of men. Others claim disproportionate numbers of women among believers in astrology (horoscopes in most "women's" but few "men's" magazines), crystals, ESP and the like. Some commentators suggest that there is something peculiarly male about skepticism. It's hard-driving, competitive, confrontational, tough-minded—whereas women, they say, are more accepting, consensus-building, and uninterested in challenging conventional wisdom. But in my experience women scientists have just as finely honed skeptical senses as their male counterparts; that's just part of being a scientist. This criticism, if that's what it is, is presented to the world in the usual ragged disguise: If you discourage women from being skeptical and don't train them in skepticism, then sure enough you may find that many women aren't skeptical. Open the doors and let them in, and they're as skeptical as anybody else.

One of the stereotyped occupations is science. Scientists are nerds, socially inept, working on incomprehensible subjects that no normal person would find in any way interesting—even if he were willing to invest the time required, which, again, no sensible person would. "Get a life," you might want to tell them.

I asked for a fleshed-out contemporary characterization of science-nerds from an expert on eleven-year-olds of my acquaintance. I should stress that she is merely reporting, not necessarily endorsing, the conventional prejudices:

Nerds wear their belts just under their rib cages. Their short-sleeve shirts are equipped with pocket protectors in which are displayed a formidable array of multicolored pens and pencils. A programmable calculator is carried in a special belt holster. They all wear thick glasses with broken nose-pieces that have been repaired with Band-Aids. They are bereft of social skills, and oblivious or indifferent to the lack. When they laugh, what comes out is a snort. They jabber at each other in an incomprehensible language. They'll jump at the opportunity to work for extra credit in all classes except gym. They look down on normal people, who in turn laugh at them. Most nerds have names like Norman. (The Norman Conquest involved a horde of high-belted, pocket-protected, calculator-carrying nerds with broken glasses invading England.) There are more boy nerds than girl nerds, but there are plenty of both. Nerds don't date. If you're a nerd you can't be cool. Also vice versa.

This of course is a stereotype. There are scientists who dress elegantly, who are devastatingly cool, who many people long to date, who do not carry concealed calculators to social events. Some you'd never guess were scientists if you invited them to your home.

But other scientists do match the stereotype, more or less. They're pretty socially inept. There may be, proportionately, many more nerds among scientists than among backhoe operators or fashion designers or highway patrol officers. Perhaps scientists are more nerdish than bartenders or surgeons or short-order cooks. Why should this be? Maybe people untalented in getting along with others find a refuge in impersonal pursuits, particularly mathematics and the physical sciences. Maybe the serious study of difficult subjects requires so much time and dedication that very little is left over for learning more than the barest social niceties. Maybe it's a combination of both.

Like the mad-scientist image to which it's closely related, the nerd-scientist stereotype is pervasive in our society. What's wrong with a little good-natured fun at the expense of scientists? If, for whatever reason, people dislike the stereotypical scientist, they are less likely to support science. Why subsidize geeks to pursue their absurd and incomprehensible little projects? Well, we know the answer to that: Science is supported because it provides spectacular benefits at all levels in the society, as I have argued earlier in this book. So those who find nerds distasteful, but at the same time crave the products of science, face a kind of dilemma. A tempting resolution is to direct the activities of the scientists. Don't give them money to go off in weird directions;

instead tell them what we need—this invention, or that process. Subsidize not the curiosity of the nerds, but what will benefit the society. It seems simple enough.

The trouble is that ordering someone to go out and make a specific invention, even if price is no object, hardly guarantees that it gets done. There may be an underpinning of knowledge that's unavailable, without which no one will ever build the contrivance you have in mind. And the history of science shows that often you can't go after the underpinnings in a directed way, either. They may emerge out of the idle musings of some lonely young person off in the boondocks. They're ignored or rejected even by other scientists, sometimes until a new generation of scientists comes along. Urging major practical inventions while discouraging curiosity-driven research would be spectacularly counterproductive.

—

Suppose: You are, by the Grace of God, Victoria, Queen of the United Kingdom of Great Britain and Ireland, and Defender of the Faith in the most prosperous and triumphant age of the British Empire. Your dominions stretch across the planet. Maps of the world are abundantly splashed with British red. You preside over the world's leading technological power. The steam engine is perfected in Great Britain, largely by Scottish engineers—who provide technical expertise on the railways and steamships that bind up the Empire.

Suppose in the year 1860 you have a visionary idea, so daring it would have been rejected by Jules Verne's publisher. You want a machine that will carry your voice, as well as moving pictures of the glory of the Empire, into every home in the kingdom. What's more, the sounds and pictures must come not through conduits or wires, but somehow out of the air—so people at work and in the field can receive instantaneous inspirational offerings designed to insure loyalty and the work ethic. The Word of God could also be conveyed by the same contrivance. Other socially desirable applications would doubtless be found.

So with the Prime Minister's support, you convene the Cabinet, the Imperial General Staff, and the leading scientists and engineers of the Empire. You will allocate a million pounds, you tell them— big money in 1860. If they need more, just ask. You don't care how they do it; just get it done. Oh, yes, it's to be called the Westminster Project.

Probably there would be some useful inventions emerging out of such an endeavor—"spin-off." There always are when you spend huge amounts of money on technology. But the Westminster Project would almost certainly fail. Why? Because the underlying science hadn't been done. By 1860 the telegraph was in existence. You could imagine at great expense telegraphy sets in every home, with people ditting and dahing messages out in Morse code. But that's not what the Queen asked for. She had radio and television in mind, but they were far out of reach.

In the real world, the physics necessary to invent radio and television would come from a direction that no one could have predicted:

James Clerk Maxwell was born in Edinburgh, Scotland, in 1831. At age two he found that he could use a tin plate to bounce an image of the Sun off the furniture and make it dance against the walls. As his parents came running he cried out, "It's the Sun! I got it with the tin plate!" In his boyhood, he was fascinated by bugs, grubs, rocks, flowers, lenses, machines. "It was humiliating," later recalled his aunt Jane, "to be asked so many questions one couldn't answer by a child like that."

Naturally, by the time he got to school he was called "Dafty"—daft being a Britishism for not quite right in the head. He was an exceptionally handsome young man, but he dressed carelessly, for comfort rather than style, and his Scottish provincialisms in speech and conduct were a cause for derision, especially by the time he reached college. And he had peculiar interests.

Maxwell was a nerd.

He fared little better with his teachers than with his fellow students. Here's a poignant couplet he wrote at the time:

Ye years roll on, and haste the expected time
When flogging boys shall be accounted crime.

Many years later, in 1872, in his inaugural lecture as professor of experimental physics at Cambridge University, he alluded to the nerdish stereotype:

It is not so long ago since any man who devoted himself to geometry, or to any science requiring continued application, was looked upon as necessarily a misanthrope, who must have abandoned all human interests, and betaken himself to abstractions so far re-

moved from all the world of life and action that he has become in-
sensible alike to the attractions of pleasure and to the claims of
duty.

I suspect that "not so long ago" was Maxwell's way of recalling the experiences of his youth. He then went on to say,

In the present day, men of science are not looked upon with the
same awe or with the same suspicion. They are supposed to be in
league with the material spirit of the age, and to form a kind of advanced Radical party among men of learning.

We no longer live in a time of untrammeled optimism about the
benefits of science and technology. We understand that there is a
downside. Circumstances today are much closer to what Maxwell remembered from his childhood.

He made enormous contributions to astronomy and physics—
from the conclusive demonstration that the rings of Saturn are composed of small particles, to the elastic properties of solids, to the
disciplines now called the kinetic theory of gases and statistical mechanics. It was he who first showed that an enormous number of tiny
molecules, moving on their own, incessantly colliding with each other
and bouncing elastically, leads not to confusion, but to precise statistical laws. The properties of such a gas can be predicted and understood. (The bell-shaped curve that describes the speeds of molecules
in a gas is now called the Maxwell-Boltzmann distribution.) He invented a mythical being, now called "Maxwell's demon," whose actions generated a paradox that took modern information theory and
quantum mechanics to resolve.

The nature of light had been a mystery since antiquity. There were
acrimonious learned debates on whether it was a particle or a wave.
Popular definitions ran to the style, "Light is darkness—lit up."
Maxwell's greatest contribution was his discovery that electricity and
magnetism, of all things, join together to become light. The now conventional understanding of the electromagnetic spectrum—running
in wavelength from gamma rays to X rays to ultraviolet light to visible
light to infrared light to radio waves—is due to Maxwell. So is radio,
television, and radar.

But Maxwell wasn't after any of this. He was interested in how electricity makes magnetism and vice versa. I want to describe what Maxwell did, but his historic accomplishment is highly mathematical. In a few pages, I can at best give you only a flavor. If you do not fully understand what I'm about to say, please bear with me. There's no way we can get a feeling for what Maxwell did without looking at a little mathematics.

Mesmer, the inventor of "mesmerism," believed he had discovered a magnetic fluid, "almost the same thing as the electric fluid," that permeated all things. On this matter as well, he was mistaken. We now know that there is no special magnetic fluid, and that all magnetism — including the power that resides in a bar or horseshoe magnet — is due to moving electricity. The Danish physicist Hans Christian Oersted had performed a little experiment in which electricity was made to flow down a wire and induce a nearby compass needle to waver and tremble. The wire and the compass were not in physical contact. The great English physicist Michael Faraday had done the complementary experiment: He made a magnetic force turn on and off and thereby generated a current of electricity in a nearby wire. Time-varying electricity had somehow reached out and generated magnetism, and time-varying magnetism had somehow reached out and generated electricity. This was called "induction" and was deeply mysterious, close to magic.

Faraday proposed that the magnet had an invisible "field" of force that extended into surrounding space, stronger close to the magnet, weaker farther away. You could track the form of the field by placing tiny iron filings on a piece of paper and waving a magnet underneath. Likewise, your hair after a good combing on a low-humidity day generates an electric field which invisibly extends out from your head, and which can even make small pieces of paper move by themselves.

The electricity in a wire, we now know, is caused by submicroscopic electrical particles, called electrons, which respond to an electric field and move. The wires are made of materials like copper which have lots of free electrons — electrons not bound within atoms, but able to move. Unlike copper, though, most materials, wood, say, are not good conductors; they are instead insulators or "dielectrics." In them, comparatively few electrons are available to move in response to the impressed electric or magnetic field. Not much of a current is produced. Of course there's some movement or "displacement" of electrons, and the bigger the electric field, the more displacement occurs.

Maxwell devised a way of writing down what was known about electricity and magnetism in his time, a method of summarizing precisely all those experiments with wires and currents and magnets. Here they are, the four Maxwell equations for the behavior of electricity and magnetism in matter:

$$\nabla \cdot \mathbf{E} = \rho/\varepsilon_0$$
$$\nabla \cdot \mathbf{B} = 0$$
$$\nabla \times \mathbf{E} = - \dot{\mathbf{B}}$$
$$\nabla \times \mathbf{B} = \mu_0 \mathbf{j} + \mu_0 \varepsilon_0 \dot{\mathbf{E}}$$

It takes a few years of university-level physics to really understand these equations. They are written using a branch of mathematics called vector calculus. A vector, written in bold-face type, is any quantity with both a magnitude and a direction. Sixty miles an hour isn't a vector, but sixty miles an hour due north on Highway 1 is. $\mathbf{E}$ and $\mathbf{B}$ represent the electric and magnetic fields. The triangle, called a nabla (because of its resemblance to a certain ancient Middle Eastern harp), expresses how the electric or magnetic fields vary in three-dimensional space. The "dot product" and the "cross product" after the nablas are statements of two different kinds of spatial variation.

$\dot{\mathbf{E}}$ and $\dot{\mathbf{B}}$ represent the time variation, the rate of change of the electric and magnetic fields. $\mathbf{j}$ stands for an electrical current. The lower-case Greek letter ρ (rho) represents the density of electrical charges, while ε_0 (pronounced "epsilon zero") and μ_0 (pronounced "mu zero") are not variables, but properties of the substance $\mathbf{E}$ and $\mathbf{B}$ are measured in, and determined by experiment. In a vacuum, ε_0 and μ_0 are constants of Nature.

Considering how many different quantities are being brought together in these equations, it's striking how simple they are. They could have gone on for pages, but they don't.

The first of the four Maxwell equations tells how an electric field due to electrical charges (electrons, for example) varies with distance (it gets weaker the farther away we go). But the greater the charge density (the more electrons, say, in a given space), the stronger the field.

The second equation tells us that there's no comparable statement in magnetism, because Mesmer's magnetic "charges" (or magnetic "monopoles") do not exist: Saw a magnet in half and you won't be holding an isolated "north" pole and an isolated "south" pole; each piece now has its own "north" and "south" pole.

The third equation tells us how a changing magnetic field induces an electric field.

The fourth describes the converse—how a changing electric field (or an electrical current) induces a magnetic field.

The four equations are essentially distillations of generations of laboratory experiments, mainly by French and British scientists. What I've described here vaguely and qualitatively, the equations describe exactly and quantitatively.

Maxwell then asked himself a strange question: What would these equations look like in empty space, in a vacuum, in a place where there were no electrical charges and no electrical currents? We might very well anticipate no electric and no magnetic fields in a vacuum. Instead, he suggested that the right form of the Maxwell equations for the behavior of electricity and magnetism in empty space is this:

$$\nabla \cdot \mathbf{E} = 0$$
$$\nabla \cdot \mathbf{B} = 0$$
$$\nabla \times \mathbf{E} = -\dot{\mathbf{B}}$$
$$\nabla \times \mathbf{B} = \mu_0 \varepsilon_0 \dot{\mathbf{E}}$$

He set ρ equal to zero, indicating that there are no electrical charges. He also set $\mathbf{j}$ equal to zero, indicating that there are no electrical currents. But he didn't discard the last term in the fourth equation, $\mu_0 \varepsilon_0 \dot{\mathbf{E}}$, the feeble displacement current in insulators.

Why not? As you can see from the equations, Maxwell's intuition preserved the symmetry between the magnetic and electric fields. Even in a vacuum, in the total absence of electricity, or even matter, a changing magnetic field, he proposed, elicits an electric field and vice versa. The equations were to represent Nature, and Nature is, Maxwell believed, beautiful and elegant. (There was also another, more technical, reason for preserving the displacement current in a vacuum, which we pass over here.) This partly esthetic judgment by a nerdish physicist, entirely unknown except to a few other academic scientists, has done more to shape our civilization than any ten recent presidents and prime ministers.

Briefly, the four Maxwell equations for a vacuum say (1) there are no electrical charges in a vacuum; (2) there are no magnetic monopoles in a vacuum; (3) a changing magnetic field generates an electrical field; and (4) vice versa.

When the equations were written down like this, Maxwell was readily able to show that **E** and **B** propagated through empty space as if they were *waves*. What's more, he could calculate the speed of the wave. It was just 1 divided by the square root of ε_0 and μ_0. But ε_0 and μ_0 had been measured in the laboratory. When you plugged in the numbers you found that the electric and magnetic fields in a vacuum ought to propagate, astonishingly, at the same speed as had already been measured for light. The agreement was too close to be accidental. Suddenly, disconcertingly, electricity and magnetism were deeply implicated in the nature of light.

Since light now appeared to behave as waves and to derive from electric and magnetic fields, Maxwell called it electromagnetic. Those obscure experiments with batteries and wires had something to do with the brightness of the Sun, with how we see, with what light is. Ruminating on Maxwell's discovery many years later, Albert Einstein wrote, "To few men in the world has such an experience been vouchsafed."

Maxwell himself was baffled by the results. The vacuum seemed to act like a dielectric. He said that it can be "electrically polarized." Living in a mechanical age, Maxwell felt obliged to offer some kind of mechanical model for the propagation of an electromagnetic wave through a perfect vacuum. So he imagined space filled with a mysterious substance he called the aether, which supported and contained the time-varying electric and magnetic fields—something like a throbbing but invisible Jell-O permeating the Universe. The quivering of the aether was the reason that light traveled through it—just as water waves propagate through water and sound waves through air.

But it had to be very odd stuff, this aether, very thin, ghostly, almost incorporeal. The Sun and the Moon, the planets and the stars had to pass through it without being slowed down, without noticing. And yet it had to be stiff enough to support all these waves propagating at prodigious speed.

The word "aether" is still, in a desultory fashion, in use—in English mainly in the adjective ethereal, residing in the aether. It has some of the same connotations as the more modern "spacey" or "spaced out." When, in the early days of radio, they would say "On the air," the aether is what they had in mind. (The Russian phrase is quite literally "on the aether," *v efir.*) But of course radio readily travels through a vacuum, one of Maxwell's main results. It doesn't need air to propagate. The presence of air is, if anything, an impediment.

The whole idea of light and matter moving through the aether was to lead in another forty years to Einstein's Special Theory of Relativity, $E=mc^2$, and a great deal else. Relativity, and experiments leading up to it, showed conclusively that there is no aether supporting the propagation of electromagnetic waves, as Einstein writes in the extract from his famous paper that I reproduced in Chapter 2. The wave goes by itself. The changing electric field generates a magnetic field; the changing magnetic field generates an electric field. They hold each other up—by their bootstraps.

Many physicists were deeply troubled by the demise of the "luminiferous" aether. They had needed some mechanical model to make the whole notion of the propagation of light in a vacuum reasonable, plausible, understandable. But this is a crutch, a symptom of our difficulties in reconnoitering realms in which common sense no longer serves. The physicist Richard Feynman described it this way:

> Today, we understand better that what counts are the equations themselves and not the model used to get them. We may only question whether the equations are true or false. This is answered by doing experiments, and untold numbers of experiments have confirmed Maxwell's equations. If we take away the scaffolding he used to build it, we find that Maxwell's beautiful edifice stands on its own.

But what *are* these time-varying electric and magnetic fields permeating all of space? What do $\dot{E}$ and $\dot{B}$ *mean*? We feel so much more comfortable with the idea of things touching and jiggling, pushing and pulling, rather than "fields" magically moving objects at a distance, or mere mathematical abstractions. But, as Feynman pointed out, our sense that at least in everyday life we can rely on solid, sensible physical contact—to explain, say, why the butter knife comes to you when you pick it up—is a misconception. What does it mean to have physical contact? What exactly is happening when you pick up a knife, or push a swing, or make a wave in a waterbed by pressing down on it periodically? When we investigate deeply, we find that there is no physical contact. Instead, the electrical charges on your hand are influencing the electrical charges on the knife or swing or waterbed, and vice versa. Despite everyday experience and common sense, even here, there is only the interaction of electric fields. Nothing is touching anything.

No physicist started out impatient with commonsense notions, eager to replace them with some mathematical abstraction that could be understood only by rarefied theoretical physics. Instead, they began, as we all do, with comfortable, standard, commonsense notions. The trouble is that Nature does not comply. If we no longer insist on our notions of how Nature *ought* to behave, but instead stand before Nature with an open and receptive mind, we find that common sense often doesn't work. Why not? Because our notions, both hereditary and learned, of how Nature works were forged in the millions of years our ancestors were hunters and gatherers. In this case common sense is a faithless guide because no hunter-gatherer's life ever depended on understanding time-variable electric and magnetic fields. There were no evolutionary penalties for ignorance of Maxwell's equations. In our time it's different.

Maxwell's equations show that a rapidly varying electric field (making $\dot{E}$ large) ought to generate electromagnetic waves. In 1888 the German physicist Heinrich Hertz did the experiment and found that he had generated a new kind of radiation, radio waves. Seven years later, British scientists in Cambridge transmitted radio signals over a distance of a kilometer. By 1901, Guglielmo Marconi of Italy was using radio waves to communicate across the Atlantic Ocean.

The linking-up of the modern world economically, culturally, and politically by broadcast towers, microwave relays, and communication satellites traces directly back to Maxwell's judgment to include the displacement current in his vacuum equations. So does television, which imperfectly instructs and entertains us; radar, which may have been the decisive element in the Battle of Britain and in the Nazi defeat in World War II (which I like to think of as "Dafty," the boy who didn't fit in, reaching into the future and saving the descendants of his tormentors); the control and navigation of airplanes, ships, and spacecraft; radio astronomy and the search for extraterrestrial intelligence; and significant aspects of the electrical power and microelectronics industries.

What's more, Faraday's and Maxwell's notion of fields has been enormously influential in understanding the atomic nucleus, quantum mechanics, and the fine structure of matter. His unification of electricity, magnetism, and light into one coherent mathematical whole is the inspiration for subsequent attempts—some successful, some still in their rudimentary stages—to unify all aspects of the physical world, including gravity and nuclear forces, into one grand theory. Maxwell may fairly be said to have ushered in the age of modern physics.

Our current view of the silent world of Maxwell's varying electric and magnetic vectors is described by Richard Feynman in these words:

> Try to imagine what the electric and magnetic fields look like at present in the space of this lecture room. First of all, there is a steady magnetic field; it comes from the currents in the interior of the earth—that is, the earth's steady magnetic field. Then there are some irregular, nearly static electric fields produced perhaps by electric charges generated by friction as various people move about in their chairs and rub their coat sleeves against the chair arms. Then there are other magnetic fields produced by oscillating currents in the electrical wiring—fields which vary at a frequency of 60 cycles per second, in synchronism with the generator at Boulder Dam. But more interesting are the electric and magnetic fields varying at much higher frequencies. For instance, as light travels from window to floor and wall to wall, there are little wiggles of the electric and magnetic fields moving along at 186,000 miles per second. Then there are also infrared waves travelling from the warm foreheads to the cold blackboard. And we have forgotten the ultraviolet light, the X rays, and the radiowaves travelling through the room.
>
> Flying across the room are electromagnetic waves which carry music of a jazz band. There are waves modulated by a series of impulses representing pictures of events going on in other parts of the world, or of imaginary aspirins dissolving in imaginary stomachs. To demonstrate the reality of these waves it is only necessary to turn on electronic equipment that converts these waves into pictures and sounds.

If we go into further detail to analyze even the smallest wiggles, there are tiny electromagnetic waves that have come into the room from enormous distances. There are now tiny oscillations of the electric field, whose crests are separated by a distance of one foot, that have come from millions of miles away, transmitted to the earth from the *Mariner* [2] space craft which has just passed Venus. Its signals carry summaries of information it has picked up about the planets (information obtained from electromagnetic waves that travelled from the planet to the space craft).

There are very tiny wiggles of the electric and magnetic fields that are waves which originated billions of light-years away—from galaxies in the remotest corners of the universe. That this is true has been found by "filling the room with wires"—by building antennas as large as this room. Such radiowaves have been detected from places in space beyond the range of the greatest optical telescopes. Even they, the optical telescopes, are simply gatherers of electromagnetic waves. What we call the stars are only inferences, inferences drawn from the only physical reality we have yet gotten from them—from a careful study of the unendingly complex undulations of the electric and magnetic fields reaching us on earth.

There is, of course, more: the fields produced by lightning miles away, the fields of the charged cosmic ray particles as they zip through the room, and more, and more. What a complicated thing is the electric field in the space around you!

If Queen Victoria had ever called an urgent meeting of her counselors, and ordered them to invent the equivalent of radio and television, it is unlikely that any of them would have imagined the path to lead through the experiments of Ampère, Biot, Oersted and Faraday, four equations of vector calculus, and the judgment to preserve the displacement current in a vacuum. They would, I think, have gotten nowhere. Meanwhile, on his own, driven only by curiosity, costing the government almost nothing, himself unaware that he was laying the ground for the Westminster Project, "Dafty" was scribbling away. It's doubtful whether the self-effacing, unsociable Mr. Maxwell would even have been thought of to perform such a study. If he had, probably the government would have been telling him what to think about and what not, impeding rather than inducing his great discovery.

Late in life, Maxwell did have one interview with Queen Victoria.

He worried about it beforehand—essentially about his ability to communicate science to a non-expert—but the Queen was distracted and the interview was short. Like the four other greatest British scientists of recent history, Michael Faraday, Charles Darwin, P.A.M. Dirac, and Francis Crick, Maxwell was never knighted (although Lyell, Kelvin, J.J. Thomson, Rutherford, Eddington, and Hoyle in the next tier were). In Maxwell's case, there was not even the excuse that he might hold opinions at variance with the Church of England: He was an absolutely conventional Christian for his time, more devout than most. Maybe it was his nerdishness.

The communications media—the instruments of education and entertainment that James Clerk Maxwell made possible—have never, so far as I know, offered even a mini-series on the life and thought of their benefactor and founder. By contrast, think of how difficult it is to grow up in America without television teaching you about, say, the life and times of Davy Crockett or Billy the Kid or Al Capone.

Maxwell married young, but the bond seems to have been passionless as well as childless. His excitement was reserved for science. This founder of the modern age died in 1879 at the age of 47. While he is almost forgotten in popular culture, radar astronomers who map other worlds have remembered: The greatest mountain range on Venus, discovered by sending radio waves from Earth, bouncing them off Venus, and detecting the faint echoes, is named for him.

———

Less than a century after Maxwell's prediction of radio waves, the first quest was initiated for signals from possible civilizations on planets of other stars. Since then there have been a number of searches, some of which I referred to earlier, for the time-varying electric and magnetic fields crossing the vast interstellar distances from possible other intelligences—biologically very different from us—who had also benefited sometime in their histories from the insights of local counterparts of James Clerk Maxwell.

In October 1992—in the Mojave Desert, and in a Puerto Rican karst valley—we initiated by far the most promising, powerful, and comprehensive search for extraterrestrial intelligence (SETI). For the first time NASA would organize and operate the program. The entire sky would be examined over a 10-year period with unprecedented sensitivity and frequency range. If, on a planet of any of the 400 billion other stars that make up the Milky Way Galaxy, anyone had been sending us a radio message, we might have had a pretty fair chance of hearing them.

Just one year later, Congress pulled the plug. SETI was not of pressing importance; its interest was limited; it was too expensive. But every civilization in human history has devoted some of its resources to investigating deep questions about the Universe, and it's hard to think of a deeper one than whether we are alone. Even if we never decrypted the message contents, the receipt of such a signal would transform our view of the Universe and ourselves. And if we could understand the message from an advanced technical civilization, the practical benefits might be unprecedented. Far from being narrowly based, the SETI program, strongly supported by the scientific community, is also embedded in popular culture. The fascination with this enterprise is broad and enduring, and for very good reason. And far from being too expensive, the program would have cost about one attack helicopter per year.

I wonder why those members of Congress concerned about pricetags don't devote greater attention to the Department of Defense—which, with the Soviet Union gone and the Cold War over, still spends, when all costs are tallied, well over $300 billion a year. (And elsewhere in government there are many programs that amount to welfare for the well-to-do.) Perhaps our descendants will look back on our time and marvel at us—possessed of the technology to detect other beings, but closing our ears because we insisted on spending the national wealth to protect us from an enemy that no longer exists.*

David Goodstein, a physicist at Cal Tech, notes that science has been growing nearly exponentially for centuries and that it cannot continue such growth—because then everybody on the planet would have to be a scientist, and *then* the growth would have to stop. He speculates that for this reason, and not because of any fundamental disaffection from science, the growth in funding of science has slowed measurably in the last few decades.

Nevertheless, I'm worried about how research funds are *distrib-*

---

* The SETI program was briefly resurrected, using private contributions, in 1995 under the appropriate name Project Phoenix.

uted. I'm worried that canceling government funds for SETI is part of a trend. The Government has been pressuring the National Science Foundation to move away from basic scientific research and to support technology, engineering, applications. Congress is suggesting doing away with the U.S. Geological Survey, and slashing support for study of the Earth's fragile environment. NASA support for research and analysis of data already obtained is increasingly constrained. Many young scientists are not only unable to find grants to support their research; they are unable to find jobs.

Industrial research and development funded by American companies has slowed across the board in recent years. Government funding for research and development has declined in the same period. (Only military research and development increased in the decade of the 1980s.) In annual expenditures, Japan is now the world's leading investor in civilian research and development. In such fields as computers, telecommunications equipment, aerospace, robotics, and scientific precision equipment, the U.S. share of global exports has been declining, while the Japanese share has been increasing. In that same period the United States lost its lead to Japan in most semiconductor technologies. It experienced severe declines in market share of color TVs, VCRs, phonographs, telephone sets, and machine tools.

Basic research is where scientists are free to pursue their curiosity and interrogate Nature, not with any short-term practical end in view, but to seek knowledge for its own sake. Scientists of course have a vested interest in basic research. It's what they like to do, in many cases why they became scientists in the first place. But it is in society's interest to support such research. This is how the major discoveries that benefit humanity are largely made. Whether a few grand and ambitious scientific projects are a better investment than a larger number of small programs is a worthwhile question.

We are rarely smart enough to set about on purpose making the discoveries that will drive our economy and safeguard our lives. Often, we lack the fundamental research. Instead, we pursue a broad range of investigations of Nature, and applications we never dreamed of emerge. Not always, of course. But often enough.

Giving money to someone like Maxwell might have seemed the most absurd encouragement of mere "curiosity-driven" science, and an imprudent judgment for practical legislators. Why grant money now, so nerdish scientists talking incomprehensible gibberish can indulge their hobbies, when there are urgent unmet national needs? From this point of view it's easy to understand the contention that science is just another lobby, another pressure group anxious to keep the grant money rolling in so the scientists don't ever have to do a hard day's work or meet a payroll.

Maxwell wasn't thinking of radio, radar, and television when he first scratched out the fundamental equations of electromagnetism; Newton wasn't dreaming of space flight or communications satellites when he first understood the motion of the Moon; Roentgen wasn't contemplating medical diagnosis when he investigated a penetrating radiation so mysterious he called it "X-rays"; Curie wasn't thinking of cancer therapy when she painstakingly extracted minute amounts of radium from tons of pitchblende; Fleming wasn't planning on saving the lives of millions with antibiotics when he noticed a circle free of bacteria around a growth of mold; Watson and Crick weren't imagining the cure of genetic diseases when they puzzled over the X-ray diffractometry of DNA; Rowland and Molina weren't planning to implicate CFCs in ozone depletion when they began studying the role of halogens in stratospheric photochemistry.

Members of Congress and other political leaders have from time to time found it irresistible to poke fun at seemingly obscure scientific research proposals that the government is asked to fund. Even as bright a senator as William Proxmire, a Harvard graduate, was given to making episodic "Golden Fleece" awards—many commemorating ostensibly useless scientific projects—including SETI. I imagine the same spirit in previous governments—a Mr. Fleming wishes to study bugs in smelly cheese; a Polish woman wishes to sift through tons of Central African ore to find minute quantities of a substance she says will glow in the dark; a Mr. Kepler wants to hear the songs the planets sing.

These discoveries and a multitude of others that grace and characterize our time, to some of which our very lives are beholden, were made ultimately by scientists given the opportunity to explore what in their opinion, under the scrutiny of their peers, were basic questions in Nature. Industrial applications, in which Japan in the last two decades has done so well, are excellent. But applications of what? Fundamental research, research into the heart of Nature, is the means by which we acquire the new knowledge that gets applied.

Scientists have an obligation, especially when asking for big money, to explain with great clarity and honesty what they're after. The Superconducting Supercollider (SSC) would have been the pre-eminent instrument on the planet for probing the fine structure of matter and the nature of the early Universe. Its price tag was $10 to $15 billion. It was canceled by Congress in 1993 after about $2 billion had been spent—a worst of both worlds outcome. But *this* debate was not, I think, mainly about declining interest in the support of science. Few in Congress understood what modern high-energy accelerators are for. They are not for weapons. They have no practical applications. They are for something that is, worrisomely from the point of view of many, called "the theory of everything." Explanations that involve entities called quarks, charm, flavor, color, etc. sound as if physicists are being cute. The whole thing has an aura, in the view of at least some Congresspeople I've talked to, of "nerds gone wild"—which I suppose is an uncharitable way of describing curiosity-based science. No one asked to pay for this had the foggiest idea of what a Higgs boson is. I've read some of the material intended to justify the SSC. At the very end, some of it wasn't too bad, but there was nothing that really addressed what the project was about on a level accessible to bright but skeptical non-physicists. If physicists are asking for 10 or 15 billion dollars to build a machine that has no practical value, at the very least they should make an extremely serious effort, with dazzling graphics, metaphors, and capable use of the English language, to justify their proposal. More than financial mismanagement, budgetary constraints, and political incompetence, I think this is the key to the failure of the SSC.

There is a growing free-market view of human knowledge, according to which basic research should compete without government support with all the other institutions and claimants in the society. If they couldn't have relied on government support, and had to compete in the free market economy of their day, it's unlikely that any of the scientists on my list would have been able to do their groundbreaking research. And the cost of basic research is substantially greater than it was in Maxwell's day—both theoretical and, especially, experimental.

But that aside, would free market forces be adequate to support basic research? Only about 10 percent of meritorious research proposals in medicine are funded today. More money is spent on quack medicine than on all of medical research. What would it be like if government opted out of medical research?

A necessary aspect of basic research is that its applications lie in the future — sometimes decades or even centuries ahead. What's more, no one knows which aspects of basic research will have practical value and which will not. If scientists cannot make such predictions, is it likely that politicians or industrialists can? If free market forces are focused only towards short-term profit — as they certainly mainly are in an America with steep declines in corporate research — is not this solution tantamount to abandoning basic research?

Cutting off fundamental, curiosity-driven science is like eating the seed corn. We may have a little more to eat next winter, but what will we plant so we and our children will have enough to get through the winters to come?

Of course there are many pressing problems facing our nation and our species. But reducing basic scientific research is not the way to solve them. Scientists do not constitute a voting bloc. They have no effective lobby. However, much of their work is in everybody's interest. Backing off from fundamental research constitutes a failure of nerve, of imagination, and of that vision thing that we still don't seem to have a handle on. It might strike one of those hypothetical extraterrestrials that we were planning not to have a future.

Of course we need literacy, education, jobs, adequate medical care and defense, protection of the environment, security in our old age, a balanced budget, and a host of other matters. But we are a rich society. Can't we also nurture the Maxwells of our time? To take one symbolic example, is it really true that we can't afford one attack helicopter's worth of seed corn to listen to the stars?

Chapter 24

SCIENCE AND WITCHCRAFT*

---

* Written with Ann Druyan. The following two chapters include more political content than elsewhere in this book. I do not wish to suggest that advocacy of science and skepticism necessarily leads to all the political or social conclusions I draw. Although skeptical thinking is invaluable in politics, politics is not a science.

Ubi dubium ibi libertas:
Where there is doubt, there is freedom.

LATIN PROVERB

The 1939 New York World's Fair—that so transfixed me as a small visitor from darkest Brooklyn—was about "The World of Tomorrow." Merely by adopting such a motif, it promised that there would *be* a world of tomorrow, and the most casual glance affirmed that it would be better than the world of 1939. Although the nuance wholly passed me by, many people longed for such a reassurance on the eve of the most brutal and calamitous war in human history. I knew at least that I would be growing up in the future. The sleek and clean "tomorrow" portrayed by the Fair was appealing and hopeful. And something called science was plainly the means by which that future would be realized.

But if things had gone a little differently, the Fair could have given me enormously more. A fierce struggle had gone on behind the scenes. The vision that prevailed was that of the Fair's president and chief spokesman, Grover Whalen—former corporate executive, New York City police chief in a time of unprecedented police brutality, and public relations innovator. It was he who had envisioned the exhibit buildings as chiefly commercial, industrial, oriented to consumer products, and he who had convinced Stalin and Mussolini to build lavish national pavilions. (He later complained about how often he had been obliged to give the fascist salute.) The level of the exhibits, as one designer described it, was pitched to the mentality of a twelve-year-old.

However, as recounted by the historian Peter Kuznick of American University, a group of prominent scientists—including Harold Urey and Albert Einstein—advocated presenting science for its own sake, not just as the route to gadgets for sale; concentrating on the way of thinking and not just the products of science. They were convinced that broad popular understanding of science was the antidote to superstition and bigotry; that, as science popularizer Watson Davis put it, "the scientific way is the democratic way." One scientist even suggested that widespread public appreciation of the methods of science might work "a final conquest of stupidity"—a worthy, but probably un-realizable, goal.

As events transpired, almost no real science was tacked on to the Fair's exhibits, despite the scientists' protests and their appeals to high principles. And yet, some of the little that was added trickled down to me and helped to transform my childhood. The corporate and con-sumer focus remained central, though, and essentially nothing ap-peared about science as a way of thinking, much less as a bulwark of a free society.

—

Exactly half a century later, in the closing years of the Soviet Union, Ann Druyan and I found ourselves at a dinner in Peredelkino, a village outside Moscow where Communist Party officials, retired generals, and a few favored intellectuals had their summer homes. The air was electric with the prospect of new freedoms—especially the right to speak your mind even if the government doesn't like what you're say-ing. The fabled revolution of rising expectations was in full flower.

But, despite *glasnost*, there were widespread doubts. Would those in power really allow their own critics to be heard? Would freedom of speech, of assembly, of the press, of religion, really be permitted? Would people inexperienced with freedom be able to bear its burdens?

Some of the Soviet citizens present at the dinner had fought—for decades and against long odds—for the freedoms that most Americans take for granted; indeed, they had been inspired by the American ex-periment, a real-world demonstration that nations, even multicultural and multiethnic nations, could survive and prosper with these free-doms reasonably intact. They went so far as to raise the possibility that prosperity was *due* to freedom—that, in an age of high technology and swift change, the two rise or fall together, that the openness of science and democracy, their willingness to be judged by experiment, were closely allied ways of thinking.

There were many toasts, as there always are at dinners in that part of the world. The most memorable was given by a world-famous So-viet novelist. He stood up, raised his glass, looked us in the eye, and said, "To the Americans. They have a little freedom." He paused a beat, and then added: "And they know how to keep it."

Do we?

The ink was barely dry on the Bill of Rights before politicians found a way to subvert it—by cashing in on fear and patriotic hysteria. In 1798, the ruling Federalist Party knew that the button to push was ethnic and cultural prejudice. Exploiting tensions between France and the U.S., and a widespread fear that French and Irish immigrants were somehow intrinsically unfit to be Americans, the Federalists passed a set of laws that have come to be known as the Alien and Sedition Acts.

One law upped the residency requirement for citizenship from five to 14 years. (Citizens of French and Irish origin usually voted for the opposition, Thomas Jefferson's Democratic-Republican Party.) The Alien Act gave President John Adams the power to deport any foreigner who aroused his suspicions. Making the President nervous, said a member of Congress, "is the new crime." Jefferson believed the Alien Act had been framed particularly to expel C. F. Volney,* the French historian and philosopher; Pierre Samuel du Pont de Nemours, patriarch of the famous chemical family; and the British scientist Joseph Priestley, the discoverer of oxygen and an intellectual antecedent of James Clerk Maxwell. In Jefferson's view, these were just the sort of people America needed.

The Sedition Act made it unlawful to publish "false or malicious" criticism of the government or to inspire opposition to any of its acts. Some two dozen arrests were made, ten people were convicted, and many more were censored or intimidated into silence. The act attempted, Jefferson said, "to crush all political opposition by making criticism of Federalist officials or policies a crime."

As soon as Jefferson was elected, indeed in the first week of his presidency in 1801, he began pardoning every victim of the Sedition

---

*A typical passage from Volney's 1791 book *Ruins*:

> You dispute, you quarrel, you fight for that which is uncertain, that of which you doubt. O men! Is this not folly? . . . We must trace a line of distinction between those that are capable of verification, and those that are not, and separate by an inviolable barrier the world of fantastical beings from the world of realities; that is to say, all civil effect must be taken away from theological and religious opinions.

Act because, he said, it was as contrary to the spirit of American freedoms as if Congress had ordered us all to fall down and worship a golden calf. By 1802, none of the Alien and Sedition Acts remained on the books.

From across two centuries, it's hard to recapture the frenzied mood that made the French and the "wild Irish" seem so grave a threat that we were willing to surrender our most precious freedoms. Giving credit for French and Irish cultural triumphs, advocating equal rights for them, was in effect decried in conservative circles as sentimental—unrealistic political correctness. But that's how it always works. It always seems an aberration later. But by then we're in the grip of the next hysteria.

Those who seek power at any price detect a societal weakness, a fear that they can ride into office. It could be ethnic differences, as it was then, perhaps different amounts of melanin in the skin; different philosophies or religions; or maybe it's drug use, violent crime, economic crisis, school prayer, or "desecrating" (literally, making unholy) the flag.

Whatever the problem, the quick fix is to shave a little freedom off the Bill of Rights. Yes, in 1942, Japanese-Americans were protected by the Bill of Rights, but we locked them up anyway—after all, there was a war on. Yes, there are Constitutional prohibitions against unreasonable search and seizure, but we have a war on drugs and violent crime is racing out of control. Yes, there's freedom of speech, but we don't want foreign authors here, spouting alien ideologies, do we? The pretexts change from year to year, but the result remains the same: concentrating more power in fewer hands and suppressing diversity of opinion—even though experience plainly shows the dangers of such a course of action.

———

If we do not know what we're capable of, we cannot appreciate measures taken to protect us from ourselves. I discussed the European witch mania in the alien abduction context; I hope the reader will forgive me for returning to it in its political context. It is an aperture to human self-knowledge. If we focus on what was considered acceptable evidence and a fair trial by the religious and secular authorities in the fifteenth-to-seventeenth-century witch hunts, many of the novel and peculiar features of the eighteenth-century U.S. Constitution and Bill of Rights become clear: including trial by jury, prohibitions against self-incrimination and against cruel and unusual punishment, freedom of speech and the press, due process, the balance of powers and the separation of church and state.

Friedrich von Spee (pronounced "Shpay") was a Jesuit priest who had the misfortune to hear the confessions of those accused of witchcraft in the German city of Würzburg (see Chapter 7). In 1631, he published *Cautio Criminalis (Precautions for Prosecutors)*, which exposed the essence of this Church/State terrorism against the innocent. Before he was punished he died of the plague—as a parish priest serving the afflicted. Here is an excerpt from his whistle-blowing book:

1. Incredibly among us Germans, and especially (I am ashamed to say) among Catholics, are popular superstitions, envy, calumnies, backbiting, insinuations, and the like, which, being neither punished nor refuted, stir up suspicion of witchcraft. No longer God or nature, but witches are responsible for everything.

2. Hence everybody sets up a clamor that the magistrates investigate the witches—whom only popular gossip has made so numerous.

3. Princes, therefore, bid their judges and counselors bring proceedings against the witches.

4. The judges hardly know where to start, since they have no evidence [*indicia*] or proof.

5. Meanwhile, the people call this delay suspicious; and the princes are persuaded by some informer or another to this effect.

6. In Germany, to offend these princes is a serious offense; even clergymen approve whatever pleases them, not caring by whom these princes (however well-intentioned) have been instigated.

7. At last, therefore, the judges yield to their wishes and contrive to begin the trials.

8. Other judges who still delay, afraid to get involved in this ticklish matter, are sent a special investigator. In this field of investigation, whatever inexperience or arrogance he brings to the job is held zeal for justice. His zeal for justice is also whetted by hopes of

profit, especially with a poor and greedy agent with a large family, when he receives as stipend so many dollars per head for each witch burned, besides the incidental fees and perquisites which investigating agents are allowed to extort at will from those they summon.

9. If a madman's ravings or some malicious and idle rumor (for no proof of the scandal is ever needed) points to some helpless old woman, she is the first to suffer.

10. Yet to avoid the appearance that she is indicted solely on the basis of rumor, without other proofs, a certain presumption of guilt is obtained by posing the following dilemma: Either she has led an evil and improper life, or she has led a good and proper one. If an evil one, then she should be guilty. On the other hand, if she has led a good life, this is just as damning; for witches dissemble and try to appear especially virtuous.

11. Therefore the old woman is put in prison. A new proof is found through a second dilemma: she is afraid or not afraid. If she is (hearing of the horrible tortures used against witches), this is sure proof; for her conscience accuses her. If she does not show fear (trusting in her innocence), this too is a proof; for witches characteristically pretend innocence and wear a bold front.

12. Lest these should be the only proofs, the investigator has his snoopers, often depraved and infamous, ferret out all her past life. This, of course, cannot be done without turning up some saying or doing of hers which men so disposed can easily twist or distort into evidence of witchcraft.

13. Any who have borne her ill now have ample opportunity to bring against her whatever accusations they please; and everyone says that the evidence is strong against her.

14. And so she is hurried to the torture, unless, as often happens, she was tortured on the very day of her arrest.

15. In these trials nobody is allowed a lawyer or any means of fair defense, for witchcraft is reckoned an exceptional crime [of such enormity that all rules of legal procedure may be suspended], and whoever ventures to defend the prisoner falls himself under suspicion of witchcraft—as well as those who dare to utter a protest in these cases and to urge the judges to exercise prudence, for they are forthwith labeled supporters of witchcraft. Thus everybody keeps quiet for fear.

16. So that it may seem that the woman has an opportunity to defend herself, she is brought into court and the indications of her guilt are read and examined—if it can be called an examination.

17. Even though she denies these charges and satisfactorily answers every accusation, no attention is paid and her replies are not even recorded; all the indictments retain their force and validity, however perfect her answers to them. She is ordered back into prison, there to consider more carefully whether she will persist in obstinacy—for, since she has already denied her guilt, she is obstinate.

18. Next day she is brought out again, and hears a decree of torture—just as if she had never refuted the charges.

19. Before torture, however, she is searched for amulets: her entire body is shaved, and even those privy parts indicating the female sex are wantonly examined.

20. What is so shocking about this? Priests are treated the same way.

21. When the woman has been shaved and searched, she is tortured to make her confess the truth—that is, to declare what they want, for naturally anything else will not and cannot be the truth.

22. They start with the first degree, i.e., the less severe torture. Although exceedingly severe, it is light compared to those tortures which follow. Wherefore if she confesses, they say the woman has confessed without torture!

23. Now, what prince can doubt her guilt when he is told she has confessed voluntarily, without torture?

24. She is therefore put to death without scruple. But she would have been executed even if she had not confessed; for when once the torture has begun, the die is already cast; she cannot escape, she has perforce to die.

25. The result is the same whether she confesses or not. If she confesses, her guilt is clear: she is executed. All recantation is in vain. If she does not confess, the torture is repeated—twice, thrice, four times. In exceptional crimes, the torture is not limited in duration, severity, or frequency.

26. If, during the torture, the old woman contorts her features with pain, they say she is laughing; if she loses consciousness, she

is sleeping or has bewitched herself into taciturnity. And if she is taciturn, she deserves to be burned alive, as lately has been done to some who, though several times tortured, would not say what the investigators wanted.

27. And even confessors and clergymen agree that she died obstinate and impenitent; that she would not be converted or desert her incubus, but kept faith with him.

28. If, however, she dies under so much torture, they say the devil broke her neck.

29. Wherefore the corpse is buried underneath the gallows.

30. On the other hand, if she does not die under torture, and if some exceptionally scrupulous judge hesitates to torture her further without fresh proofs or to burn her without her confession, she is kept in prison and more harshly chained, there to rot until she yields, even if it take a whole year.

31. She can never clear herself. The investigating committee would feel disgraced if it acquitted a woman; once arrested and in chains, she has to be guilty, by fair means or foul.

32. Meanwhile, ignorant and headstrong priests harass the wretched creature so that, whether truly or not, she will confess herself guilty; unless she does so, they say, she cannot be saved or partake of the sacraments.

33. More understanding or learned priests cannot visit her in prison lest they counsel her or inform the princes what goes on. Nothing is more dreaded than that something be brought to light to prove the innocence of the accused. Persons who try to do so are labeled troublemakers.

34. While she is kept in prison and tortured, the judges invent clever devices to build up new proofs of guilt to convict her to her face, so that, when reviewing the trial, some university faculty can confirm her burning alive.

35. Some judges, to appear ultrascrupulous, have the woman exorcized, transferred elsewhere, and tortured all over again, to break her taciturnity; if she maintains silence, then at last they can burn her. Now, in Heaven's name, I would like to know, since she who confesses and she who does not both perish alike, how can anybody, no matter how innocent, escape? O unhappy woman, why have you rashly hoped? Why did you not, on first entering prison, admit whatever they wanted? Why, foolish and crazy

woman, did you wish to die so many times when you might have died but once? Follow my counsel, and, before undergoing all these pains, say you are guilty and die. You will not escape, for this were a catastrophic disgrace to the zeal of Germany.

36. When, under stress of pain, the witch has confessed, her plight is indescribable. Not only cannot she escape herself, but she is also compelled to accuse others whom she does not know, whose names are frequently put into her mouth by the investigators or suggested by the executioner, or of whom she has heard as suspected or accused. These in turn are forced to accuse others, and these still others, and so it goes on: who can help seeing that it must go on and on?

37. The judges must either suspend these trials (and so impute their validity) or else burn their own folk, themselves, and everybody else; for all sooner or later are falsely accused and, if tortured, all are proved guilty.

38. Thus eventually those who at first clamored most loudly to feed the flames are themselves involved, for they rashly failed to see that their turn too would come. Thus Heaven justly punishes those who with their pestilent tongues created so many witches and sent so many innocent to the stake...

Von Spee is not explicit about the sickening methods of torture employed. Here is an excerpt from an invaluable compilation, *The Encyclopedia of Witchcraft and Demonology*, by Rossell Hope Robbins (1959):

One might glance at some of the special tortures at Bamberg, for example, such as the forcible feeding of the accused on herrings cooked in salt, followed by denial of water—a sophisticated method which went side by side with immersion of the accused in baths of scalding water to which lime had been added. Other ways with witches included the wooden horse, various kinds of racks, the heated iron chair, leg vises [Spanish boots], and large boots of leather or metal into which (with the feet in them, of course) was poured boiling water or molten lead. In the water torture, the *question de l'eau*, water was poured down the throat of the accused, along with a soft cloth to cause choking. The cloth was pulled out quickly so that the entrails would be torn. The thumbscrews [*grésillons*] were a vise designed to compress the thumbs or the big toes to the root of the nails, so that the crushing of the digit would cause excruciating pain.

In addition, and more routinely applied, were the strappado and squassation and still more ghastly tortures that I will avoid describing. After torture, and with the instruments of torture in plain view, the victim was asked to sign a statement. This was then described as a "free confession," voluntarily admitted to.

At great personal risk, von Spee protested the witch mania. So did a few others, mainly Catholic and Protestant clergy who had witnessed these crimes at first hand—including Gianfrancesco Ponzinibio in Italy, Cornelius Loos in Germany, and Reginald Scot in Britain in the sixteenth century; as well as Johann Mayfurth ["Listen, you money-hungry judges and bloodthirsty prosecutors, the apparitions of the Devil are all lies"] in Germany and Alonzo Salazar de Frias in Spain in the seventeenth century. Along with von Spee and the Quakers generally, they are heroes of our species. Why are they not better known?

In *A Candle in the Dark* (1656), Thomas Ady addressed a key question:

Some again will object and say, If Witches cannot kill, and do many strange things by Witchcraft, why have many confessed that they have done such Murthers, and other strange matters, whereof they have been accused?

To this I answer, If Adam and Eve in their innocency were so easily overcome, and tempted to sin, how much more may poor Creatures now after the Fall, by perswasions, promises, and threatenings, by keeping from sleep, and continual torture, be brought to confess that which is false and impossible, and contrary to the faith of a Christian to believe?

It was not until the eighteenth century that the possibility of hallucination as a component in the persecution of witches was seriously entertained; Bishop Francis Hutchinson, in his *Historical Essay Concerning Witchcraft* (1718), wrote

Many a man hath verily believed he hath seen a spirit externally before him, when it hath been only an internal image dancing in his own brain.

Because of the courage of these opponents of the witch mania, its extension to the privileged classes, the danger it posed to the growing institution of capitalism, and especially the spread of the ideas of the European Enlightenment, witch burnings eventually disappeared. The last execution for witchcraft in Holland, cradle of the Enlightenment, was in 1610; in England, 1684; America, 1692; France, 1745; Germany, 1775; and Poland, 1793. In Italy, the Inquisition was condemning people to death until the end of the eighteenth century, and inquisitorial torture was not abolished in the Catholic Church until 1816. The last bastion of support for the reality of witchcraft and the necessity of punishment has been the Christian churches.

The witch mania is shameful. How could we do it? How could we be so ignorant about ourselves and our weaknesses? How could it have happened in the most "advanced," the most "civilized" nations then on Earth? Why was it resolutely supported by conservatives, monarchists, and religious fundamentalists? Why opposed by liberals, Quakers and followers of the Enlightenment? If we're absolutely sure that our beliefs are right, and those of others wrong; that we are motivated by good, and others by evil; that the King of the Universe speaks to us, and not to adherents of very different faiths; that it is wicked to challenge conventional doctrines or to ask searching questions; that our main job is to believe and obey—then the witch mania will recur in its infinite variations down to the time of the last man. Note Friedrich von Spee's very first point, and the implication that improved public understanding of superstition and skepticism might have helped to short-circuit the whole train of causality. If we fail to understand how it worked in the last round, we will not recognize it as it emerges in the next.

—

"It is the absolute right of the state to supervise the formation of public opinion," said Josef Goebbels, the Nazi propaganda minister. In George Orwell's novel 1984, the "Big Brother" state employs an army of bureaucrats whose only job is to alter the records of the past so they conform to the interests of those currently in power. 1984 was not just an engaging political fantasy; it was based on the Stalinist Soviet Union, where the rewriting of history was institutionalized. Soon after Stalin took power, pictures of his rival Leon Trotsky—a monumental figure in the 1905 and 1917 revolutions—began to disappear. Heroic and wholly anhistoric paintings of Stalin and Lenin together directing the Bolshevik Revolution took their place, with Trotsky, the founder of the Red Army, nowhere in evidence. These images became icons of the state. You could see them in every office building, on outdoor advertising signs sometimes ten stories high, in museums, on postage stamps.

New generations grew up believing that *was* their history. Older generations began to feel that they remembered something of the sort, a kind of political false-memory syndrome. Those who made the accommodation between their real memories and what the leadership wished them to believe exercised what Orwell described as "doublethink." Those who did not, those old Bolsheviks who could recall the peripheral role of Stalin in the Revolution and the central role of Trotsky, were denounced as traitors or unreconstructed bourgeoisie or "Trotskyites" or "Trotsky-fascists," and were imprisoned, tortured, made to confess their treason in public, and then executed. It *is* possible—given absolute control over the media and the police—to rewrite the memories of hundreds of millions of people, if you have a generation to accomplish it in. Almost always, this is done to improve the hold that the powerful have on power, or to serve the narcissism or megalomania or paranoia of national leaders. It throws a monkey wrench into the error-correcting machinery. It works to erase public memory of profound political mistakes, and thus to guarantee their eventual repetition.

In our time, with total fabrication of realistic stills, motion pictures, and videotapes technologically within reach, with television in every home, and with critical thinking in decline, restructuring societal memories even without much attention from the secret police seems possible. What I'm imagining here is not that each of us has a budget of memories implanted in special therapeutic sessions by state-appointed psychiatrists, but rather that small numbers of people will have so much control over news stories, history books, and deeply affecting images as to work major changes in collective attitudes.

We saw a pale echo of what is now possible in 1990–1991, when Saddam Hussein, the autocrat of Iraq, made a sudden transition in the American consciousness from an obscure near-ally—granted commodities, high technology, weaponry, and even satellite intelligence data—to a slavering monster menacing the world. I am not myself an admirer of Mr. Hussein, but it was striking how quickly he could be brought from someone almost no American had heard of into the incarnation of evil. These days the apparatus for generating indignation is busy elsewhere. How confident are we that the power to drive and determine public opinion will always reside in responsible hands?

Another contemporary example is the "war" on drugs—where the government and munificently funded civic groups systematically distort and even invent scientific evidence of adverse effects (especially of marijuana), and in which no public official is permitted even to raise the topic for open discussion.

But it's hard to keep potent historical truths bottled up forever. New data repositories are uncovered. New, less ideological, generations of historians grow up. In the late 1980s and before, Ann Druyan and I would routinely smuggle copies of Trotsky's *History of the Russian Revolution* into the USSR—so our colleagues could know a little about their own political beginnings. By the fiftieth anniversary of the murder of Trotsky (Stalin's assassin had cracked Trotsky's head open with a hammer), *Izvestia* could extol Trotsky as "a great and irreproachable* revolutionary," and a German Communist publication went so far as to describe him as

> fight[ing] for all of us who love human civilization, for whom this civilization is our nationality. His murderer . . . tried, in killing him, to kill this civilization . . . [This] was a man who had in his head the most valuable and best-organized brain that was ever crushed by a hammer.

Trends working at least marginally towards the implantation of a very narrow range of attitudes, memories, and opinions include con-

---

* Suggesting that the authorities have learned nothing from their history, except substituting one historical figure for another on the list of Irreproachables.

trol of major television networks and newspapers by a small number of similarly motivated powerful corporations and individuals, the disappearance of competitive daily newspapers in many cities, the replacement of substantive debate by sleaze in political campaigns, and episodic erosion of the principle of the separation of powers. It is estimated (by the American media expert Ben Bagdikian) that fewer than two dozen corporations control more than half "of the global business in daily newspapers, magazines, television, books and movies." The proliferation of cable television channels, cheap long-distance telephone calls, fax machines, computer bulletin boards and networks, inexpensive computer self-publishing, and surviving instances of the traditional liberal arts university curriculum are trends that might work in the opposite direction.

It's hard to tell how it's going to turn out.

The business of skepticism is to be dangerous. Skepticism challenges established institutions. If we teach everybody, including, say, high school students, habits of skeptical thought, they will probably not restrict their skepticism to UFOs, aspirin commercials, and 35,000-year-old channelees. Maybe they'll start asking awkward questions about economic, or social, or political, or religious institutions. Perhaps they'll challenge the opinions of those in power. Then where would we be?

—

Ethnocentrism, xenophobia, and nationalism are these days rife in many parts of the world. Government repression of unpopular views is still widespread. False or misleading memories are inculcated. For the defenders of such attitudes, science is disturbing. It claims access to truths that are largely independent of ethnic or cultural biases. By its very nature, science transcends national boundaries. Put scientists working in the same field of study together in a room and even if they share no common spoken language, they will find a way to communicate. Science itself is a transnational language. Scientists are naturally cosmopolitan in attitude and are more likely to see through efforts to divide the human family into many small and warring factions. "There is no national science," said the Russian playwright Anton Chekhov, "just as there is no national multiplication table." (Likewise, for many, there is no such thing as a national religion, although the religion of nationalism has millions of adherents.)

In disproportionate numbers, scientists are found in the ranks of social critics (or, less charitably, "dissidents"), challenging the policies and myths of their own nations. The heroic names of the physicists Andrei Sahkarov* in the former USSR, Albert Einstein and Leo Szilard in the United States, and Fang Li-zhu in China spring readily enough to mind—the first and last risking their lives. Especially in the aftermath of the invention of nuclear weapons, scientists have been portrayed as ethical cretins. This is an injustice, considering all those who, sometimes at considerable personal peril, have spoken out against their own countries' misapplications of science and technology.

For example, the chemist Linus Pauling (1901–1994) was, more than any other person, responsible for the Limited Test Ban Treaty of 1963, which halted aboveground explosions of nuclear weapons by the United States, the Soviet Union, and the United Kingdom. He mounted a blistering campaign of moral outrage and scientific data, made more credible by the fact that he was a Nobel laureate. In the American press, he was generally vilified for his troubles, and in the 1950s the State Department canceled his passport because he had been insufficiently anti-communist. His Nobel Prize was awarded for the application of quantum mechanical insights—resonances, and what is called hybridization of orbitals—to explain the nature of the chemical bond that joins atoms together into molecules. These ideas are now the bread and butter of modern chemistry. But in the Soviet Union, Pauling's work on structural chemistry was denounced as incompatible with dialectical materialism and declared off-limits to Soviet chemists.

Undaunted by this criticism, East and West—indeed, not even slowed down—he went on to do monumental work on how anesthetics work, identified the cause of sickle cell anemia (a single nucleotide substitution in DNA), and showed how the evolutionary history of life might be read by comparing the DNAs of various organisms. He was

---

* As a much-decorated "Hero" of the Soviet Union, and privy to its nuclear secrets, Sakharov in the Cold War year 1968 boldly wrote—in a book published in the West and widely distributed in samizdat in the USSR—"Freedom of thought is the only guarantee against an infection of peoples by the mass myths, which, in the hands of treacherous hypocrites and demagogues, can be transformed into bloody dictatorships." He was thinking of both East and West. I would add that free thought is a necessary, but not a sufficient, condition for democracy.

hot on the trail of the structure of DNA; Watson and Crick were consciously rushing to get there before Pauling. The verdict on his assessment of Vitamin C is apparently still out. "That man is a real genius" was Albert Einstein's assessment.

In all this time he continued to work for peace and amity. When Ann and I once asked Pauling about the roots of his dedication to social issues, he gave a memorable reply: "I did it to be worthy of the respect of my wife," Helen Ava Pauling. He won a second Nobel Prize, this one in peace, for his work on the nuclear test ban, becoming the only person in history to win two unshared Nobel Prizes.

There were some who saw Pauling as a troublemaker. Those unhappy about social change may be tempted to view science itself with suspicion. Technology is safe, they tend to think, readily guided and controlled by industry and government. But pure science, science for its own sake, science as curiosity, science that might lead anywhere and challenge anything, that's another story. Certain areas of pure science are the unique pathway to future technologies—true enough—but the attitudes of science, if applied broadly, can be perceived as dangerous. Through salaries, social pressures, and the distribution of prestige and awards, societies try to herd scientists into some reasonably safe middle ground—between too little long-term technological progress and too much short-term social criticism.

Unlike Pauling, many scientists consider their job to be science, narrowly defined, and believe that engaging in politics or social criticism is not just a distraction from but antithetical to the scientific life. As mentioned earlier, during the Manhattan Project, the successful World War II U.S. effort to build nuclear weapons before the Nazis did, certain participating scientists began to have reservations—the more so when it became clear how immensely powerful these weapons were. Some, such as Leo Szilard, James Franck, Harold Urey, and Robert R. Wilson, tried to call the attention of political leaders and the public (especially after the Nazis were defeated) to the dangers of the forthcoming arms race, which they foresaw very well, with the Soviet Union. Others argued that policy matters were outside their jurisdiction. "I was put on Earth to make certain discoveries," said Enrico Fermi, "and what the political leaders do with them is not my business." But even so, Fermi was so appalled by the dangers of the thermonuclear weapon Edward Teller was advocating that he coauthored a famous document urging the United States not to build it, calling it "evil."

Jeremy Stone, the president of the Federation of American Scientists, has described Teller—whose efforts to justify thermonuclear weapons I recounted in a previous chapter—in these words:

> Edward Teller . . . insisted, at first for personal intellectual reasons and later for geopolitical reasons, that a hydrogen bomb be built. Using tactics of exaggeration and even smear, he successfully manipulated the policy-making process for five decades, denouncing all manner of arms control measures and promoting arms-race-escalating programs of many kinds.
>
> The Soviet Union, hearing of his H-bomb project, built its own H-bomb. As a direct consequence of the unusual personality of this particular individual and of the power of the H-bomb, the world may have risked a level of annihilation that might not otherwise have transpired, or might have come later and under better political controls.
>
> If so, no scientist has ever had more influence on the risks that humanity has run than Edward Teller, and Teller's general behavior throughout the arms race was reprehensible. . .
>
> Edward Teller's fixation on the H-bomb may have led him to do more to imperil life on this planet than any other individual in our species. . .
>
> Compared to Teller, the leaders of Western atomic science were frequently babes in the political woods—their leadership having been determined by their professional skills rather than by, in this case, their political skills.

My purpose here is not to castigate a scientist for succumbing to very human passions, but to reiterate that new imperative: The unprecedented powers that science now makes available must be accompanied by unprecedented levels of ethical focus and concern by the scientific community—as well as the most broadly based public education into the importance of science and democracy.

Chapter 25

REAL
PATRIOTS
ASK
QUESTIONS*

———

* Written with Ann Druyan.

It is not the function of our government to keep
the citizen from falling into error; it is the function
of the citizen to keep the government from falling
into error.

U.S. SUPREME COURT JUSTICE
ROBERT H. JACKSON,
1950

It is a fact of life on our beleaguered little planet that widespread torture, famine, and governmental criminal irresponsibility are much more likely to be found in tyrannical than in democratic governments. Why? Because the rulers of the former are much less likely to be thrown out of office for their misdeeds than the rulers of the latter. This is error-correcting machinery in politics.

The methods of science—with all its imperfections—can be used to improve social, political, and economic systems, and this is, I think, true no matter what criterion of improvement is adopted. How is this possible if science is based on experiment? Humans are not electrons or laboratory rats. But every act of Congress, every Supreme Court decision, every Presidential National Security Directive, every change in the Prime Rate is an experiment. Every shift in economic policy, every increase or decrease in funding for Head Start, every toughening of criminal sentences is an experiment. Exchanging needles, making condoms freely available, or decriminalizing marijuana are all experiments. Doing nothing to help Abyssinia against Italy, or to prevent Nazi Germany from invading the Rhineland, was an experiment. Communism in Eastern Europe, the Soviet Union, and China was an experiment. Privatizing mental health care or prisons is an experiment. Japan and West Germany investing a great deal in science and technology and next to nothing on defense—and finding that their economies boomed—was an experiment. Handguns are available for self-protection in Seattle, but not in nearby Vancouver, Canada; handgun killings are five times more common and the handgun suicide rate is ten times greater in Seattle. Guns make impulsive killing easy. This is also an experiment. In almost all of these cases, adequate control experiments are not performed, or variables are insufficiently separated. Nevertheless, to a certain and often useful degree, policy ideas can be tested. The great waste would be to ignore the results of social experiments because they seem to be ideologically unpalatable.

There is no nation on Earth today optimized for the middle of the twenty-first century. We face an abundance of subtle and complex problems. We need therefore subtle and complex solutions. Since there is no deductive theory of social organization, our only recourse is scientific experiment—trying out sometimes on small scales (community, city, and state level, say) a wide range of alternatives. One of the perquisites of power on becoming prime minister in China in the fifth century B.C. was that you got to construct a model state in your home district or province. It was Confucius' chief life failing, he lamented, that he never got to try.

Even a casual scrutiny of history reveals that we humans have a sad tendency to make the same mistakes again and again. We're afraid of strangers or anybody who's a little different from us. When we get scared, we start pushing people around. We have readily accessible buttons that release powerful emotions when pressed. We can be manipulated into utter senselessness by clever politicians. Give us the right kind of leader and, like the most suggestible subjects of the hypnotherapists, we'll gladly do just about anything he wants—even things we know to be wrong. The framers of the Constitution were students of history. In recognition of the human condition, they sought to invent a means that would keep us free in spite of ourselves.

Some of the opponents of the U.S. Constitution insisted that it would never work; that a republican form of government spanning a land with "such dissimilar climates, economies, morals, politics, and peoples," as Governor George Clinton of New York said, was impossible; that such a government and such a Constitution, as Patrick Henry of Virginia declared, "contradicts all the experience of the world." The experiment was tried anyway.

Scientific findings and attitudes were common in those who invented the United States. The supreme authority, outranking any personal opinion, any book, any revelation, was—as the Declaration of Independence puts it—"the laws of nature and of nature's GOD." Benjamin Franklin was revered in Europe and America as the founder of the new field of electrical physics. At the Constitutional Convention of 1789 John Adams repeatedly appealed to the analogy of mechanical balance in machines; others to William Harvey's discovery of the circulation of the blood. Late in life Adams wrote, "All mankind are chemists from their cradles to their graves. . . The Material Universe is a chemical experiment." James Madison used chemical and biological metaphors in *The Federalist Papers*. The American revolutionaries were creatures of the European Enlightenment which provides an essential background for understanding the origins and purpose of the United States.

"Science and its philosophical corollaries," wrote the American historian Clinton Rossiter,

> were perhaps the most important intellectual force shaping the destiny of eighteenth-century America . . . Franklin was only one of a number of forward-looking colonists who recognized the kinship of scientific method and democratic procedure. Free inquiry, free exchange of information, optimism, self-criticism, pragmatism, objectivity—all these ingredients of the coming republic were already active in the republic of science that flourished in the eighteenth century.

—

Thomas Jefferson was a scientist. That's how he described himself. When you visit his home at Monticello, Virginia, the moment you enter its portals you find ample evidence of his scientific interests—not just in his immense and varied library, but in copying machines, automatic doors, telescopes, and other instruments, some at the cutting edge of early nineteenth-century technology. Some he invented, some he copied, some he purchased. He compared the plants and animals of America with Europe's, uncovered fossils, used the calculus in the design of a new plow. He mastered Newtonian physics. Nature destined him, he said, to be a scientist, but there were no opportunities for scientists in pre-revolutionary Virginia. Other, more urgent, needs took precedence. He threw himself into the historic events that were transpiring around him. Once independence was won, he said, later generations could devote themselves to science and scholarship.

Jefferson was an early hero of mine, not because of his scientific interests (although they very much helped to mold his political philosophy), but because he, almost more than anyone else, was responsible for the spread of democracy throughout the world. The idea—breathtaking, radical, and revolutionary at the time (in many places in the world, it still is)—is that not kings, not priests, not big-city bosses, not dictators, not a military cabal, not a *de facto* conspiracy of the wealthy, but ordinary people, working together, are to rule the nations. Not only was Jefferson a leading theoretician of this cause; he was also involved in the most practical way, helping to bring about the great American political experiment that has, all over the world, been admired and emulated since.

He died at Monticello on July 4, 1826, fifty years to the day after the colonies issued that stirring document, written by Jefferson, called the Declaration of Independence. It was denounced by conservatives worldwide: Monarchy, aristocracy, and state-supported religion—that's what conservatives were defending then. In a letter composed a few days before his death, he wrote that it was the "light of science" that had demonstrated that "the mass of mankind has not been born with saddles on their backs," nor were a favored few born "booted and spurred." He had written in the Declaration of Independence that we all must have the same opportunities, the same "unalienable" rights. And if the definition of "all" was disgracefully incomplete in 1776, the spirit of the Declaration was generous enough that today "all" is far more inclusive.

Jefferson was a student of history—not just the compliant and safe history that praises our own time or country or ethnic group, but the real history of real humans, our weaknesses as well as our strengths. History taught him that the rich and powerful will steal and oppress if given half a chance. He described the governments of Europe, which he saw at first hand as the American ambassador to France. Under the pretense of government, he said, they had divided their nations into two classes: wolves and sheep. Jefferson taught that every government degenerates when it is left to the rulers alone, because rulers—by the very act of ruling—misuse the public trust. The people themselves, he said, are the only prudent repository of power.

But he worried that the people—and the argument goes back to Thucydides and Aristotle—are easily misled. So he advocated safeguards, insurance policies. One was the constitutional separation of powers; accordingly, various groups, some pursuing their own selfish interests, balance one another, preventing any one of them from running away with the country: the executive, legislative and judicial branches; the House and the Senate; the States and the Federal Government. He also stressed, passionately and repeatedly, that it was essential for the people to understand the risks and benefits of government, to educate themselves, and to involve themselves in the political process. Without that, he said, the wolves will take over. Here's how he put it in *Notes on Virginia*, stressing how the powerful and unscrupulous find zones of vulnerability they can exploit:

> In every government on earth is some trace of human weakness, some germ of corruption and degeneracy, which cunning will discover and wickedness insensibly open, cultivate and improve. Every government degenerates when trusted to the rulers of the people alone. The people themselves therefore are its only safe depositories. And to render even them safe, their minds must be improved . . .

Jefferson had little to do with the actual writing of the U.S. Constitution; as it was being formulated, he was serving as American minister to France. When he read its provisions, he was pleased, but with two reservations. One deficiency: no limit was provided on the number of terms the President could serve. This, Jefferson feared, was a way for a President to become a king, in fact if not in law. The other major deficiency was the absence of a bill of rights. The citizen—the average person—was insufficiently protected, Jefferson thought, from the inevitable abuses of those in power.

He advocated freedom of speech, in part so that even wildly unpopular views could be expressed, so that deviations from the conventional wisdom could be offered for consideration. Personally he was an extremely amiable man, reluctant to criticize even his sworn enemies. He displayed a bust of his arch-adversary Alexander Hamilton in the vestibule at Monticello. Nevertheless, he believed that the habit of skepticism is an essential prerequisite for responsible citizenship. He argued that the cost of education is trivial compared to the cost of ignorance, of leaving the government to the wolves. He taught that the country is safe only when the people rule.

Part of the duty of citizenship is not to be intimidated into conformity. I wish that the oath of citizenship taken by recent immigrants, and the pledge that students routinely recite, included something like "I promise to question everything my leaders tell me." That would be really to Thomas Jefferson's point. "I promise to use my critical faculties. I promise to develop my independence of thought. I promise to educate myself so I can make my own judgments."

I also wish that the Pledge of Allegiance were directed at the Constitution and the Bill of Rights, as it is when the President takes his oath of office, rather than to the flag and the nation.

When we consider the founders of our nation—Jefferson, Washington, Samuel and John Adams, Madison and Monroe, Benjamin Franklin, Tom Paine and many others—we have before us a list of at least ten and maybe even dozens of great political leaders. They were well-educated. Products of the European Enlightenment, they were students of history. They knew human fallibility and weakness and corruptibility. They were fluent in the English language. They wrote their own speeches. They were realistic and practical, and at the same time motivated by high principles. They were not checking the pollsters on what to think this week. They knew what to think. They were comfortable with long-term thinking, planning even further ahead than the next election. They were self-sufficient, not requiring careers as politicians or lobbyists to make a living. They were able to bring out the best in us. They were interested in and, at least two of them, fluent in science. They attempted to set a course for the United States into the far future—not so much by establishing laws as by setting limits on what kinds of laws could be passed.

The Constitution and its Bill of Rights have done remarkably well, constituting, despite human weaknesses, a machine able, more often than not, to correct its own trajectory.

At that time, there were only about two and a half million citizens of the United States. Today there are about a hundred times more. So if there were ten people of the caliber of Thomas Jefferson then, there ought to be 10 x 100 = 1,000 Thomas Jeffersons today.

Where are they?

—

One reason the Constitution is a daring and courageous document is that it allows for continuing change, even of the form of government itself, if the people so wish. Because no one is wise enough to foresee which ideas may answer urgent societal needs—even if they're counterintuitive and have been troubling in the past—this document tries to guarantee the fullest and freest expression of views.

There is, of course, a price. Most of us are for ~~~
sion when there's a danger that our own views wi.
We're not all that upset, though, when views we desp
little censorship here and there. But within certain narrowly ~
scribed limits—Justice Oliver Wendell Holmes's famous example was
causing panic by falsely crying "fire" in a crowded theater—great liberties are permitted in America:

- Gun collectors are free to use portraits of the Chief Justice, the
Speaker of the House, or the Director of the FBI for target practice;
outraged civic-minded citizens are free to burn in effigy the President
of the United States.
- Even if they mock Judeo-Christian-Islamic values, even if they
ridicule everything most of us hold dear, devil-worshipers (if there are
any) are entitled to practice their religion, so long as they break no
constitutionally valid law.
- A purported scientific article or popular book asserting the "superiority" of one race over another may not be censored by the government, no matter how pernicious it is; the cure for a fallacious
argument is a better argument, not the suppression of ideas.
- Individuals or groups are free to argue that a Jewish or Masonic
conspiracy is taking over the world, or that the Federal government is
in league with the Devil.
- Individuals may, if they wish, praise the lives and politics of such
undisputed mass murderers as Adolf Hitler, Josef Stalin, and Mao Zedong. Even detestable opinions have a right to be heard.

The system founded by Jefferson, Madison, and their colleagues
offers means of expression to those who do not understand its origins
and wish to replace it by something very different. For example, Tom
Clark, Attorney General and therefore chief law enforcement officer
of the United States, in 1948 offered this suggestion: "Those who do
not believe in the ideology of the United States shall not be allowed to
stay in the United States." But if there is one key and characteristic
U.S. ideology, it is that there are no mandatory and no forbidden ideologies. Some more recent 1990s cases: John Brockhoeft, in jail for
bombing an abortion clinic in Cincinnati, wrote, in a "pro-life"
newsletter:

ı a very narrow-minded, intolerant, reactionary, Bible-thump-
ing fundamentalist . . . a zealot and fanatic . . . The reason the
United States was once a great nation, besides being blessed by
God, is because she was founded on truth, justice, and narrow-
mindedness.

Randall Terry, founder of "Operation Rescue," an organization that
blockades abortion clinics, told a congregation in August 1993:

Let a wave of intolerance wash over you . . . Yes, hate is good . . .
Our goal is a Christian nation . . . We are called by God to con-
quer this country . . . We don't want pluralism.

The expression of such views is protected, and properly so, under the
Bill of Rights, even if those protected would abolish the Bill of Rights
if they got the chance. The protection for the rest of us is to use that
same Bill of Rights to get across to every citizen the indispensability of
the Bill of Rights.

What means to protect themselves against human fallibility, what
error-protection machinery do these alternative doctrines and institutions offer? An infallible leader? Race? Nationalism? Wholesale disen-
gagement from civilization, except for explosives and automatic
weapons? How can they be *sure*—especially in the darkness of the
twentieth century? Don't they need candles?

In his celebrated little book *On Liberty*, the English philosopher
John Stuart Mill argued that silencing an opinion is "a peculiar evil."
If the opinion is right, we are robbed of the "opportunity of exchanging error for truth"; and if it's wrong, we are deprived of a deeper
understanding of the truth in "its collision with error." If we know
only our own side of the argument, we hardly know even that; it becomes stale, soon learned only by rote, untested, a pallid and lifeless
truth.

Mill also wrote, "If society lets any considerable number of its
members grow up as mere children, incapable of being acted on by rational consideration of distant motives, society has itself to blame." Jefferson made the same point even more strongly: "If a nation expects to
be both ignorant and free in a state of civilization, it expects what
never was and never will be." In a letter to Madison, he continued the

thought: "A society that will trade a little liberty for a little order will lose both, and deserve neither."

When permitted to listen to alternative opinions and engage in substantive debate, people have been known to change their minds. It can happen. For example, Hugo Black, in his youth, was a member of the Ku Klux Klan; he later became a Supreme Court justice and was one of the leaders in the historic Supreme Court decisions, partly based on the 14th Amendment to the Constitution, that affirmed the civil rights of all Americans: It was said that when he was a young man, he dressed up in white robes and scared black folks; when he got older, he dressed up in black robes and scared white folks.

In matters of criminal justice, the Bill of Rights recognizes the temptation that may be felt by police, prosecutors, and the judiciary to intimidate witnesses and expedite punishment. The criminal-justice system is fallible: Innocent people might be punished for crimes they did not commit; governments are perfectly capable of framing those who, for reasons unconnected with the purported crime, they do not like. So the Bill of Rights protects defendants. A kind of cost-benefit analysis is made. The guilty may on occasion be set free so that the innocent will not be punished. This is not only a moral virtue; it also inhibits the misuse of the criminal-justice system to suppress unpopular opinions or despised minorities. It is part of the error-correction machinery.

—

New ideas, invention, and creativity in general, always spearhead a kind of freedom—a breaking out from hobbling constraints. Freedom is a prerequisite for continuing the delicate experiment of science—which is one reason the Soviet Union could not remain a totalitarian state and be technologically competitive. At the same time, science—or rather its delicate mix of openness and skepticism, and its encouragement of diversity and debate—is a prerequisite for continuing the delicate experiment of freedom in an industrial and highly technological society.

Once you questioned the religious insistence on the prevailing view that the Earth was at the center of the Universe, why should you accept the repeated and confident assertions by religious leaders that God sent kings to rule over us? In the seventeenth century, it was easy to whip English and Colonial juries into a frenzy over this impiety or that heresy. They were willing to torture people to death for their beliefs. By the late eighteenth century, they weren't so sure.

Rossiter again (from *Seedtime of the Republic*, 1953):

> Under the pressure of the American environment, Christianity grew more humanistic and temperate—more tolerant with the struggle of the sects, more liberal with the growth of optimism and rationalism, more experimental with the rise of science, more individualistic with the advent of democracy. Equally important, increasing numbers of colonists, as a legion of preachers loudly lamented, were turning secular in curiosity and skeptical in attitude.

The Bill of Rights decoupled religion from the state, in part because so many religions were steeped in an absolutist frame of mind—each convinced that it alone had a monopoly on the truth and therefore eager for the state to impose this truth on others. Often, the leaders and practitioners of absolutist religions were unable to perceive any middle ground or recognize that the truth might draw upon and embrace apparently contradictory doctrines.

The framers of the Bill of Rights had before them the example of England, where the ecclesiastical crime of heresy and the secular crime of treason had become nearly indistinguishable. Many of the early Colonists had come to America fleeing religious persecution, although some of them were perfectly happy to persecute other people for *their* beliefs. The Founders of our nation recognized that a close relation between the government and any of the quarrelsome religions would be fatal to freedom—*and* injurious to religion. Justice Black (in the Supreme Court decision *Engel* v. *Vitale*, 1962) described the Establishment Clause of the First Amendment this way:

> Its first and most immediate purpose rested on the belief that a union of government and religion tends to destroy government and degrade religion.

Moreover, here too the separation of powers works. Each sect and cult, as Walter Savage Landor once noted, is a moral check on the others:

"Competition is as wholesome in religion as in commerce." But the price is high: This competition is an impediment to religious bodies acting in concert to address the common good.

Rossiter concludes:

> The twin doctrines of separation of church and state and liberty of individual conscience are the marrow of our democracy, if not indeed America's most magnificent contribution to the freeing of Western man.

Now it's no good to have such rights if they're not used—a right of free speech when no one contradicts the government, freedom of the press when no one is willing to ask the tough questions, a right of assembly when there are no protests, universal suffrage when less than half the electorate votes, separation of church and state when the wall of separation is not regularly repaired. Through disuse they can become no more than votive objects, patriotic lip-service. Rights and freedoms: Use 'em or lose 'em.

Due to the foresight of the framers of the Bill of Rights—and even more so to all those who, at considerable personal risk, insisted on exercising those rights—it's hard now to bottle up free speech. School library committees, the immigration service, the police, the FBI—or the ambitious politician looking to score cheap votes—may attempt it from time to time, but sooner or later the cork pops. The Constitution is, after all, the law of the land, public officials are sworn to uphold it, and activists and the courts episodically hold their feet to the fire.

However, through lowered educational standards, declining intellectual competence, diminished zest for substantive debate, and social sanctions against skepticism, our liberties can be slowly eroded and our rights subverted. The Founders understood this well: "The time for fixing every essential right on a legal basis is while our rulers are honest, and ourselves united," said Thomas Jefferson.

> From the conclusion of this [Revolutionary] war we shall be going downhill. It will not then be necessary to resort every moment to the people for support. They will be forgotten, therefore, and their rights disregarded. They will forget themselves but in the sole faculty of making money, and will never think of uniting to effect a

due respect for their rights. The shackles, therefore, which shall not be knocked off at the conclusion of this war will remain on us long, will be made heavier and heavier, 'til our rights shall revive or expire in a convulsion.

—

Education on the value of free speech and the other freedoms reserved by the Bill of Rights, about what happens when you don't have them, and about how to exercise and protect them, should be an essential prerequisite for being an American citizen—or indeed a citizen of any nation, the more so to the degree that such rights remain unprotected. If we can't think for ourselves, if we're unwilling to question authority, then we're just putty in the hands of those in power. But if the citizens are educated and form their own opinions, then those in power work for *us*. In every country, we should be teaching our children the scientific method and the reasons for a Bill of Rights. With it comes a certain decency, humility and community spirit. In the demon-haunted world that we inhabit by virtue of being human, this may be all that stands between us and the enveloping darkness.

Acknowledgments

It has been my great pleasure over many years to teach a Senior Seminar on Critical Thinking at Cornell University. I've been able to select students from all over the University on the basis both of ability, and of cultural and disciplinary diversity. We stress written assignments and oral argumentation. Towards the end of the course, students select a range of wildly controversial social issues in which they have major emotional investments. Paired two-by-two they prepare for a succession of end-of-semester oral debates. A few weeks before the debates, however, they are informed that it is the task of each to present the point of view of the opponent in a way that's satisfactory to the opponent—so the opponent will say, "Yes, that's a fair presentation of my views." In the joint written debate they explore their differences, but also how the debate process has helped them to better understand the opposing point of view. Some of the topics in this book were first presented to these students; I have learned much from their reception and criticism of my ideas, and want to thank them here. I'm also grateful to Cornell's Department of Astronomy, and its Chair, Yervant Terzian, for permitting me to teach the course, which—although labeled Astronomy 490—presents only a little astronomy.

Some of this book has also been presented in *Parade* magazine, a supplement to Sunday newspapers all over North America, with some 83 million readers each week. The vigorous feedback I've received from *Parade* readers has greatly enhanced my understanding of the issues described in this book and the variety of public attitudes. I have in several places excerpted some of my mail from *Parade* readers which, it seems to me, has provided a kind of finger on the pulse of the citizenry of the United States. The Editor-in-Chief of *Parade*, Walter Anderson, and the Senior Editor, David Currier, as well as the editorial and research staff of this remarkable magazine, have in many cases greatly improved my presentation. They also have permitted opinions to be expressed that might not have made it into print in mass market publications less dedicated to the First Amendment of the U.S. Constitution. Some portions of the text first appeared in *The Washington Post* and *The New York Times.* The last chapter is based in part on an address I had the pleasure of delivering on July 4, 1992 from the East Portico at Monticello—the "back of the nickel"—on the occasion of the induction to U.S. citizenship of people from 31 other nations.

My opinions on democracy, the method of science, and public education have been influenced by enormous numbers of people over the years, many of whom I mention in the body of the text. But I would like to single out here the inspiration I have received from Martin Gardner, Isaac Asimov, Philip Morrison, and Henry Steele Commager. There is not room to thank the many others who have helped provide understanding and lucid examples, or who have corrected errors of omission or commission, but I want them all to know how deeply grateful I am to them. I must however explicitly thank the following friends and colleagues for critically reviewing all or part of earlier drafts of this book: Bill Aldridge; Susan Blackmore; William Cromer; Fred Frankel; Kendrick Frazier; Martin Gardner; Ira Glasser; Fred Golden; Kurt Gottfried; Lester Grinspoon; Philip Klass; Paul Kurtz; Elizabeth Loftus; David Morrison; Richard Ofshe; Jay Orear; Albert Pennybacker; Frank Press; James Randi; Theodore Roszak; Dorion Sagan; David Saperstein; Robert Seiple; Steven Soter; Jeremy Stone; Peter Sturrock; and Yervant Terzian.

I also am very grateful to my literary agent, Morton Janklow, and members of his staff for wise counsel; Ann Godoff and others responsible for the production process at Random House—Enrica Gadler, J. K. Lambert, and Kathy Rosenbloom; William Barnett for ushering the manuscript through its final phases; Andrea Barnett, Laurel Parker, Karenn Gobrecht, Cindi Vita Vogel, Ginny Ryan, and Christopher Ruser for their assistance; and the Cornell Library system, including the rare books collection on mysticism and superstition originally compiled by the University's first president, Andrew Dickson White.

Parts of four of the chapters in this book were written with my wife and long-time collaborator, Ann Druyan, who is also the elected Secretary of the Federation of American Scientists—an organization founded in 1945 by the original Manhattan Project scientists to monitor the ethical use of science and high technology. She has also provided enormously helpful guidance, suggestions and criticism on content and style throughout the book and at every stage of writing it over the course of nearly a decade. I have learned from her more than I can say. I know how lucky I am to find in the same person someone whose advice and judgment, sense of humor and courageous vision I so much admire, who is also the love of my life.

References

(a few citations and suggestions for further reading)

Chapter 1, The Most Precious Thing

Martin Gardner, "Doug Henning and the Giggling Guru," *Skeptical Inquirer*, May/June 1995, pp. 9–11, 54.

Daniel Kahneman and Amos Tversky, "The Psychology of Preferences," *Scientific American*, vol. 246 (1982), pp. 160–173.

Ernest Mandel, *Trotsky as Alternative* (London: Verso, 1995), p. 110.

Maureen O'Hara, "Of Myths and Monkeys: A Critical Look at Critical Mass," in Ted Schultz, ed., *The Fringes of Reason* (see below), pp. 182–186.

Max Perutz, *Is Science Necessary?: Essays on Science and Scientists* (Oxford: Oxford University Press, 1991).

Ted Schultz, ed., *The Fringes of Reason: A Whole Earth Catalog: A Field Guide to New Age Frontiers, Unusual Beliefs & Eccentric Sciences* (New York: Harmony, 1989).

Xianghong Wu, "Paranormal in China," *Skeptical Briefs*, vol. 5 (1995), no. 1, pp. 1–3, 14.

J. Peder Zane, "Soothsayers as Business Advisers," *The New York Times*, September 11, 1994, sec. 4, p. 2.

Chapter 2, Science and Hope

Albert Einstein, "On the Electrodynamics of Moving Bodies," pp. 35–65 (originally published as "Zur Elektrodynamik bewegter Körper," *Annalen der Physik* 17 [1905], pp. 891–921), in H. Lorentz, A. Einstein, H. Minkowski, and H. Weyl, *The Principle of Relativity: A Collection of Original Memoirs on the*

Special and General Theory of Relativity (New York: Dover, 1923).

Harry Houdini, *Miracle Mongers and Their Methods* (Buffalo, NY: Prometheus Books, 1981).

Chapter 3, The Man in the Moon and the Face on Mars

John Michell, *Natural Likeness: Faces and Figures in Nature* (New York: E. P. Dutton, 1979).

Carl Sagan and Paul Fox, "The Canals of Mars: An Assessment after *Mariner 9*," *Icarus*, vol. 25 (1972), pp. 601–612.

Chapter 4, Aliens

E. U. Condon, *Scientific Study of Unidentified Flying Objects* (New York: Bantam Books, 1969).

Philip J. Klass, *Skeptics UFO Newsletter*, Washington, D.C., various issues. (Address: 404 "N" St. SW, Washington, D.C. 20024.)

Charles Mackay, *Extraordinary Popular Delusions and the Madness of Crowds* (first edition published in 1841) (New York: Farrar, Straus and Giroux, 1932, 1974) (also, New York: Gordon Press, 1991).

Curtis Peebles, *Watch the Skies!: A Chronicle of the Flying Saucer Myth* (Washington and London: Smithsonian Institution Press, 1994).

Donald B. Rice, "No Such Thing as 'Aurora,'" *The Washington Post*, December 27, 1992, p. 10.

Carl Sagan and Thornton Page, eds., *UFO's—A Scientific Debate* (Ithaca, NY: Cornell University Press, 1972.)

Jim Schnabel, *Round in Circles: Physicists, Poltergeists, Pranksters and the Secret History of the Cropwatchers* (London: Penguin Books, 1994) (first published in Great Britain by Hamish Hamilton in 1993).

Chapter 6, Hallucinations

K. Dewhurst and A. W. Beard, "Sudden Religious Conversions in Temporal Lobe Epilepsy," *British Journal of Psychiatry*, vol. 117 (1970), pp. 497–507.

Michael A. Persinger, "Geophysical Variables and Behavior: LV. Predicting the Details of Visitor Experiences and the Personality of Experients: The Temporal Lobe Factor," *Perceptual and Motor Skills*, vol. 68 (1989), pp. 55–65.

R. K. Siegel and L. J. West, eds., *Hallucinations: Behavior, Experience and Theory* (New York: Wiley, 1975).

Chapter 7, The Demon-Haunted World

Katherine Mary Briggs, *An Encyclopedia of Fairies, Hobgoblins, Brownies, Bogies, and Other Supernatural Creatures* (New York: Pantheon, 1976), pp. 239–242.

Thomas E. Bullard, "UFO Abduction Reports: The Supernatural Kidnap Narrative Returns in Technological Guise," *Journal of American Folklore*, vol. 102, no. 404 (April–June 1989), pp. 147–170.

Norman Cohn, *Europe's Inner Demons* (New York: Basic Books, 1975)

Ted Daniel, *Millennial Prophecy Report*, The Millennium Watch Institute, P.O. Box 34201, Philadelphia, PA 19101–4021, various issues.

Edward Gibbon, *The Decline and Fall of the Roman Empire*, Volume I, 180 A.D.–395 A.D. (New York: Modern Library, n.d.), pp. 410, 361, 432.

Martin S. Kottmeyer, "Entirely Unpredisposed," *Magonia*, January 1990.

Martin S. Kottmeyer, "Gauche Encounters: Badfilms and the UFO Mythos" (unpublished manuscript).

John E. Mack, *Abduction: Human Encounters with Aliens* (New York: Scribner, 1994).

John E. Mack, *Nightmares and Human Conflict* (Boston: Little Brown, 1970), pp. 227, 228.

Annemarie de Waal Malefijt, *Religion and Culture: An Introduction to Anthropology of Religion* (Prospect Heights, IL: Waveland Press, 1989) (originally published in 1968 by Macmillan), pp. 286 ff.

Jacques Vallee, *Passport to Magonia* (Chicago: Henry Regnery, 1969).

Chapter 8, On the Distinction Between True and False Visions

S. Ceci, M. L. Huffman, E. Smith, and E. Loftus, "Repeatedly Thinking About a Non-Event: Source Misattributions Among Pre-Schoolers," *Consciousness and Cognition*, Vol 3, 1994, pp. 388-407.

William A. Christian, Jr., *Apparitions in Late Medieval and Renaissance Spain* (Princeton, NJ: Princeton University Press, 1981).

Chapter 9, Therapy

Anonymous, "Trial in Woman's Blinding Offers Chilling Glimpse of Hoodoo," *The New York Times*, September 25, 1994, p. 23.

Ellen Bass and Laura Davis, *The Courage to Heal: A Guide for Women Survivors of Child Sexual Abuse* (New York: Perennial Library, 1988) (second and third editions, 1993 and 1994).

Richard J. Boylan and Lee K. Boylan, *Close Extraterrestrial Encounters: Positive Experiences with Mysterious Visitors* (Tigard, OR: Wild Flower Press, 1994).

Gail S. Goodman, Jianjian Qin, Bette L. Bottoms, and Philip R. Shaver, "Characteristics and Sources of Allegations of Ritualistic Child Abuse," Final Report, Grant 90CA1405, to the National Center on Child Abuse and Neglect, 1994.

David M. Jacobs, *Secret Life: First-Hand Accounts of UFO Abductions* (New York: Simon and Schuster, 1992), p. 293.

Carl Gustav Jung, Introduction to *The Unobstructed Universe*, by Stewart Edward White (New York: E. P. Dutton, 1941).

Kenneth V. Lanning, "Investigator's Guide to Allegations of 'Ritual' Child Abuse" (Washington: FBI, January 1992).

Elizabeth Loftus and Katherine Ketcham, *The Myth of Repressed Memory* (New York: St. Martin's, 1994)

Mike Males, "Recovered Memory, Child Abuse, and Media Escapism," *Extra!* September/October 1994, pp. 10, 11.

Ulric Neisser, keynote address, "Memory with a Grain of Salt," *Memory and Reality: Emerging Crisis* conference, Valley Forge, PA, as reported by *FMS Foundation Newsletter*, (Philadelphia, PA) vol. 2, no. 4 (May 3, 1993), p. 1.

Richard Ofshe and Ethan Watters. *Making Monsters* (New York: Scribner, 1994)

Nicholas P. Spanos, Patricia A. Cross, Kirby Dixon, and Susan C. DuBreuil, "Close Encounters: An Examination of UFO Experiences," *Journal of Abnormal Psychology*, vol. 102 (1993), pp. 624–632.

Rose E. Waterhouse, "Government Inquiry Decides Satanic Abuse Does Not Exist," *Independent on Sunday*, London, April 24, 1994.

Lawrence Wright, *Remembering Satan: A Case of Recovered Memory and the Shattering of an American Family* (New York: Knopf, 1994).

Michael D. Yapko, *True and False Memories of Childhood Sexual Trauma: Suggestions of Abuse* (New York: Simon and Schuster, 1994).

Chapter 10, The Dragon in My Garage

Thomas J. Flotte, Norman Michaud, and David Pritchard, in *Alien Discussions*, Andrea Pritchard et al., eds., pp. 279–295 (Cambridge, MA: North Cambridge Press, 1994).

Richard L. Franklin, *Overcoming the Myth of Self-Worth: Reason and Fallacy in What You Say to Yourself* (Appleton, WI: R. L. Franklin, 1994).

Robert Lindner, "The Jet-Propelled Couch," in *The Fifty-Minute Hour: A Collection of True Psychoanalytic Tales* (New York and Toronto: Rinehart, 1954).

James Willwerth, "The Man from Outer Space," *Time*, April 25, 1994.

Chapter 12, The Fine Art of Baloney Detection

George O. Abell and Barry Singer, eds., *Science and the Paranormal: Probing the Existence of the Supernatural* (New York: Scribner's, 1981).

Robert Basil, ed., *Not Necessarily the New Age* (Buffalo: Prometheus, 1988)

Susan Blackmore, "Confessions of a Parapsychologist," in Ted Schultz, ed., *The Fringes of Reason*, pp. 70–74.

Russell Chandler, *Understanding the New Age* (Dallas: Word, 1988)

T. Edward Damer, *Attacking Faulty Reasoning*, second edition (Belmont, CA: Wadsworth, 1987).

Kendrick Frazier, ed., *Paranormal Borderlands of Science* (Buffalo, NY: Prometheus, 1981).

Martin Gardner, *The New Age: Notes of a Fringe Watcher* (Buffalo, NY: Prometheus, 1991).

Daniel Goleman, "Study Finds Jurors Often Hear Evidence with a Closed Mind," *The New York Times*, November 29, 1994, pp. C-1, C-12.

J.B.S. Haldane, *Fact and Faith* (London: Watts & Co., 1934).

Philip J. Hilts, "Grim Findings on Tobacco Made the 70's a Decade of Frustration" (including box, p. 12, "Top Scientists for Companies Saw the Perils"), *The New York Times*, June 18, 1994, pp. 1, 12.

Philip J. Hilts, "Danger of Tobacco Smoke Is Said to be Underplayed," *New York Times*, December 21, 1994, D23.

Howard Kahane, *Logic and Contemporary Rhetoric: The Use of Reason in Everyday Life*, 7th edition (Belmont, CA: Wadsworth, 1992).

Noel Brooke Moore and Richard Parker, *Critical Thinking* (Palo Alto, CA: Mayfield, 1991).

Graham Reed, *The Psychology of Anomalous Experience* (Buffalo, NY: Prometheus, 1988).

Theodore Schick, Jr., and Lewis Vaughn, *How to Think About Weird Things: Critical Thinking for a New Age* (Mountain View, CA: Mayfield, 1995).

Leonard Zusne and Warren H. Jones, *Anomalistic Psychology* (Hillsdale, NJ: Lawrence Erlbaum, 1982).

Chapter 13, Obsessed with Reality

Alvar Nuñez Cabeza de Vaca, *Castaways*, translated by Frances M. López-Morillas (Berkeley: University of California Press, 1993).

"Faith Healing: Miracle or Fraud," special issue of *Free Inquiry*, vol. 6, no. 2 (Spring 1986).

Paul Kurtz, *The New Skepticism: Inquiry and Reliable Knowledge* (Buffalo, NY: Prometheus Books, 1992).

William A. Nolen, M.D., *Healing: A Doctor in Search of a Miracle* (New York: Random House, 1974).

David P. Phillips and Daniel G. Smith, "Postponement of Death Until Symbolically Meaningful Occasions," *Journal of the American Medical Association*, vol. 263 (1990), pp 1947–1951.

James Randi, *The Faith Healers* (Buffalo, NY: Prometheus Books, 1989).

James Randi, *Flimflam!: The Truth About Unicorns, Parapsychology & Other Delusions* (Buffalo, NY: Prometheus Books, 1982).

David Spiegel, "Psychosocial Treatment and Cancer Survival," *The Harvard Mental Health Letter*, vol. 7 (1991), no. 7, pp. 4–6.

Charles Whitfield, *Healing the Child Within* (Deerfield Beach, FL: Health Communications, Inc., 1987).

Chapter 14, Antiscience

Joyce Appleby, Lynn Hunt, and Margaret Jacob, *Telling the Truth About History* (New York: W. W. Norton, 1994).

Morris R. Cohen, *Reason and Nature: An Essay on the Meaning of Scientific Method* (New York: Dover, 1978) (first edition published by Harcourt Brace in 1931).

Gerald Holton, *Science and Anti-Science* (Cambridge: Harvard University Press, 1993), chs. 5 and 6.

John Keane, *Tom Paine: A Political Life* (Boston: Little, Brown, 1995).

Michael Krause, *Relativism: Interpretation and Confrontation* (South Bend, IN: University of Notre Dame, 1989).

Harvey Siegel, *Relativism Refuted* (Dordrecht, Netherlands: D. Reidel, 1987).

Chapter 15, Newton's Sleep

Henry Gordon, *Channeling into the New Age* (Buffalo: Prometheus, 1988)

Charles T. Tart, "The Science of Spirituality," in Ted Schultz, ed., *The Fringes of Reason*, p. 67.

Chapter 16, When Scientists Know Sin

William Broad, *Teller's War: The Top-Secret Story Behind the Star Wars Deception* (New York: Simon and Schuster, 1992).

David Holloway, *Stalin and the Bomb* (New Haven: Yale University Press, 1994).

John Passmore, *Science and Its Critics* (London: Duckworth, 1978).

Stockholm International Peace Research Institute, *SIPRI Yearbook 1994* (Oxford: Oxford University Press, 1994), p. 378.

Carl Sagan, *Pale Blue Dot: A Vision of the Human Future in Space* (New York: Random House, 1994).

Carl Sagan and Richard Turco, *A Path Where No Man Thought: Nuclear Winter and the End of the Arms Race* (New York: Random House, 1990).

Chapter 17, The Marriage of Skepticism and Wonder

R. B. Culver and P. A. Ianna, *The Gemini Syndrome: A Scientific Explanation of Astrology* (Buffalo, NY: Prometheus, 1984).

David J. Hess, *Science in the New Age: The Paranormal, Its Defenders and Debunkers, and American Culture* (Madison, WI: The University of Wisconsin Press, 1993).

Carl Sagan, "Objections to Astrology" (letter to the editor), *The Humanist*, vol. 36, no. 1 (January/February 1976), p. 2.

Robert Anton Wilson, *The New Inquisition: Irrational Rationalism and the Citadel of Science* (Phoenix: Falcon Press, 1986).

Chapter 18, The Wind Makes Dust

Alan Cromer, *Uncommon Sense: The Heretical Nature of Science* (New York: Oxford University Press, 1993).

Richard Borshay Lee, *The !Kung San: Men, Women, and Work in a Foraging Society* (Cambridge, UK: Cambridge University Press, 1979).

Chapter 19, No Such Thing as a Dumb Question

Youssef M. Ibrahim, "Muslim Edicts Take on New Force," *The New York Times*, February 12, 1995, p. A14.

Catherine S. Manegold, "U.S. Schools Misuse Time, Study Asserts," *The New York Times*, May 5, 1994, p. A21.

"The Competitive Strength of U.S. Industrial Science and Technology: Strategic Issues," a report of the National Science Board Committee on Industrial Support for R&D, National Science Foundation, Washington, D.C., August 1992.

Chapter 21, *The Path to Freedom*

Walter R. Adam and Joseph O. Jewell, "African-American Education Since *An American Dilemma*," *Daedalus* 124, 77–100, 1995.

J. Larry Brown, ed., "The Link Between Nutrition and Cognitive Development in Children," Center on Hunger, Poverty and Nutrition Policy, School of Nutrition, Tufts University, Medford, MA, 1993, and references given there.

Gerald S. Coles, "For Whom the Bell Curves," *The Bookpress* 5 (1), 8–9, 15, February 1995.

Frederick Douglass, *Autobiographies: Narrative of a Life, My Bondage & My Freedom, Life and Times*, Henry L. Gates, Jr., ed. (New York: Library of America, 1994).

Leon J. Kamin, "Behind the Bell Curve," *Scientific American*, February 1995, pp. 99–103.

Tom McIver, "The Protocols of Creationism: Racism, Anti-Semitism and White Supremacy in Christian Fundamentalism," *Skeptic*, vol. 2, no. 4 (1994), pp. 76–87.

Chapter 22, *Significance Junkies*

Tom Gilovich, *How We Know What Isn't So: The Fallibility of Human Reason in Everyday Life* (New York: Free Press, 1991).

"O. J. Who?" *New York*, October 17, 1994, p. 19.

Chapter 23, *Maxwell and the Nerds*

Richard P. Feynman, Robert B. Leighton, and Matthew Sands, *The Feynman Lectures on Physics*, Volume II, *The Electromagnetic Field* (Reading, MA: Addison-Wesley, 1964). [Passages quoted appear on pp. 18–2, 20–8, and 20–9.]

Ivan Tolstoy, *James Clerk Maxwell: A Biography* (Chicago: University of Chicago Press, 1982) (originally published by Canongate Publishing Ltd., Edinburgh, 1981).

Chapter 24, Science and Witchcraft

William Glaberson, "The Press: Bought and Sold and Grey All Over," *The New York Times*, July 30, 1995, Section 4, pp. 1, 6.

Peter Kuznick, "Losing the World of Tomorrow: The Battle Over the Presentation of Science at the 1939 World's Fair," *American Quarterly*, vol. 46, no. 3 (September 1994), pp. 341–373.

Ernest Mandel, *Trotsky as Alternative*.

Rossell Hope Robbins, *The Encyclopedia of Witchcraft and Demonology* (New York: Crown, 1959).

Jeremy J. Stone, "Conscience, Arrogation and the Atomic Scientists" and "Edward Teller: A Scientific Arrogator of the Right," *F.A.S.* [Federation of American Scientists] *Public Interest Report*, vol. 47, no. 4 (July/August 1994), pp. 1, 11.

Chapter 25, Real Patriots Ask Questions

I. Bernard Cohen, *Science and the Founding Fathers* (Cambridge: Harvard University Press, 1995).

Clinton Rossiter, *Seedtime of the Republic* (New York: Harcourt Brace, 1953). Excerpted in Rossiter, *The First American Revolution* (San Diego: Harvest).

J. H. Sloan, F. P. Rivera, D. T. Reay, J.A.J. Ferris, M.R.C. Path, and A. L. Kellerman, "Firearm Regulations and Rates of Suicide: A Comparison of Two Metropolitan Areas," *New England Journal of Medicine*, vol. 311 (1990), pp. 369–373.

"Post Script," *Conscience*, vol. 15, no. 1 (Spring 1994), p. 77.

Index

Robert Reichert

ABOUT THE AUTHOR

CARL SAGAN served as the David Duncan Professor of Astronomy and Space Sciences and Director of the Laboratory for Planetary Studies at Cornell University. He played a leading role in the Mariner, Viking, Voyager, and Galileo spacecraft expeditions to the planets for which he received the NASA Medals for Exceptional Scientific Achievement and (twice) for Distinguished Public Service.

His Emmy and Peabody Award–winning television series, *Cosmos*, became the most widely watched series in the history of American public television. The accompanying book, also called *Cosmos*, is one of the bestselling science books ever published in the English language. Dr. Sagan received the Pulitzer Prize, the Oersted Medal, and many other awards—including twenty honorary degrees from American colleges and universities—for his contributions to science, literature, education, and the preservation of the environment.

Dr. Sagan died on December 20, 1996.